# Ecological Studies, Vol. 140

Analysis and Synthesis

Ecological Studies

Volumes published since 1993 are listed at the end of this book.

Springer
*Tokyo*
*Berlin*
*Heidelberg*
*New York*
*Barcelona*
*Hong Kong*
*London*
*Milan*
*Paris*
*Singapore*

Edi Guhardja Mansur Fatawi Maman Sutisna
Tokunori Mori Seiichi Ohta (Eds.)

# Rainforest Ecosystems of East Kalimantan

## El Niño, Drought, Fire and Human Impacts

With 146 Figures, 23 in Color, and 47 Tables

Springer

Dr. Edi Guhardja
Director, Master of Management in Agribusiness
Bogor Agricultural University
P.O. Box 2019, Bogor Timur 16144, Indonesia

Dr. Mansur Fatawi
Director of PUSREHUT (Tropical Rain Forest Research Center)
Mulawarman University, Kampus Gunung Kelua
P.O. Box 1165, Samarinda, East Kalimantan, Indonesia

Dr. Maman Sutisna
Lecturer, Faculty of Forestry, Mulawarman University
Kampus Gunung Kelua, Samarinda, East Kalimantan, Indonesia

Dr. Tokunori Mori
Project Team Leader of JICA
PUSREHUT (Tropical Rain Forest Research Center)
Mulawarman University, Kampus Gunung Kelua
P.O. Box 1165, Samarinda, East Kalimantan, Indonesia

Dr. Seiichi Ohta
Section Chief of Forest Soil
Forestry and Forest Product Research Institute (FFPRI)
P.O. Box 16, Tsukuba Norin-Danchi, Ibaraki 305-8687, Japan

ISSN 0070-8356
ISBN 4-431-70272-5 Springer-Verlag Tokyo Berlin Heidelberg New York

Library of Congress Cataloging-in-Publication Data

Rainforest ecosystems of East Kalimantan : El Niño, drought, fire, and human impacts /
Edi Guhardja...[et al.].
p.cm. — (Ecological studies ; v.140)
Includes bibliographical references.
ISBN 4431702725 (hardcover : alk. paper)
1. Forests and forestry—Environmental aspects—Indonesia—Kalimantan Timur. 2. Rain forests—Indonesia—Kalimantan Timur—Management. 3. Rain forest ecology—Indonesia—Kalimantan Timur. I. Guhardja, Edi, 1933-II. Series.

SD387.E58 R35 2000
333.75'153'095983—dc21

99-049504

Printed on acid-free paper



Printing and binding: Obun, Japan
SPIN: 10745474

# Foreword

Kalimantan, which covers more than 70% of Borneo, has one of the most important tropical forests, not only in Indonesia or Southeast Asia but also in the world. Although in the past the island was almost covered by tropical rainforest, the forest area has decreased rapidly with expanding human activities and with ecological and environmental changes. Almost every year, especially during the dry years, forest fire is among the greatest threats. The forest fire in 1982–83, when approximately 3.5 million ha of forest were destroyed, is still remembered today. The most recent fires were in 1997–98, which coincided with the exceptionally dry years of the El Niño–Southern Oscillation phenomenon.

In the past, forest utilization practices were designed to produce a certain quantity of trees with high economic value and were based on sustained yield principles (SYP). Some of the yield regulation methods in classical forest management science at that time were created on the basis of economic principles and forest land capacity in order to produce trees in an even-aged forest.

With the worsening of environmental problems and the increase of human exploitation of the forest, SYP is no longer adequate. At present, the principle followed is that of sustained forest management (SFM). SFM means that forest management activity should achieve the maximum benefit by optimizing production (economy), the environment (ecology), and the sociocultural function of sustained forest ecosystems.

SFM should be adequately applied in Indonesian forests, where the majority of the forest is tropical rainforest, and should take into account the local flora and fauna and the sociocultural characteristics of the local community. The tropical rainforest ecosystems are known for their rich biodiversity and high value—not only in conventional economic terms (market price) but also in terms of ecological and sociocultural values. Based on this fact, the survival of tropical rainforest ecosystems should be maintained by applying SFM principles.

One of the important requirements when applying SFM in forest management is the availability of data with high levels of accuracy and precision. Some data that are needed for this purpose concern the characteristics of forest ecosystems and the sociocultural characteristics of the local community. At present, such data for each ecosystem unit of the tropical rainforests in Indonesia are still very limited.

Mulawarman University (Universitas Mulawarman; UNMUL) at Samarinda, the capital of East Kalimantan, is one of the National Public Universities developed in the early 1970s. A priority research field at the university is rehabilitation of tropical rainforest. The Japan International Cooperation Agency (JICA) became interested in cooperating with UNMUL in the 1970s, and in 1979 founded the Tropical Rain Forest Research Center (Pusat Studi Reboisasi Hutan Tropica Humida; PUSREHUT) in UNMUL, together with a field research station in Bukit Soeharto. PUSREHUT has become a center for studies on forest rehabilitation for many scientists from a number of countries and also has become a field station for many university students.

This book was compiled as one of the activities of the cooperative research project *Tropical Rain Forest Research*, which was started at PUSREHUT in 1985 with cooperation between the Ministry of Education and Culture in Indonesia and JICA. Information collected from the collaboration involving UNMUL in Samarinda, Bogor Agricultural University (IPB, Bogor) and the University of Gadjah Mada (UGM) in Jogyakarta, and JICA researchers is very important academically and socially, as well as culturally. It is a great pleasure for me to contribute the foreword for this book, which contains information on the tropical rainforest ecosystem based on the activities of PUSREHUT. I welcome its publication. The data and information presented in this book provide an invaluable source of reference material, especially for scientists and students interested in ecology, tropical biology, forestry, socioeconomics, and forest policy in the tropics.

Finally, to the authors and editors for their work in preparing the manuscript and to Springer-Verlag, the publisher, I am very grateful and I hope that this book will be useful not only for the development of forestry science but also for sustainable tropical rainforest management.

Prof. Dr. Ir. Jajah Koswara
Director
Directorate Research and Community
Service Development
Directorate General of Higher Education
Ministry of Education and Culture
Indonesia

**Tropical Rain Forest Research Project**

**1985-1999**

**The Indonesian Ministry of Education and Culture**

**Japan International Cooperation Agency**

PUSREHUT
**(Tropical Rain Forest Research Center)**

# Preface

Since the late 1960s, the human impact of commercial logging, large-scale agricultural exploitation, and immigration from other islands has been increasingly evident in the rainforest of East Kalimantan, and the rapid changes in the landscape have decreased the biodiversity of the area. In addition, prolonged droughts linked to particularly strong El Niño–Southern Oscillation (ENSO) events occurred in 1982–83 and 1997–98 in East Borneo. Large human-induced wildfires occurred during the ENSO events and burned large areas of the lowlands in East Kalimantan. It is possible that prolonged droughts and large wildfires in rainforest areas may affect global carbon and water cycles. However, we have little ecological or socioeconomic understanding of the changing ecosystems in the lowland dipterocarp forests of Southeast Asia.

Our research has spanned a 15-year cycle of severe ENSO events that took place in this area, and has revealed possible interactive effects of drought, fire, and human activity on the rainforest ecosystems. In presenting such long-term research, this book addresses the current condition of the rainforest ecosystem and the regeneration techniques that can be used to develop sustainable forest management in East Kalimantan. Part 1 introduces developmental, historical, and climatological information, and discusses the large forest fires in East Kalimantan. Part 2 examines the forest soil conditions that are linked with changes in the vegetation caused by human activity. Part 3 reveals the effects of drought and forest fires on the function and dynamics of forest vegetation and insect communities. Part 4 deals with the human impact on forest vegetation and on the Bornean gibbon community. Part 5 presents a variety of topics on biotic impacts, such as the effects of mycorrhiza and plant disease on seedling growth as ways of developing artificial and natural regeneration, and rehabilitation techniques for degraded lowland dipterocarp forests.

This volume is a compilation of the results of a cooperative research project between Indonesia and Japan. The project was conceived in 1985 at PUSREHUT (Tropical Rain Forest Research Center), which was established by a Japanese Aid Program in Samarinda, East Kalimantan in 1981, called the Tropical Rain Forest Research Project. The project involves technical cooperation between the Indonesian Ministry of Education and Culture (MoEC) and the Japan International Cooperation Agency (JICA). The project has continued successfully for 15 years, through Phases I to III, thanks to the endeavors and enthusiasm of a great many people.

It would not have been possible to publish this volume without the marvelous cooperation, understanding, and consideration of many individuals and institutions in both Indonesia and Japan. We would like to express our sincerest gratitude to the people of the Directorate General of Higher Education, MoEC, and the Forestry and Fisheries Development Cooperation Department, JICA, and to the Rectors of Mulawarman University, Bogor Agricultural University, and the University of Gadjah Mada. We wish to extend our sincerest thanks to all the partners, assistants, and staff members of PUSREHUT and Bukit Soeharto Education Forest, and to our research colleagues in Japan and Indonesia. We thank the Forest Fire Prevention Management Project (Indonesian Ministry of Forestry and Estate Crops and JICA) in Bogor for kindly providing the hot-spot data and satellite image photographs. Finally, we are also deeply grateful to Drs. Atsushi Ishida and Takeshi Toma for their earnest and devoted assistance in planning and editing this volume and to Dr. Kazuhito Morisada and Mr. Kazuhiro Matsuzawa for their overall services.

THE EDITORS

# Contents

## Part 2: Soil Characteristics and Human Impacts

# Contributors

AKSA, DARUL
Tropical Rain Forest Research Center (PUSREHUT), Mulawarman University, P. O. Box 1165, Samarinda 75117, East Kalimantan, Indonesia

EFFENDI, SYARIF
Tropical Rain Forest Research Center (PUSREHUT), Mulawarman University, P. O. Box 1165, Samarinda 75117, East Kalimantan, Indonesia

FATAWI, MANSUR
Tropical Rain Forest Research Center (PUSREHUT), Mulawarman University, P. O. Box 1165, Samarinda 75117, East Kalimantan, Indonesia

GHOZALI, DADANG I.
Tropical Rain Forest Research Center (PUSREHUT), Mulawarman University, P. O. Box 1165, Samarinda 75117, East Kalimantan, Indonesia

GUHARDJA, EDI
Faculty of Mathematics and Natural Science, Bogor Agricultural University, Gedung MMA-IPB, Jl. Raya Pajajaran, Bogor, Indonesia

HASTANIAH
Tropical Rain Forest Research Center (PUSREHUT), Mulawarman University, P. O. Box 1165, Samarinda 75117, East Kalimantan, Indonesia

INOUE, MAKOTO
Division of Forest Science, Graduate School of Agricultural and Life Sciences, The University of Tokyo, Yayoi, Bunkyo, Tokyo 113-8657, Japan

ISHIDA, ATSUSHI
Forest Environment Division, Forestry and Forest Products Research Institute (FFPRI), P. O. Box 16, Tsukuba Norin Danchi, Tsukuba, Ibaraki 305-8687, Japan

ISKANDAR, ECEP
Tropical Rain Forest Research Center (PUSREHUT), Mulawarman University, P. O. Box 1165, Samarinda 75117, East Kalimantan, Indonesia

ITO, SATOSHI
Forest Science Division, Faculty of Agriculture, Miyazaki University, Gakuen-Kibanadai-Nishi, Miyazaki 889-2192, Japan

KINUURA, HARUO
Tohoku Research Center, Forestry and Forest Products Research Institute (FFPRI), Shimokuriya-gawa, Morioka, Iwate 020-0123, Japan

KIYONO, YOSHIYUKI
Kansai Research Center, Forestry and Forest Products Research Institute (FFPRI), Momoyama, Fushimi, Kyoto 612-0855, Japan

KOSWARA, JAJAH
Research and Community Service Development, Directrate General of Higher Education, Ministry of Education and Culture, Jl. Pintu 1, Senayan, Jakarta, Indonesia

KUSTIWAN, WAWAN
Faculty of Forestry, Mulawarman University, P. O. Box 1013, Samarinda 75117, East Kalimantan, Indonesia

MAKIHARA, HIROSHI
Tropical Rain Forest Research Center (PUSREHUT), Mulawarman University, P. O. Box 1165, Samarinda 75117, East Kalimantan, Indonesia
Present Address: Forest Biology Division, Forestry and Forest Products Research Institute (FFPRI), P. O. Box 16, Tsukuba Norin Danchi, Tsukuba, Ibaraki 305-8687, Japan

MARDJI, DJUMALI
Faculty of Forestry, Mulawarman University, P. O. Box 1013, Samarinda 75117, East Kalimantan, Indonesia

MARJENAH
Tropical Rain Forest Research Center (PUSREHUT), Mulawarman University, P. O. Box 1165, Samarinda 75117, East Kalimantan, Indonesia

MATIUS, PAULUS
Faculty of Forestry, Mulawarman University, P. O. Box 1013, Samarinda 75117, East Kalimantan, Indonesia

MIURA, SATORU
Forest Environment Division, Forestry and Forest Products Research Institute (FFPRI), P. O. Box 16, Tsukuba Norin Danchi, Tsukuba, Ibaraki 305-8687, Japan

MORI, SHIGETA
Tohoku Research Center, Forestry and Forest Products Research Institute (FFPRI), Shimokuriya-gawa, Morioka, Iwate 020-0123, Japan

MORI, TOKUNORI
Tropical Rain Forest Research Center (PUSREHUT), Mulawarman University, P. O. Box 1165, Samarinda 75117, East Kalimantan, Indonesia
Present Address: Bio-resources Technology Division, Forestry and Forest Products Research Institute (FFPRI), P. O. Box 16, Tsukuba Norin Danchi, Tsukuba, Ibaraki 305-8687, Japan

MORISADA, KAZUHITO
Forest Environment Division, Forestry and Forest Products Research Institute (FFPRI), P. O. Box 16, Tsukuba Norin Danchi, Tsukuba, Ibaraki 305-8687, Japan

NISHIYAMA, YOSHIHIKO
Tama Forest Science Garden, Forestry and Forest Products Research Institute (FFPRI), Todori, Hachiohji, Tokyo 193-0843, Japan

OHTA, SEIICHI
Forest Environment Division, Forestry and Forest Products Research Institute (FFPRI), P. O. Box 16, Tsukuba Norin Danchi, Tsukuba, Ibaraki 305-8687, Japan

OKA, TERUKI
Primate Research Institute, Kyoto University, Inuyama, Aichi 484-8506, Japan

OKIMORI, YASUYUKI
Biological Environment Institute, Kansai Environmental Engineering Center Co., Ltd., Ujimatafuri, Uji, Kyoto 611-0021, Japan

RUSLIM, YOSEF
Faculty of Forestry, Mulawarman University, P. O. Box 1013, Samarinda 75117, East Kalimantan, Indonesia

SUHARDI
Faculty of Forestry, The University of Gajah Mada, Bulaksumur, Yogyakarta 55261, Indonesia

SUTISNA, MAMAN
Faculty of Forestry, Mulawarman University, P. O. Box 1013, Samarinda 75117, East Kalimantan, Indonesia

SOEYAMTO, CHRISTOPHORUS
Faculty of Forestry, Mulawarman University, P. O. Box 1013, Samarinda 75117, East Kalimantan, Indonesia

TANAKA, NAGAHARU
Hokkaido Research Center, Forestry and Forest Products Research Institute (FFPRI), Hitsujigaoka, Toyohira, Sapporo 062-0045, Japan

TOMA, TAKESHI
Tropical Rain Forest Research Center (PUSREHUT), Mulawarman University, P. O. Box 1165, Samarinda 75117, East Kalimantan, Indonesia
Present Address: Forest Environment Division, Forestry and Forest Products Research Institute (FFPRI), P. O. Box 16, Tsukuba Norin Danchi, Tsukuba, Ibaraki 305-8687, Japan

WATANABE, RYUICHI
Faculty of Education, Shinshu University, Nishi-Nagano, Nagano 380-8544, Japan

YAHATA, HISASHI
Division of Land-Water Resources and Environmental Conservation, Institute of Tropical Agriculture, Kyushu University, Higashi, Fukuoka 812-8581, Japan

YAHIRO, KATSURO
Lake Biwa Museum, Kusatsu, Shiga 525-0001, Japan

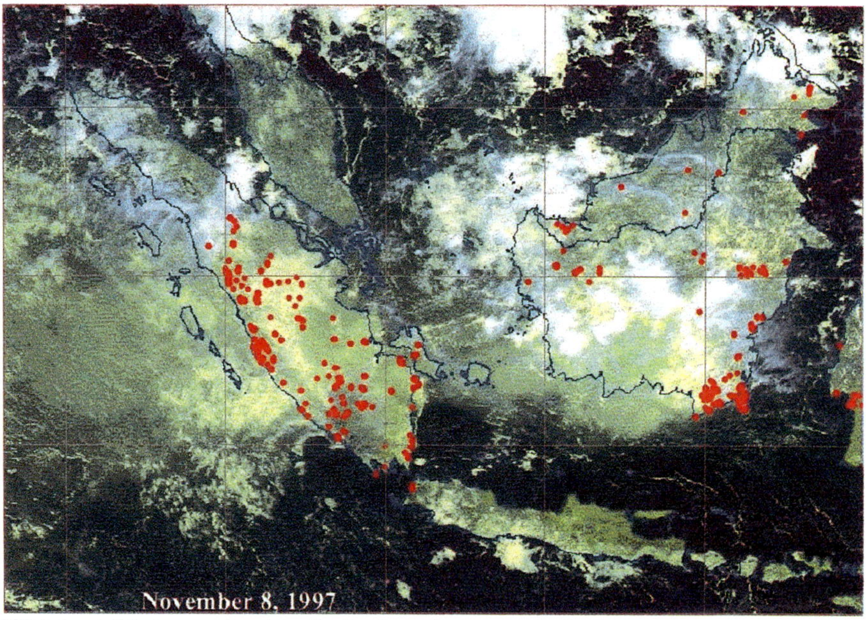


NOAA satellite image showing distribution of hot spots (red points) and haze and clouds in South East Asia on November 8, 1997, during an ENSO (El Niño Southern-Oscillation) event

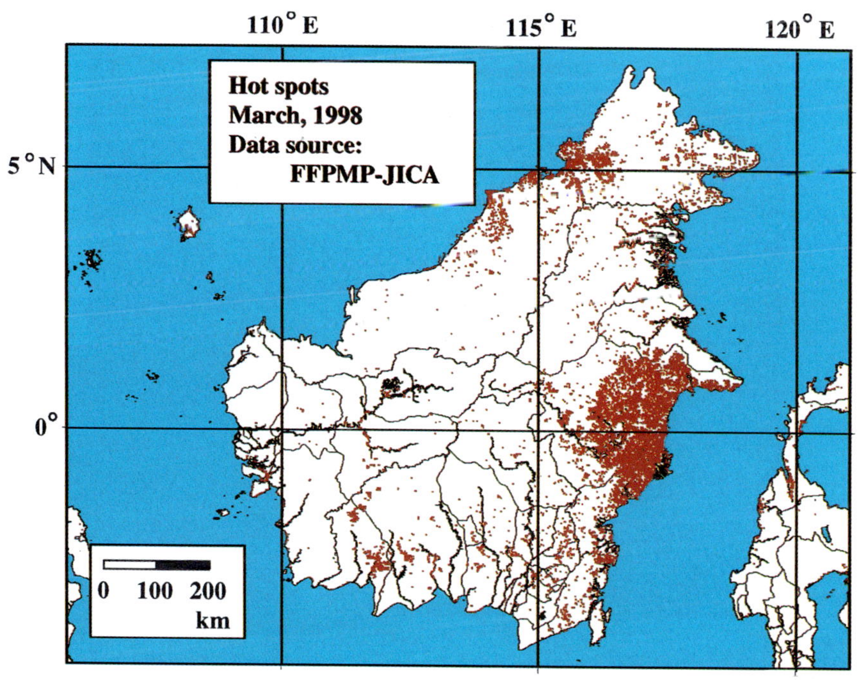


Distribution of hot spots (red points) on Borneo island for the entire month of March 1998, during an ENSO event (Data from the Forest Fire Prevention Management Project, Bogor, Indonesia)

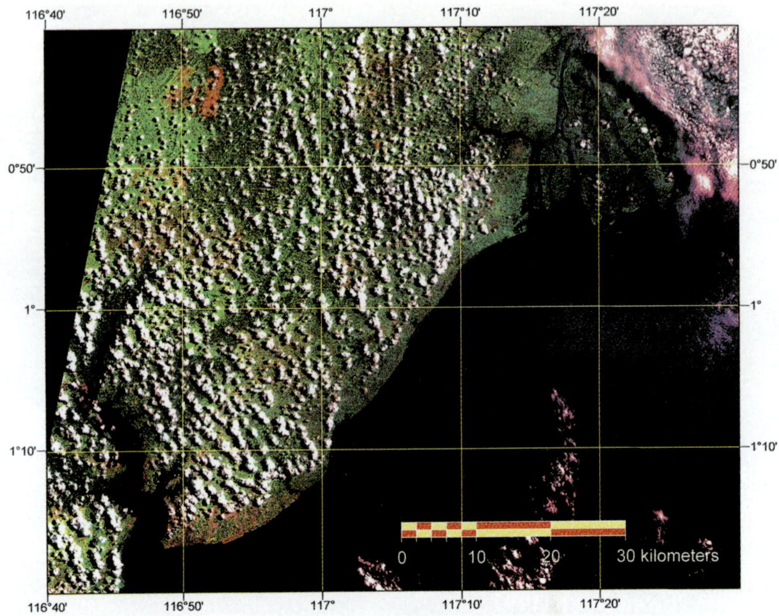


Landsat TM image (D116-061) April 13, 1997, before drought and wildfires around the study area between Balikpapan and Samarinda

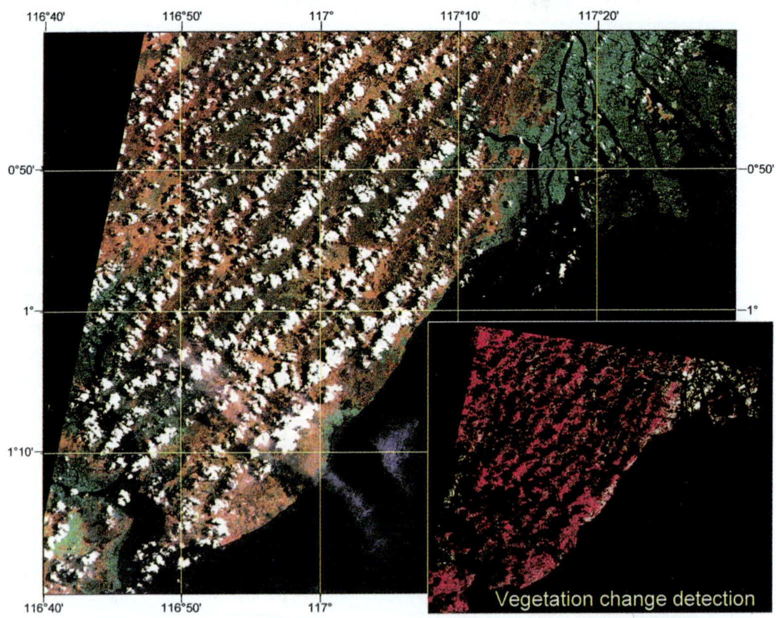


Landosat TM image March 31, 1998, just after wildfires associated with a 1997-98 ENSO event around the study area. *Inset:* detected decrease in vegetation (*pink areas*) between April 1997 and March 1998, using Change Vector Analysis (CVA) examined by Satoshi Tsuyuki

A grassland fire during a prolonged drought associated with an ENSO event. Photographed February 1998

A surface fire on the understory of a near-primary forest during a prolonged drought associated with an ENSO event. Photographed February 1998

Mixed dipterocarp forest after the forest fire of March 1998. Photographed April 1998

A near-primary forest before the forest fires of 1998. Photographed July 1996

A near-primary forest burning. Photographed February 1998. The smoke is from forest fires during a prolonged drought associated with an ENSO event

A near-primary forest after the forest fires of February 1998. Photographed July 1998

A secondary forest consisting of *Macaranga* trees developed after the forest fires of 1983 associated with an ENSO event. Photographed August 1997. Dead standing trees are from the forest fires of 1983

A secondary forest, affected by the prolonged drought associated with a 1997-98 ENSO event, before forest fires. Photographed February 1998. The smoke is from forest fires

A secondary forest after the forest fires of February 1998. Photographed May 1998

Land use by pepper, banana, and other plantations. A typical landscape in East Kalimantan

An alang-alang (*Imperata cylindrica*) grassland developed after the abandonment of a pepper plantation (woody stakes in *background*), and burning coal on an outcrop (*foreground*). A typical landscape (savannization) in East Kalimantan

Mosaic of vegetation produced by land use by migratory farmers. Photographed November 1996

Just after a fire in slash-and-burn agriculture

*Dipterocarpus cornutus* fruits at the top of a canopy. Flowering occurred during a 1997-98 ENSO event. Photographed January 1998

The fruit body of *Scleroderma columnare*, a typical ectomycorrhizal fungus, in a potted dipterocarp (Shorea smithiana) seedling **(left)**. *Xylotrechus australis* (longicorn beetle) on *Mikania micratha* flowers **(right)**

Vertical profiles of Typic Paleudults under a degraded secondary forest **(left)** and an *Imperata* grassland **(middle)**, which are the major soil types of the lowland dipterocarp forest area in East Kalimantan, and a profile of Rhodic Kandiperox **(right)** derived from serpentinite found in the southern part of East Kalimantan

# Part 1

# Introduction

# 1 Description of Forests and Forestry in East Kalimantan

MANSUR FATAWI and TOKUNORI MORI

## 1.1 Introduction

Kalimantan, the Indonesian part of Borneo, occupies 73% of the land area of the island, which is the third largest island in the world. Kalimantan is subdivided into four Indonesian states. One of these states, East Kalimantan, includes the southeastern parts of the central mountain ranges of the island and lies between approximately 4° north latitude and 2° south latitude (Fig. 1.1). It has an area of 21 144 $km^2$ which occupies about 11% of the whole country. Despite its large area, the population of East Kalimantan, which is about 2.3 million, is only about 1% of the population of Indonesia. Nevertheless, this state has played a key role in Indonesia's economic development and is a major earner of foreign revenue. This wealth is not due to industrial production or to agricultural plantations but to the rich reserves of natural resources: forests, oil, gas, coal, and other minerals. The exploitation of these resources, which is the main measure of the development of East Kalimantan, increased rapidly after the late 1960s. This chapter will provide the background information for this volume.

With respect to the geological features of East Kalimantan, melange and old volcanic rocks are found in the central mountain area (Chapter 8, this volume) and alluvium deposits and peat are found in the northern coastal plain of the Bulungan district, the lake area along the Mahakam River and the Mahakam River delta. Most lowlands and hills of the state are composed of sedimentary rocks. Limestone is found in the Sambaliung mountains in the central part of the state. Ultramafic rocks are found in the Meratus mountain area (see Chapter 7, this volume).

Different soil types occur in narrow belts from the northeast to the southwest. The major soils are Dystropepts in the mountain area, Tropudults (acid soils) in the plain area, Tropohemists (peats) or Tropaquepts in the swamp area, and Rendolls in the limestone area. The physical and chemical properties of the soils of the lowland dipterocarp forest, which is the main forest in East Kalimantan, have been reported in detail by Ohta et al. (1992ab, 1993, Chapter 4, this volume). Our study site (Bukit Soeharto Education Forest: BSEF) is located on hills near the east coast and is covered by Ultisols. The hills, which have an altitudinal difference of about 100 m, were elevated at the end of the Tertiary and subsequently acquired an undulating topogra-

Ecological Studies, Vol. 140
Guhardja et al. (eds.) Rainforest Ecosystems of East Kalimantan: El Niño, Drought, Fire and Human Impacts

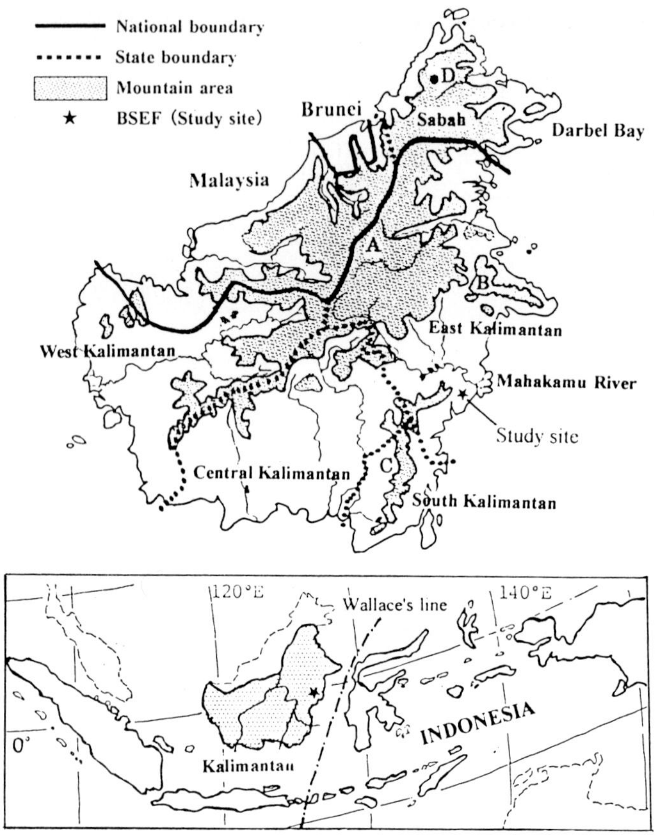


**Fig. 1.1.** Map of Indonesia and Borneo showing mountain ranges and main rivers. *A*, Central mountain ridge; *B*, Sambaliung Mts.; *C*, Meratus Mts.; *D*, Mt. Kinabalu. *Inset*: Indonesia

phy by peneplanation (Kawana 1984).

The climate of Borneo is typical of the equatorial tropics, and air temperatures are relatively constant at around 28±1°C throughout the year. The main climatic variable in Borneo is rainfall. The precipitation is higher in the central mountains and in the northwest parts than in the northeast and southeast parts of Borneo. The driest parts are the coastal lowlands in the central and southern parts of East Kalimantan and around Darvel Bay and in the south valley of Mt. Kinabalu in Sabah, Malaysia (Woods and Bower 1982, Kawana 1984). These parts have a seasonally weak drier period from August to October (during which precipitation is about 80–150 mm month) and are sometimes visited by prolonged droughts linked to El Niño Southern Oscillation events. The periodical droughts seem to play an important role in dipterocarp forest dynamics and tree growth in this area (see Chapters 3, 10, 11, and 12, this volume). The recent climatic conditions of our study site are described in detail by Toma et al. in Chapter 2 of this volume.

## 1.2 Forest Flora and Fauna in East Kalimantan

The distribution of plants, animals, and vegetation have been strongly affected by the geological history (mainly continental drift) and climatic history (mainly climatic fluctuations) in the area. These events have been described in detail (e.g., Whitmore 1990, MacKinnon et al. 1996). A brief description is given here. The flora of the Indonesia archipelago are divided by Wallace's line into Asian (Laurasian) and Australasian (Gondwanan) genera (Fig. 1.1). This line is a sharp boundary between western and eastern flora and fauna in Indonesia. Borneo is on the western side of Wallace's line. This biogeographical location of Borneo largely determines the characteristics of flora and fauna in East Kalimantan.

When all of the flora is considered, the flora of Borneo is the richest in Malesia (the equatorial area from the Malay peninsula to Melanesia). It consists of about 10000 to 15000 plant species, compared with 8500 species for the Malay peninsula, 8000 species for Sumatra, 7000 species for the Philippines, and 9000 species for New Guinea (Jacobs 1988). Thus, Borneo is often called the center of Laurasian flora in Malesia (Whitmore 1990).

Borneo has five altitudinal floristic zones (Whitmore 1984). These are the lowland dipterocarp zone (less than 300 m), the hill dipterocarp zone (300–800 m), the upper dipterocarp zone (800–1200 m), the oak-chestnut zone (1200–1500 m), and the montane ericaceous zone (more than 1500 m). Forest types in East Kalimantan include mangrove forest, peat and fresh water swamp forest, heath forest, lowland dipterocarp forest, montane forest, and limestone forest (MacKinnon and MacKinnon 1986). Most areas of East Kalimantan are covered by lowland dipterocarp forest (lowland and hill dipterocarp zones) as shown in Fig. 1.2. Our study site is located in a lowland dipterocarp forest.

The dipterocarp forest has had a long and relatively stable history. The earliest evidence of dipterocarps in Borneo is from more than 30 million years ago according to a study of fossil pollen in Sarawak (Muller 1970). The long history of the dipterocarp forest has allowed a great diversity of plants to evolve. Borneo is the center of the distribution of dipterocarps with 267 species, 60% of which are endemic (Ashton 1982). There may be as many as 200 different tree species within one hectare of East Kalimantan lowland forest. According to Kartawinata et al. (1981), the undisturbed primary forest at Wanariset (which is located near Balikpapan in the same hills as our study site) is one of the most species-rich lowland forests in South East Asia. Although the forest of our study site was partially disturbed by selective logging and wildfire, the number of tree species per ha was more than 150 in 1997 (see Chapters 9 and 10, this volume). In the richest forest in our study site, about 10% of all trees and about 80% of all emergent trees are dipterocarps, and about 50% of the basal area is occupied by dipterocarps.

Borneo's fauna, like its flora, also reflects the geological and climatic history of Borneo. The fauna is divided more sharply by Wallaces line than is the flora. The rich fauna in Borneo is mostly of Asian origin, although a few species are also found in the islands to the east of Wallace's line, where the fauna is dominated by species of Australian origin (MacKinnon et al 1996.). For example, 55 species of mammals

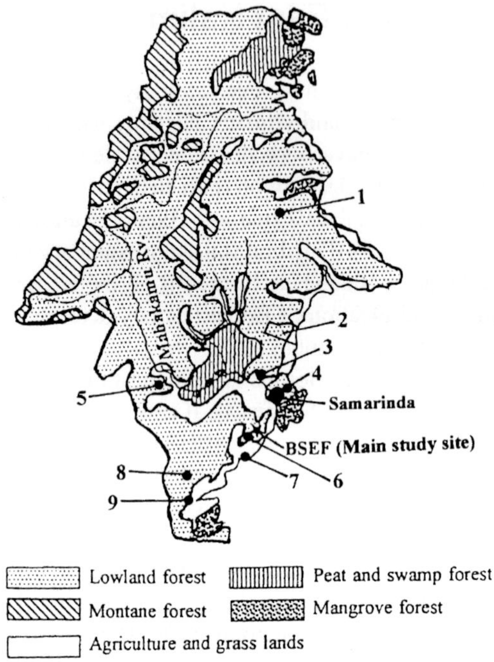


**Fig. 1.2.** Forest types in East Kalimantan and main sites referred to in this volume. 1: Berau (Chapters 3, 9, 10, 19), 2: Kutai National Park (Chapters 1, 3), 3: Sebulu (Chapters 1, 3, 9, 10), 4: Lempake (Chapters 3, 9, 10), 5: Mencimai, Barong Tongkok (Chapters 6, 8, 16), 6: Wanariset (Chapters 1, 3, 9, 10), 7: Balikpapan (Chapters 1, 2, 17), 8: Northern Pasir (Chapter 18), 9: Kuaro (Chapter7) (Source of Forest Types: ODA/Ministry of Transmigration, 1990)

are common between Borneo and Sumatra, which is located about 800 km west of Borneo, but only 9 species of mammals are common between Borneo and Sulawesi, which is located about 200 km east of Borneo. Faunal distribution in Borneo is also affected by altitude, habitat types, and geographical barriers such as mountain ranges and rivers. For example, orangutans naturally inhabit only the area on the north side of the Mahakam River in East Kalimantan.

In our study site, there are about 100 species of mammals (Yasuma 1994). This is about 45% of the 222 species that are known to exist in all of Borneo. Nine of the 13 primates of Borneo have been confirmed to inhabit our study site. These include gibbons (see Chapter 20, this volume), pig-tailed macaques, long-tailed macaques, slow lorises, etc. At least 150 species of birds have been identified in this area. The insects are also diverse. For example, more than 550 species of the family Cerambycidae (longicorn beetles) were found in a restricted area (about 1000 ha) of our study site (Makihara 1999). This number is more than half of the total number of Cerambycid species in all of North America. The species richness in tropical rain forests is due to the fact that various animals can share relatively small spaces horizontally and vertically and can be active at different times of the day and the year.

## 1.3 Forestry Activity

The Indonesian government enacted the Basic Forestry Law and Foreign Capital Investment Law in 1967 and the Domestic Capital Investment Law in 1968 (Riswan and Hartanti 1995). These laws aimed at state ownership of all forest lands and acceleration of investment for forestry activity. In 1968, forests covered an estimated 77% of Kalimantan (41.5 million ha), which was about 34% of the total forest area of Indonesia at that time (Ave and King 1986). By 1990, when more accurate baseline data were obtained, forests were estimated to cover only 63% of Kalimantan (34.7 million ha) (Ministry of Forestry (MoF)/FAO 1991). This represents a loss of 7 million ha of forest over 20 years.

From the late 1960s, large-scale forest exploitation of East Kalimantan has been continuing. The interest in exploiting the natural forests of East Kalimantan was based not only on their great extent but also on the relative accessibility of the hinterland. The number of forest concessions was 13 in East Kalimantan in 1969, but increased to over 100 in 1980 and reached 118 in 1996. In 1997, the forest area granted for the concessions was 13.4 million ha, about 68% of the total forest lands and 92% of the lands in non-convertible forest (i.e., non-exploitable forest in the government plan), in East Kalimantan (Table 1.1, MoF 1998).

Log production in East Kalimantan has been about 3–5 million $m^3$ $year^{-1}$ during the past 20 years, which accounts for about 20% of the entire production in Indonesia (Table 1.2). The seventh Development Plan (for the years 2000 to 2004) has set a log production target of 57.2 million $m^3$ $year^{-1}$ for the whole of Indonesia. This is 1.5 times the production in the present plan and about twice the production in 1996. According to the plan, most of this large increase will come from industrial forests (22.1 million $m^3$ $year^{-1}$), and from natural forests (18.3 million $m^3$ $year^{-1}$). Although the plan will be reassessed because of the economic crisis of 1997–98, the produc-

**Table 1.1.** Land utilization forest plan (1000 ha)

| | E. Kal. | Indonesia |
|---|---|---|
| Total land area | 21 144 | 192 739 |
| Protection forest | 2 935 | 34 630 |
| Reservation forest | 2 166 | 19 982 |
| Limited production forest | 4 755 | 23 863 |
| Non-convertible forest | 4 727 | 34 765 |
| Sub-total | 14 585 | 113 241 |
| Convertible forest | 5 151 | 42 639 |
| Total forest land | 19 736 | 155 880 |

E. Kal.: East Kalimantan (Source: MoF 1998).

**Table 1.2.** Log production over a 5-year period (1000 $cm^3$)

| | 1992 | 1993 | 1994 | 1995 | 1996 |
|---|---|---|---|---|---|
| E. Kal. | 5 578 | 4 482 | 4 700 | 4 591 | 4 837 |
| Indonesia | 28 267 | 26 848 | 24 027 | 24 850 | 29 069 |

E. Kal.: East Kalimantan (Source:MoF, 1998).

tive capacity of timber processing mills and upgrading the living standards of the country will require more log production than the present amount of about 25 million $m^3$ $year^{-1}$. This means that deforestation of natural forests will continue.

The Indonesian government has recommended the application of a polycyclic or selective cutting system which is based on repeated logging on a cutting cycle of 25 to 40 years. In principle, only fully matured trees of commercial value are felled and extracted and smaller trees (less than 50 cm in DBH in a normal production forest and 60 cm in a limited production forest) are left to mature for the next harvest. This selective cutting system is designed on the assumption that sufficient numbers of good-sized trees remain so that natural regeneration can take place. It is doubtful that the cutting cycle of 25 to 40 years is long enough to allow for regeneration of commercially valuable trees. It may take 100 years or more for some canopy trees to mature in the forests that have been disturbed or cut selectively (Riswan et al 1985, Toma et al 1997, Chapters 18 and 19, this volume).

Moreover, the system is rarely regulated properly and there is always short-term financial pressure to carry out repeated logging before the end of the cutting cycle that is needed for regeneration (Riswan and Hartanti 1995). Forests that were logged on a short cutting period became degraded and regeneration was poor. The government regulations make it possible to conduct clear cutting and to plant fast-growing tree species in these low productive forests. Presently, large plantation areas of young Acacia trees can be seen in the timber concession areas.

The government program for making industrial plantations of forest trees (HTI: Indonesian acronym for an industrial plantation) started in 1984 and reached about 100 000 ha $year^{-1}$ during the 4th Development Plan (1989 to 1994) but it presently remains at one third of the maximum level. In the future, HTI is expected to be the main source of log production. HTI consists of three categories: there is an HTI for pulp wood, which is based mainly on *Acacia mangium*, an HTI for non-pulp timber, which is based on native timber species, and an HTI for improving the lives of transmigrants by employing them as forestry workers. The area of HTI for pulp wood is the largest. The reforested area during the last 5 years has been about 1000 to 2000 ha $year^{-1}$ in East Kalimantan. In the *Acacia* plantations, leaching of basic cations from the soil is much higher than in dipterocarp forests (see Chapter 5, this volume) and rapid reduction of soil fertility is feared.

Planting of fast growing species such as those in the genera *Acacia*, *Eucalyptus*, *Gmelina, Paraserianthes*, etc. and important timber species such as teak (*Tectona grandis*) and *Pinus merkusiana* can be successfully conducted on open land after the clear cutting of forests. However, in degraded forests, rehabilitation techniques using many indigenous species including dipterocarps have not yet been implemented, except for a few species such as *Shorea leprosula, Peronema canescens*, etc. Intensive studies of the silvicultural characteristics of dipterocarps (e.g., Chapters 13 and 24, this volume) and various planting trials have been conducted by many researchers and teams to develop improved reforestation techniques (e.g., part V, this volume). One of the findings of these studies is that growth of dipterocarp seedlings is largely stimulated by ectomycorrhiza, and as a result, inoculation of the mycorrhiza has commonly been practiced in East Kalimantan (see Chapters 21 and 22, this volume).

## 1.4 Forest Conversion and Human Activity

East Kalimantan has a population of about 2.3 million that is mainly distributed in the coastal lowlands, the fertile Mahakam valley and upland river valleys. The population density in East Kalimantan (11 people $km^{-2}$) is very low compared with that in Java (more than 800 people $km^{-2}$). Therefore, the Indonesian government designated East Kalimantan as one of the prime targets for exploitation of the land. Since East Kalimantan has rich natural resources as mentioned above, exploitation of natural reserves has been accelerated during the last 30 years (Ministry of Transmigration (MoT) 1990). Accordingly, rapid deforestation occurred as a result of logging, land clearance for plantations, and human settlement by government-sponsored and spontaneous migrants.

Before the time of extensive development, the traditional peoples of Borneo, such as the Dayaks, had long practiced swidden (slash-and-burn) agriculture, an agriculture regime that can be sustained on poor soils when practiced at low densities and with long fallow periods (Ave and King 1986, Chapters 15 and 16, this volume). However, recent population growth and extensive forest exploitation has resulted in decreasing forest resources and land areas for farming. These conditions then forced the local people to extend the cultivation period and shorten the fallow period beyond tolerable levels. The soils in Kalimantan, unlike those of Java, are generally poor and unsuitable for intensive agriculture.

In addition, large-scale and government-sponsored transmigration began in 1950. In the whole country, half of all the transmigrants until the present moved during the 3rd Development Plan (1979–84) when almost 366 000 families (about 1.8 million people) moved with government assistance from densely populated islands to sparsely populated ones such as Sumatra and Borneo (Riswan and Hartanti 1995). Although transmigration has not been wholly successful, it is still going on. The current (1994–99) Development Plant for East Kalimantan calls for the transmigration of 24 800 families and clearing of 55 560 ha.

Furthermore, it has been estimated that, in East Kalimantan, the number of transmigrants who have come without government assistance is two or three times greater than the number of government-sponsored transmigrants. The expanding populations of the preexisting peoples and the transmigrants have resulted in con-

**Table 1.3.** Forest fires in East Kalimantan (ha)

| Year | E. Kal. | Indonesia | Year | E. Kal. | Indonesia |
|---|---|---|---|---|---|
| 1983 | - | - | 1991 | 4 693 | 118 881 |
| 1984 | 0 | 15 079 | 1992 | 3 526 | 14 531 |
| 1985 | 0 | 42 570 | 1993 | 350 | 40 897 |
| 1986 | 0 | 22 038 | 1994 | 3 963 | 161 798 |
| 1987 | 233 | 49 323 | 1995 | 34 | 6 705 |
| 1988 | 0 | 17 661 | 1996 | 301 | 10 356 |
| 1989 | 0 | 15 885 | 1997 | 21 483 | 263 991 |
| 1990 | 1 000 | 25 573 | 1998 | 51 9761[a] | - |

E. Kal.: East Kalimantan. a: Until May 1998.
Source: MoF 1997 and Forestry Department of E. Kalimantan.

version of virgin forest to agricultural lands. Shifting cultivation without long fallow periods and growing inappropriate crops has caused a rapid loss of soil fertility (see Chapter 6, this volume). As a result, the farmers eventually abandoned their degraded fields. Consequently, bushlands and alang-alang (*Imperata cylindrica*) grasslands are increasing by 10 000 ha each year (MacKinnon et al. 1996, Chapter 17, this volume).

Other types of forest conversion include the development of oil palm, rubber, and cocoa plantations. In East Kalimantan, these developments started in the 1980s and presently there are 52000 ha of oil palm plantations and 55000 ha of rubber plantations. These areas currently account for about 1/10 of the plantation areas in southern Sumatra, but oil palm plantations are rapidly increasing and are proposed to expand to 142 000 ha during the 7th Development Plan (2000–2004). During the last few years, the rate of conversion of forest land to agricultural plantations has been 30 000–45 000 ha year$^{-1}$. Similarly, the rate of conversion to farmlands for transmigrated farmers has been 12 000–14 000 ha year$^{-1}$. The current forest land use plan is shown in Table 1.1.

As a result of the marked increase in logging and agricultural lands, the frequency with which fire is used, and the areas in which fire is used, have greatly increased because land preparation by fire is the most efficient in cost and labor and the ash also supplies mineral elements. This situation has resulted in large wildfires during droughts, especially during prolonged droughts which are usually connected with El Niño Southern Oscillation events. There have been several large-scale wildfires in East Kalimantan since the early 1980s (Table 1.3). These include the fires of 1982–83, 1987, 1991–92, 1994, and 1997–98, which have resulted in significant economic losses and severe damage to ecosystems (Ministry of Environment (MoE) and UNDP 1998). The burned forest areas in East Kalimantan during the last 15 years are shown in Table 1.3. The burned forest area in 1982–83, which was one of the largest fires this century, was not recorded, but it was estimated that the area affected by the fire was about 3.6 million ha in East Kalimantan. Of the total area,

0.8 million ha was in primary rain forest, 1.4 million ha was in logged-over forests (mainly in the vicinity of settlement areas), 0.75 million ha was in secondary forests, and 0.55 million ha was in peat swamp biomes (Lennertz and Panzer 1984).

Another large fire occurred in 1997–98. The burned forests of East Kalimantan in 1998 that were reported to the Ministry of Forestry had a total area of 520 000 ha. Of this area, 85 000 ha was in national park and protected forests, 423 000 ha was in productive forests (of which, 315 000 ha was in logging forests and 108 000 ha was in industrial plantations), and 11 000 ha was in bushlands and grasslands. The government records are restricted to state forests and do not include private lands. The estimation of the total burned area in East Kalimantan must rely on satellite information. The distributed area of hot spots that were detected by NOAA satellites in 1997–98 (FFPMP-JICA 1999) was almost the same as the area assumed to be burned in the 1982–83 fires (3.6 million ha in East Kalimantan). These fires have severely damaged the forest ecosystems in plant communities (see Chapters 3 and 10, this volume) and in insect communities (see Chapter 14, this volume).

## 1.5 Concluding Remark

About 90% of East Kalimantan is still classified into forest lands according to the present Land Utilization Plan (Table 1.1, MoF 1998). In this plan, non-convertible forest is 76% of the total forest area. However, most forests in the coastal lowlands are presently degraded secondary forests and forest degradation is accelerating due to recent and frequent wildfires. About 7 million ha of undisturbed forests were lost in Kalimantan between 1968 and 1990 (Ave and King 1986, MoF/FAO 1991). The decrease of forest area is due to active transmigration and conversion of forest to agricultural land during this period. Rapid exploitation has resulted in the extinction of species that might otherwise have economic value in the future, the loss of genetic resources, increase in erosion and soil compaction, the loss of water-reservoirs, and impacts on the local and global climates (Riswan and Hartanti 1995). Partitioning the forest with roads, farmlands, and mono-culture plantations is also threatening the survival of many forest animals. Furthermore, a lack of proper reforestation/rehabilitation technology has resulted in a delay in the recovery of the degraded forests.

It is essential that we understand the function and dynamics of ecosystems and the roles that they play at many scales, from local communities to the global environment. Such an understanding will allow sustainable development and management of forest ecosystems.

## References

Ashton PS (1982) Dipterocarpaceae. Flora Malaya Ser I 9:237–552
Ave J B, King V T (1986) Borneo: The people of the weeping forest; tradition and change in Borneo. National Museum of Ethnol, Leiden

FFPMP-JICA (Forest Fires Prevention Management Project-JICA) (1999) Annual report of the FFPMP. Bogor, Indonesia (Unpublished)

Jacobs M (1988) The tropical rainforest. A first encounter. Springer, Berlin, Heidelberg

Kartawinata K, Abdulhadi R, Partomihardjo T (1981) Composition and structure of a lowland dipterocarp forest at Wanariset, East Kalimantan. Malay For 44:397–406

Kawana A (ed) (1984) Relationship between the tropical rain forest and the people (Final report of joint research) JICA (EXF, JR, 84–08) (Unpublished)

Lennertz R, Panzer KF (1984) Preliminary assessment of the drought and forest fire damage in Kalimantan Timur. Report by DFS German For Inven Serv Ltd for GTZ, Samarinda

MacKinnon J, Artha MB (1981) National conservation plan for Indonesia Vol 5: Kalimantan. FAO, Bogor

MacKinnon J, MacKinnon K (1986) Review of the protected areas system in the Indo-Malayan Realm. IUCN, Gland

MacKinnon K, Hatta G, Halim H, Mangali A (1996) The ecology of Kalimantan. Periplus Editions, Singapore

Makihara H (1999) Atlas of longicorn beetles in Bukit Soeharto Education Forest, Mulawarman University, East Kalimantan, Indonesia. PUSREHUT Special Publication No. 7, Samarinda, Indonesia

MoE, UNDP (1998) Forest and land fires in Indonesia. Vol 1, 2. Ministry of Environment and UNDP, Jakarta

MoF (1998) Forestry statistics of Indonesia, 1996/97. Bureau of Planing, Ministry of Forestry, Jakarta

MoF/FAO (1991) Indonesia national forestry action plan (country brief). Ministry of Forestry, Jakarta

MoT (1990) The land resources on Indonesia. ODA/Ministry of Transmigration, Jakarta

Muller J (1970) Palynological evidence on early differentiation of angiosperms. Biol Rev 45:417–450

Ohta S, Effindi S (1992a) Ultisols of "lowland dipterocarp forest" in East Kalimantan, Indonesia I. Morphology and physical properties. Soil Sci Plant Nutr 38:197–206

Ohta S, Effindi S (1992b) Ultisols of "lowland dipterocarp forest" in East Kalimantan, Indonesia II. Status of carbon, nitrogen, and phosphorus. Soil Sci Plant Nutr 38:207–216

Ohta S, Effindi S, Tanaka N, Miura, S (1993) Ultisols of "lowland dipterocarp forest" in East Kalimantan, Indonesia III. Clay minerals, free oxides, and exchangeable cations. Soil Sci Plant Nutr 39:1–12

Riswan S, Hartanti L (1995) Human impacts on tropical forest dynamics. Vegetatio 121:41–52

Riswan S, Kenworthy JB, Kartawinata K (1985) The estimation of temporal processes in the tropical rain forest: a study of primary mixed dipterocarp forest in Indonesia. J Trop Ecol 1:171–182

Toma T, Matius P, Hastaniah, Kiyono Y, Watanabe R, Okimori Y (1997) Floristic composition and stand dynamics in logged-over and burnt lowland dipterocarp forests in Bukit Soeharto, East Kalimantan. In: Suhartoyo H, Fatawi M (Eds) Proc 2nd Intern Symp Asean Trop For Manag, PUSREHUT Univ Mulawarman, Samarinda, pp 198–206

Woods TN, Bower RP (1982) Rainfall records, East Kalimantan: Analysis summaries and histograms. Transmigration Area Development Project, GTZ, Samarinda

Whitmore TC (1984) Tropical rain forests of the far east, 2nd edn. Clarendon, Oxford

Whitmore TC (1990) An introduction to tropical rain forest. Clarendon, Oxford

Yasuma S (1994) An invitation to the mammals of East Kalimantan. PUSREHUT special publication No.3, Mulawarman University, Samarinda

# 2 Climate in Bukit Soeharto, East Kalimantan

TAKESHI TOMA, MARJENAH, and HASTANIAH

## 2.1 Introduction

The coastal area of East Kalimantan, where the Bukit Soeharto Education Forest (BSEF) of Mulawarman University is situated, has a tropical rainforest climate and is known to be drier than the other parts of the island of Borneo (Whitmore 1984, Mackinnon et al. 1996, Walsh 1996b). The description of the climate in East Kalimantan is mainly based on the rainfall data recorded at several meteorological stations and there is little published information about climatic characteristics for East Kalimantan other than rainfall. In this chapter, we present some climatic characteristics, such as rainfall, solar radiation, air temperature and air humidity in the BSEF. Our data includes severe drought periods related to a pronounced El Niño Southern-Oscillation (ENSO) event in 1997–98.

The BSEF has suffered two exceptionally catastrophic events over the last 15 years. In East Kalimantan, an unusually prolonged and severe drought lasted from June 1982 to April 1983 and was linked to a particularly pronounced ENSO event (e.g., Goldammer et al. 1996). During the ENSO event, there were two rainless periods from July to October in 1982 and January to April in 1983 (Fig. 2.1), with large wild fires occurring during these rainless periods. About 3.5 million ha of rainforest in East Kalimantan were severely damaged by the drought and drought-linked fire (Leighton and Wirawan 1986). Fourteen years after the drought and fire events in 1982–83, another severe ENSO event occurred, lasting from 1997 to 1998. This ENSO event was longer and more pronounced than the ENSO event in 1982–83. Two rainless periods, July to October 1997 and January to April 1998, occurred in East Kalimantan during this ENSO event. The occurrence of two rainless periods during this ENSO event was quite similar to those during the pronounced ENSO event in 1982–83. During the rainless periods in 1997 and 1998, huge areas of rainforests in East Kalimantan were affected by drought and drought-linked fire (Chapter 3, this volume).

Meteorological observations in the BSEF have been conducted since 1988. There is no detailed information about climatic characteristics during the drought and fire events in 1982–83. Here we present detailed information about consecutive changes of the microclimatic characteristics before, during, and after the extraordinary drought and fire events in 1997–98 in the BSEF.

Ecological Studies, Vol. 140
Guhardja et al. (eds.) Rainforest Ecosystems of East Kalimantan: El Niño, Drought, Fire and Human Impacts

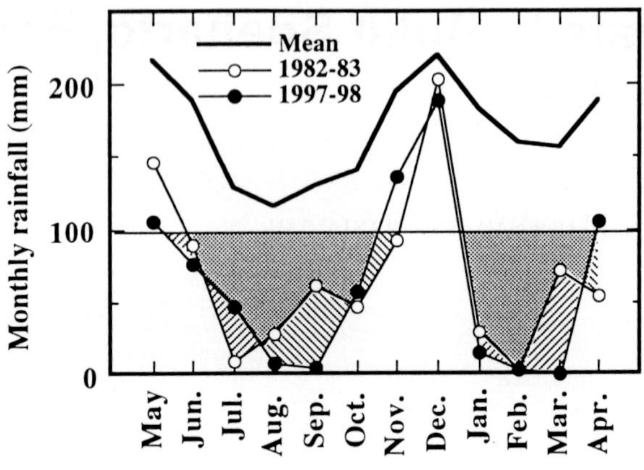


**Fig. 2.1.** Monthly rainfall at Samarinda, East Kalimantan, Indonesia (Meteorological Station at Temindung Airport). The *thick line* represents average monthly rainfall from 1978 to 1998. The *lines with open and closed circles* represent monthly rainfall in 1982–83 and 1997–98, respectively. *Shaded areas* represent the periods when monthly rainfall was less than 100 mm. 100 mm of monthly rainfall is the threshold between wet and dry months

## 2.2 General Characteristics

In a general sense, the climate in the BSEF is characteristic of a tropical rainforest region: hot and wet throughout the year (Table 2.1). From 1988 to 1998, the average annual rainfall was 2002 mm. The monthly distribution of rainfall was bimodal, with peaks of over 200 mm per month occurring both in May and December. The minimum and maximum average monthly rainfalls were 85 mm (September) and 248 mm (December), respectively. Average annual total evaporation from 1988 to 1991 measured by an evaporation pan (20 cm in diameter) was 1,273 mm per year, and the minimum and maximum average monthly evaporation were 75 mm (November) and 145 mm (February), respectively. The yearly mean air temperatures were 29.9°C for daily maxima and 21.4°C for daily minima, and the mean air temperature differed little between months. Diurnal ranges of air temperature were about 8.5°C and the changes in daily maximum and daily minimum temperatures between months were much smaller than the daily range. The yearly mean in relative air humidity was 93.2% for daily maxima and 58.5% for daily minima. The average daily range of relative humidity was larger than that between months.

**Table 2.1.** Average of monthly rainfall and evaporation, monthly means of daily maximum and minimum air temperature, and monthly means of daily maximum and minimum relative humidity at BSEF Research station, from 1988 to 1998[a]

| | Month | | | | | | | | | | | | |
|---|---|---|---|---|---|---|---|---|---|---|---|---|---|
| | Jan. | Feb. | Mar. | Apr. | May | Jun. | Jul. | Aug. | Sep. | Oct. | Nov. | Dec. | |
| | | | | | | | | | | | | | total |
| Rainfall (mm) | 171 | 172 | 185 | 183 | 229 | 142 | 141 | 108 | 85 | 140 | 199 | 248 | 2002 |
| Evaporation[b] (mm) | 121 | 87 | 145 | 119 | 121 | 90 | 104 | 113 | 116 | 82 | 75 | 101 | 1273 |
| Temperature(°C) | | | | | | | | | | | | | mean |
| Maximum | 29.5 | 29.3 | 30.3 | 30.6 | 29.9 | 29.6 | 29.1 | 30.0 | 30.7 | 30.6 | 29.9 | 28.8 | 29.9 |
| Minimum | 21.7 | 21.3 | 21.9 | 21.8 | 21.4 | 21.1 | 21.0 | 21.0 | 20.7 | 21.7 | 21.3 | 21.5 | 21.4 |
| Relative Humidity (%) | | | | | | | | | | | | | mean |
| Maximum | 95.4 | 94.9 | 94.7 | 93.0 | 93.5 | 93.3 | 92.1 | 93.2 | 92.5 | 91.0 | 91.6 | 92.8 | 93.2 |
| Minimum | 63.0 | 59.7 | 57.7 | 55.1 | 62.7 | 62.0 | 61.5 | 56.2 | 53.7 | 53.7 | 57.7 | 59.0 | 58.5 |

[a]Periods and sources: 1988–90, Okimori (1991); 1991, Hastaniah and Morisada (1992); 1992, Hastaniah (1993); 1993–1994, Kiyono and Hastaniah (1996); 1995–1998, the authors.
[b]Evaporation was measured from 1988 to 1991.

**Table 2.2.** Monthly total rainfall values (mm) at BSEF research station[a]

| | Month | | | | | | | | | | | | |
|---|---|---|---|---|---|---|---|---|---|---|---|---|---|
| | Jan. | Feb. | Mar. | Apr. | May | Jun. | Jul. | Aug. | Sep. | Oct. | Nov. | Dec. | total |
| 1988 | 239 | 170 | 103 | 148 | 132 | 102 | 143 | 285 | 189 | 98 | 281 | 295 | 2185 |
| 1989 | 273 | 298 | 354 | 253 | 149 | 195 | 133 | 159 | 161 | 195 | 370 | 338 | 2878 |
| 1990 | 187 | 87 | 248 | 288 | 366 | 76 | 269 | 61 | 68 | 77 | 68 | 337 | 2132 |
| 1991 | 141 | 125 | 237 | 186 | 274 | 116 | 25 | 4 | 22 | 51 | 145 | 208 | 1533 |
| 1992 | 189 | 44 | 183 | 147 | 355 | 137 | 211 | 64 | 97 | 130 | 271 | 162 | 1990 |
| 1993 | 75 | 138 | 149 | 169 | 247 | 177 | 44 | 33 | 50 | 84 | 65 | 113 | 1345 |
| 1994 | 106 | 144 | 209 | 118 | 203 | 82 | 12 | 30 | 0 | 148 | 136 | 211 | 1398 |
| 1995[b] | 108 | 172 | 140 | 296 | 178 | 331 | 168 | 298 | 119 | 144 | 184 | 209 | 2346 |
| 1996 | 197 | 473 | 114 | 206 | 156 | 182 | 130 | 158 | 90 | 316 | 207 | 243 | 2471 |
| 1997 | 297 | 244 | 256 | 176 | 186 | 51 | 105 | 9 | 1 | 51 | 190 | 243 | 1808 |
| 1998 | 1 | 0 | 0 | 137 | 218 | 302 | 335 | 278 | 171 | 250 | 256 | 333 | 2281 |
| mean | 171 | 172 | 185 | 183 | 229 | 142 | 141 | 108 | 85 | 140 | 199 | 248 | 2002 |
| sd | 92 | 137 | 99 | 53 | 82 | 74 | 105 | 106 | 70 | 89 | 98 | 78 | 491 |

[a]Sources: 1988–90, Okimori (1991); 1991, Hastaniah and Morisada (1992); 1992, Hastaniah (1993); 1993–1994, Kiyono and Hastaniah (1996); 1995–1998, the authors.
[b]Data in March to July were taken at Samarinda airport, and excluded from mean calculation.

## 2.3 Rainfall

### 2.3.1 Variations of Monthly Rainfall

Mohr (1933, cited in Whitmore 1984) developed a scheme to characterize monthly rainfall for Indonesia. He divided the monthly rainfall during the year into three categories; the wet, the moist and the dry months. Under his system, if monthly rainfall is more than 100 mm, precipitation predominates over evaporation, so the month is classified as wet. With monthly rainfall of less than 60 mm, evaporation from the ground exceeds the amount of precipitation, so the month is classified as dry. With monthly rainfall between 60 mm and 100 mm, there will be equilibrium between evaporation and precipitation, so the month is classified as moist. In this paper, we classified monthly rainfall into Mohr's three categories: wet, moist, and dry.

In Bukit Soeharto, the total monthly rainfall changed from year to year as did the annual rainfall (Table 2.2). From 1988 to 1998, the minimum and maximum value of the annual precipitation recorded in the BSEF was 1345 mm in 1993 and 2878 mm in 1989, respectively. Since the coefficient of variation of annual rainfall in the BSEF (24.5%) was larger than those (ca. 20%) in wet tropics described in Walsh (1996b), the BSEF was subject to frequent occasional drought compared to other wet tropical regions. Moist and/or dry months were unexpectedly found in some years in the BSEF. Furthermore, there were consecutive dry months in 1991, 1993, 1994, 1997 and 1998. The rainfall type in the BSEF belongs to Type B (slightly seasonal) according to Whitmore (1984) and to the Tropical Wet (relatively frequent short dry periods or very short dry season) according to Walsh (1996b).

### 2.3.2 Drought Periods

Because the BSEF has been subjected to occasional droughts during the last 11 years, we examined the drought period by counting the consecutive rainless days, and by calculating the running 30-day rainfall totals adopted by Brünig (1969). The running 30-day rainfall total on a day is the sum of the precipitation of the preceding 30 days. We divided the various running 30-day rainfall totals into three categories: wet, moist and dry days. If the running 30-day rainfall total of a day is more than 100 mm the day is wet. If the running 30-day rainfall total of a day is less than 60 mm, the day is dry. A running 30-day rainfall total of between 60 mm and 100 mm indicates a moist day.

Half (50%) of consecutive rainless periods (total 655) from 1988 to 1998 were only 1 day, and 94% of consecutive rainless periods were shorter than one week. At the BSEF the average (± 1 SD) duration of consecutive rainless periods was 2.9 (± 5.6) days for the monitoring period. There were three exceptionally long consecutive rainless periods lasting longer than 5 weeks between 1988 and 1998: August to September in 1994, July to August in 1997 and January to April in 1998. Such long rainless periods were rare events in the BSEF. The longest rainless period was 105 days, lasting from 2 January to 15 April 1998.

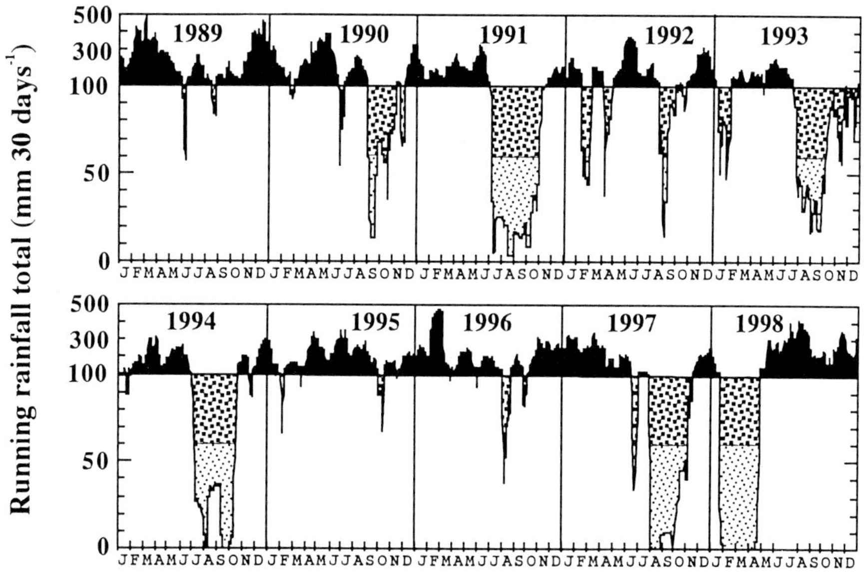


**Fig. 2.2.** Running 30-day rainfall totals in the Bukit Soeharto Education Forest (BSEF) from 1989 to 1998. The running 30-day rainfall total on a day is the sum of the precipitation of the preceding 30 days. The 30-day rainfall totals exceeding 100 mm are shown at 1/10 scale. The *solid, large and small dotted areas* represent consecutive periods when the 30-day rainfall totals were more than 100 mm (wet), between 60 mm to 100 mm (moist), and less than 60 mm (dry), respectively. The figures in each *box* represent the calendar year

Consecutive dry days, according to the running 30-day rainfall totals (Fig. 2.2) occurred even in wet months. Thus, occasional droughts occurred even in wet months. Consecutive dry days lasting longer than one month occurred in 1991, 1993, 1994, 1997 and 1998. Days with 0 mm in the running 30-day rainfall totals occurred in August and September 1994, in August and September 1997 and from January to April 1998. High mortality in *Macaranga* spp. trees was observed during the droughts in 1991, 1994 and 1997 (Chapter 10, this volume). During the prolonged drought periods in 1991, 1994, 1997 and 1998, wild fires occurred frequently in East Kalimantan.

The long and severe droughts in 1991, 1997, and 1998 in the BSEF corresponded to ENSO events. An ENSO event tends to commence in March to May and last for a year and some times longer (Walsh 1996a). Droughts would occur over the whole of Indonesia during the first year of ENSO events. A pronounced ENSO event would result in a severe drought early in the second year. In case of the pronounced 1982–83 and 1997–98 ENSO events, there were long and severe droughts both in the first and second year in East Kalimantan. There was only temporal rainfall between the droughts in the first and second years (Fig. 2.1).

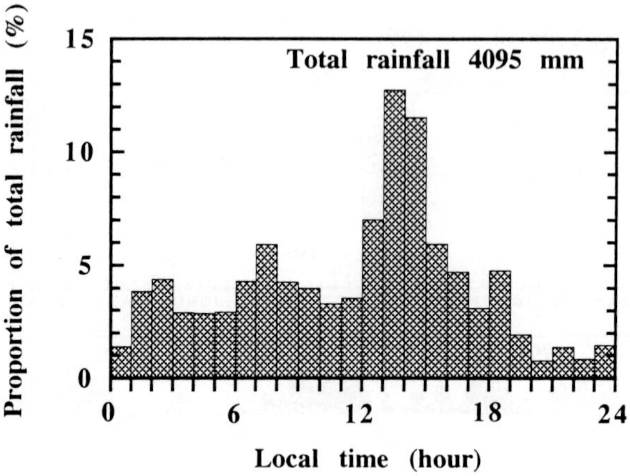


**Fig. 2.3.** Hourly precipitation distribution pattern in the BSEF from September 1996 to December 1998

## 2.3.3 Diurnal Rainfall Variations

The diurnal distribution of rain can be of ecological significance through its influence on solar radiation, sunshine, light, temperature and interception losses (Walsh 1996b). Solar radiation tends to decline with the presence of clouds associated with rainfall, leading to reduced evaporation in forests and lower photosynthetic and transpiration rates in plants. Thus, the diurnal distribution of rainfall has significant ecological implications. Figure 2.3 shows the hourly rainfall distribution from September 1996 to December 1998. Seasonal changes in the diurnal rainfall pattern were not clear through the monitoring period (data not shown). During day time (0600–1800 h), 71% of total rainfall occurred, with 45% of total rainfall occurring within the six hours of the afternoon (1200–1800 h). The highest rainfall peaks were observed during the early afternoon (1300–1400 h), the result of diurnal heating over land areas as a result of sunlight exposure. Convectional storms are caused by differential solar heating of the ground and the lower air layers, which occurs during the afternoon when warm, moist air covers an area. Much of the afternoon rainfall in the BSEF would be due to convectional storms.

## 2.3.4 Daily Rainfall Variations

A high proportion of tropical rainfall occurs during large storms with high intensity (Walsh 1996b). In the BSEF, the occurrence of high daily rainfall was less frequent, though its contribution to the total precipitation was large (Fig. 2.4). The days assigned to the low rainfall class (1–5 mm $day^{-1}$) constituted 44% of the rainy days from 1988 to 1998, but these days produced only 7% of the total precipitation for the same period. On the other hand, the rainy days belonging to the high daily rainfall class (> 25 mm $day^{-1}$) constituted only 14% of the total rainy days for the period, but these days produced 47% of the total precipitation for the period. Similar rainfall

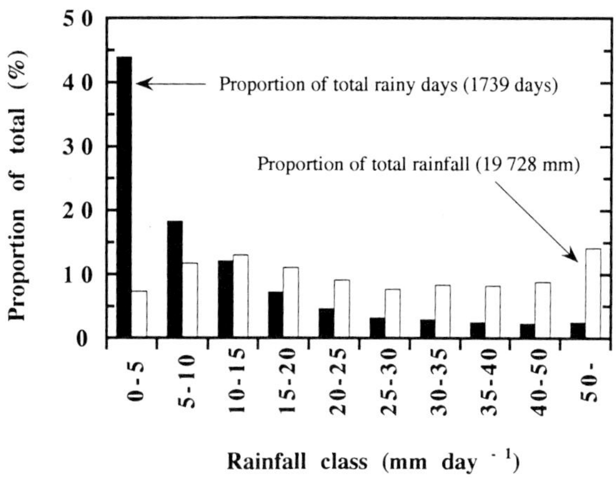


**Fig. 2.4.** Proportion of precipitation in each daily rainfall size class compared to the total precipitation (*open bars*), proportion of rainy days in each daily rainfall size class compared to the total number of rainy days (*solid bars*). From January 1988 to December 1998, in total 1739 days and 19 728 mm of total rainfall were recorded

characteristics to the BSEF were reported at Bukit Tarek in Peninsular Malaysia (Noguchi et al. 1996).

Intense rainfall as well as low vegetation cover would result in high soil erosion and nutrient leaching from soils (Longman and Jeník 1987). When the BSEF suffered surface fire during the drought of 1998, the litter accumulated on the forest floor was entirely burnt to ash and then the burnt area was covered with the ash. Many trees shed their leaves and some individual trees died after the surface fire. At the end of the drought in 1998, ash and new litter accumulated on the forest floor and the canopy of the BSEF had occasional openings compared to the canopy before the surface fire. When the drought ended in late April 1998, intense rainfall recommenced and washed away the ash. Because the intense rainfall was not intercepted by canopy leaves and reached directly to the forest floor, this lead to surface erosion. Thus, nutrient leaching and soil erosion in the BSEF were probably accelerated by the fire events and the intense rainfall after the fire events.

## 2.4 Microclimate

Two scaffolding towers were installed in the BSEF in 1996. A tower 60 m in height (60 m tower) was installed in a near-primary dipterocarp forest stand on a ridge. Top canopy height of the stand was 45 m. Another tower 30 m in height (30 m tower) was installed in a secondary forest stand on a flat river terrace. The top canopy height of the secondary stand was 20 m in 1996. The secondary stand was dominated by a pioneer tree species, *Macaranga gigantea*, regenerated after the forest fire in 1983 (Chapter 10, this volume). We have continued the hourly measurement of global radiation (GR; 400–1100 nm), photosynthetic photon flux density (PPFD), air temperature (T), and relative humidity (RH) at the top of the towers since September 1996. In addition to the measurements at the top of the towers, we measured the diurnal and vertical variations in T and RH at 15 m height intervals up the towers for a week in August 1996, April 1998 and September 1998. When the pronounced ENSO event occurred in 1997–98, there were two drought periods, from July to October 1997 and from January to April 1998. Hereafter, the first drought and the second drought referred to in this chapter represent the drought events of July to October 1997 and January to April 1998, respectively. Both the near-primary and secondary stands where the towers were installed suffered surface fire during the second drought, at the end of February 1998.

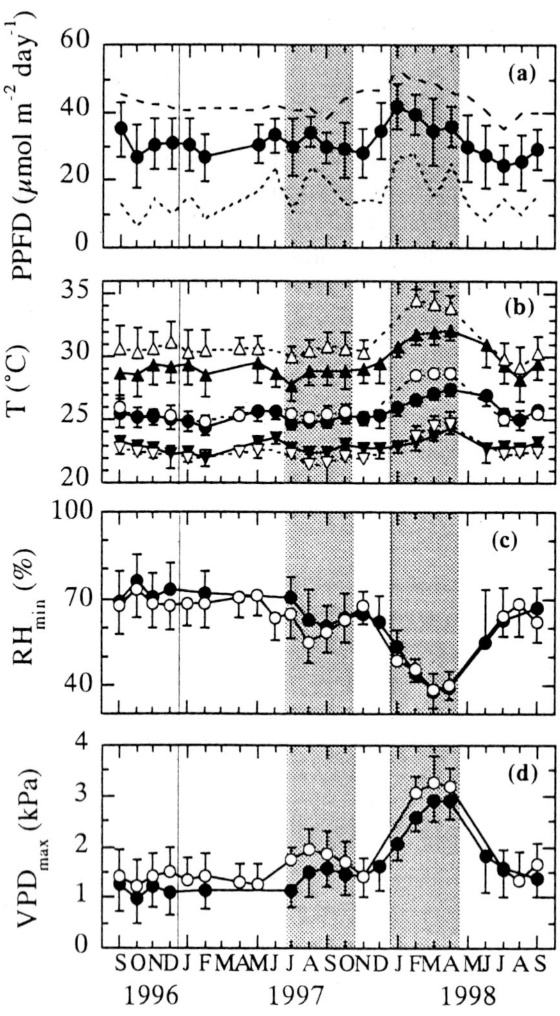


**Fig. 2.5a–d.** Monthly average values at the top of the towers. *Solid* and *open symbols* represent the values at the 60 m tower and the 30 m tower, respectively. *Error bars* indicate 1 ± SD. *Shaded areas* represent the first and second drought periods. **a** Daily totals of photosynthetic photon flux density (PPFD); *dotted lines* represent the monthly maximum and minimum values. **b** Daily maximum and minimum (*triangles*) and mean air temperature (*circles*) (T). **c** Daily minimum relative humidity ($RH_{min}$). **d** Daily maximum vapor pressure deficit ($VPD_{max}$)

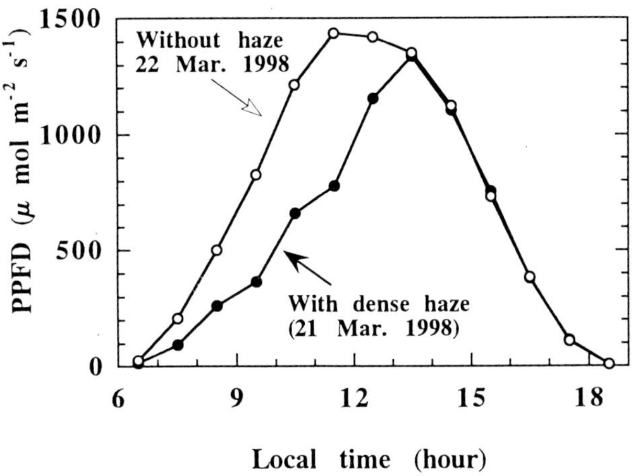


**Fig. 2.6.** Diurnal courses of PPFD on a day with dense haze in the morning to noon (*closed circles*, 21 March 1998) and on a relatively clear day (*open circles*, 22 March 1998). The values of PPFD are means of 1 h period

### 2.4.1 Radiation Above Canopy

No clear annual cycle of daily total values in GR and PPFD was recorded from September 1986 to September 1998 (Fig 2.5a). Because daily totals of PPFD and GR were well correlated (GR = 0.076 + 0.479 PPFD, $r^2 = 0.997$, GR in MJ $m^{-2}$ $day^{-1}$, PPFD in mol $m^{-2}$ $day^{-1}$), only the PPFD data is shown in Fig 2.5a. The maximum monthly mean values (± 1SD.) of daily total GR and PPFD values were 20.0 (± 3.1) MJ $^{-2}$ $day^{-1}$ and 41.6 (± 6.52) mol $m^{-2}$ $day^{-1}$, respectively, from September 1986 to September 1998. These maximum values were observed in January 1998, when the first drought started. The minimum monthly mean values of daily total GR and PPFD values were observed between June and July 1998—one to two months after the second drought. The minimum monthly mean values (± 1SD.) of daily total values of GR and PPFD were 12.6 (± 4.6) MJ $^{-2}$ $day^{-1}$ and 24.5 (± 5.9) mol $m^{-2}$ $day^{-1}$, respectively.

Day-to-day changes in solar radiation were greater than the month-to-month changes (see the monthly maximum and minimum values of daily PPFD in Fig. 2.5a). We estimated that daily total values of GR and PPFD on an entirely cloudless day were 31.2 MJ $m^{-2}$ $day^{-1}$ and 65 mol $m^{-2}$ $day^{-1}$, respectively. These values were calculated by the maximum of instantaneous values measured at every hour during six months (January, February, and from May to August in 1997). Since the mean values of daily total of GR and PPFD were 15.6 MJ $m^{-2}$ $day^{-1}$ and 31.4 mol $m^{-2}$ $day^{-1}$, respectively, from September 1996 to September 1998, some 50% of solar radiation was intercepted by clouds. The maximum and minimum daily total of GR was 24.9 and 2.7 MJ $^{-2}$ $day^{-1}$, and the maximum and minimum daily total of PPFD was 52.2 and 5.9 mol $m^{-2}$ $day^{-1}$, respectively, from September 1996 to September 1998. Thus even on the observed clearest day about 20% of solar radiation was intercepted by cloud, and over 90% of solar radiation could be intercepted by cloud on a cloudy day when the minimum values of PPFD and GR were recorded. Thus, the daily total values of GR and PPFD depended on cloud cover rather than on seasonal changes in the solar declination.

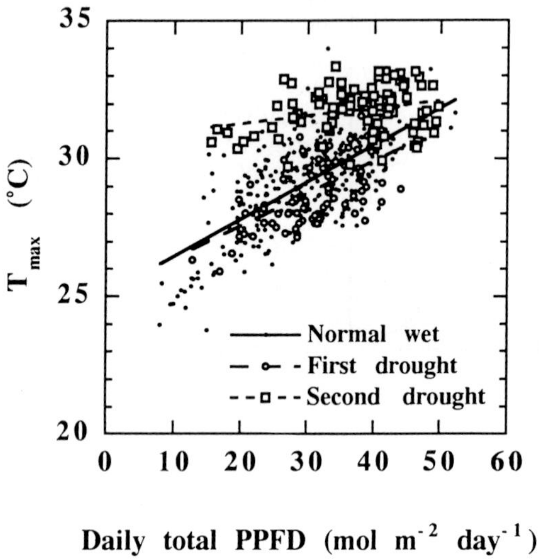


**Fig. 2.7.** Relationships between daily total PPFD and daily maximum temperature ($T_{max}$) during normal wet conditions (*dots*), the first drought (*circles*) and the second drought (*squares*). The first drought lasted from July to October 1997 and the second drought lasted from January to April 1998, related to the pronounced El Niño Southern-Oscillation (ENSO) event of 1997–98

The amount of solar radiation would be reduced not only by clouds but also by dense haze. During the first and second droughts, mainly from September to October in 1997 and February to April in 1998, large areas of East Kalimantan were affected by dense haze from wild fires and slash-and-burn land preparation. Fig. 2.6 shows reduction of hourly PPFD on a day with dense haze compare to relatively clear day. Tang et al. (1996) reported that the total daily PPFD on a day with dense haze was about 50% less than the days without visible haze in Peninsular Malaysia.

## 2.4.2 Temperature at the Top of the Towers

From September 1996 to September 1998, day-to-day variations in air temperature were larger than month-to-month variations, except for the second drought (Fig. 2.5b). There was no clear annual cycle in air temperature. Although there was no conspicuous increase in air temperature during the first drought, an increase in air temperature was found during the second drought. The monthly means of daily maximum air temperature ($T_{max}$) at the top of the 60 m tower were about 28–29°C from September 1996 to December 1997. In April 1998, the monthly mean of $T_{max}$ at the 60 m tower exceeded 32°C.

Monthly means of daily total PPFD and GR were higher during the second drought than those during the first drought and the normally wet periods (Fig. 2.5a; data not shown for GR). However, there were positive relationships between the daily totals of PPFD and $T_{max}$ during the normally wet, the first and second drought periods. At a given daily total PPFD, $T_{max}$ in the second drought period was higher than in the normally wet and the first drought periods (Fig. 2.7). Increase in $T_{max}$ (ca. 3°C) from average was also observed during the drought in February and March 1983 in Kutai National Park, East Kalimantan (Walsh 1996b). It was thought that the interaction between the foehn phenomenon and a high pressure air mass due to the strong ENSO resulted in the high temperature during the second drought (January to April) in the lowlands of Kutai, East Kalimantan (Chapter 3, this volume).

### 2.4.3 Moisture at the Top of Towers

From September 1996 to September 1998, day-to-day variations in relative air humidity were larger than month-to-month variations, except for the first and the second drought periods (Fig. 2.5c). There was no clear annual cycle in relative air humidity. The day-to-day variation in daily minimum values of relative air humidity ($RH_{min}$) was larger than that of daily maximum values of relative air humidity ($RH_{max}$). The monthly mean of $RH_{max}$ was 98% from September 1996 to September 1998, except for the second drought period when it was 89% at the top of the two towers. The monthly mean values of $RH_{min}$ at the top of the two towers were about 70% from September 1996 to September 1998, except for the first and the second drought period; during the second drought period the value was 39%. The month-to-month changes in vapor pressure deficit (VPD) were inversely related to the month-to-month changes in RH. In April 1998, the monthly mean value of $VPD_{max}$ at the 30 m and 60 m towers was 3.0 and 2.9 kPa, respectively. In April 1998, the monthly mean value of air vapor pressure was 1.9 kPa at the top of the 60 m tower. After the second drought (June and July 1998), the monthly mean value of air vapor pressure was 2.7 kPa at the top of the 60 m tower.

Because high VPD decreases photosynthetic activity due to the effect on stomatal closure, VPD is ecologically important for plants. Both net photosynthetic rate and stomatal conductance in the top canopy leaves of *Macaranga conifera* began to decrease when leaf-to-air VPD exceeded 1.5 kPa in the normally wet conditions (Ishida et al. 1999a; Chapter 12, this volume). When the upper canopy leaves are exposed to direct solar radiation, the temperature of the upper canopy leaves increased by about 7°C from the ambient air temperature for climax dipterocarp and pioneer species (Ishida et al. 1996, 1999a). For an RH value of 70%, T of 28°C, and a leaf temperature of 35°C, the air VPD and leaf-to-air VPD are 1.1 kPa and 1.7 kPa, respectively. These values of RH and T were usually found at midday at the top of the two towers even during a wet period. Thus midday depression of stomatal conductance due to high VPD would usually occur even in a wet period. During the second drought, the stomatal conductance and net photosynthetic rate decreased in the top canopy leaves of pioneer and late successional tree species even in the morning (Ishida et al. 1999b; Chapter 12, this volume).

### 2.4.4 Diurnal and Vertical Variations in Temperature and Moisture

When top canopy leaves are exposed to direct sunlight, the air temperature at the top canopy generally rises. The diurnal and vertical variations in temperature and mois-

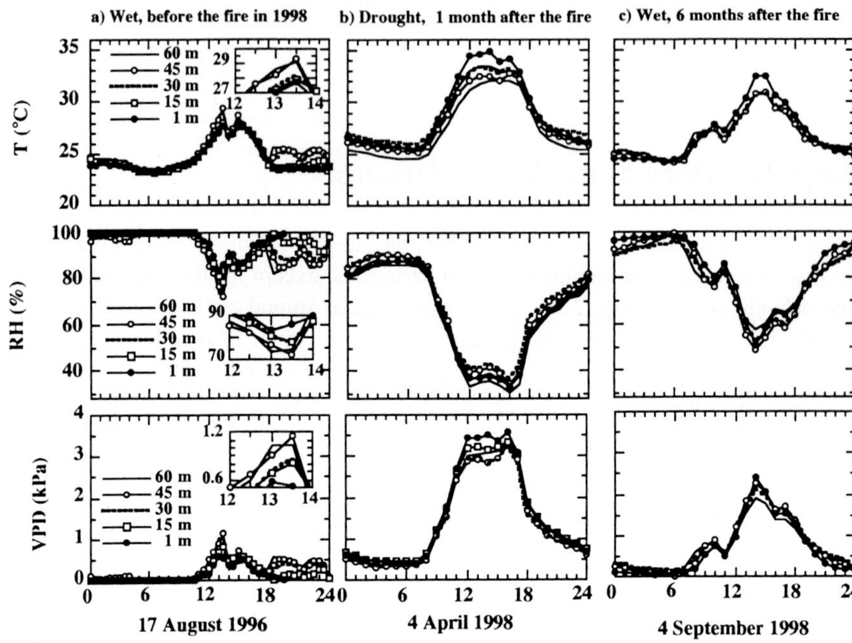


**Fig. 2.8a–c.** Daily tracks of air temperature (T), relative humidity (RH) and vapor pressure deficit (VPD) at the 60 m tower on fine days. **a** On 17 August 1996 at the 60 m tower. **b** On 4 April 1998 at the 60 m tower. **c** On 4 September 1998 at the 60 m tower. August 1996 and September 1998 were normal wet months. April 1998 was the end of the second drought. April and September 1998 were one and six months after the fire disturbance in 1998, respectively

ture were then larger on fine days than on rainy or cloudy days, because clouds reduce solar radiation and rain increases moisture. The daily tracks of T, RH and VPD on fine days at the 60 m tower are shown in Fig. 2.8. The $T_{min}$ was recorded just before dawn and T was found to increase as PPFD increased, and the $T_{max}$ was recorded in the afternoon. In contrast, RH started to decrease after sunrise and the $RH_{min}$ was recorded at the same time as the $T_{max}$ was recorded. The diurnal changes in VPD were similar to those in T, and the inverse of RH. The $VPD_{max}$ was also recorded at the same time as the $T_{max}$ and $RH_{min}$ were observed. Those time trends were similar to the 30 m tower and were not different before and after the fire.

On 17 August 1996, during a wet period, the daily variations of T, RH and VPD at or near the forest floor were smaller than those at or above the top canopy, both in the near-primary and the secondary forest stand (Fig. 2.8a; data not shown for the secondary stand). The $T_{max}$ and $VPD_{max}$ were lower and the $RH_{min}$ were higher at or near forest floor compared to those at or above top canopy level. It is well known that the daily variations in T and RH near the floor of tropical rainforests are smaller than those at the top canopy layer. T is lower and RH is higher near the forest floor than at the top canopy layer, because variations in T and RH are caused by direct solar radiation at the top canopy surface (e.g., Whitmore 1984, Kira and Yoda 1989, Walsh 1996c).

The vertical variations of T, RH, and VPD changed dramatically before and after the forest fire in 1998, especially at the forest floor. The diurnal variations in T, RH and VPD at the forest floor increased after the forest fire in 1998 (Fig. 2.8b). Because of the leaf shedding and death of individual trees which was induced by the drought and fire in 1998, the forest floor was exposed to direct solar radiation. On 4 April 1998, one month after the fire disturbance, the highest $T_{max}$ and lowest $RH_{min}$ were found at a height of 1 m, $T_{max}$ reached 35°C, $RH_{min}$ decreased to under 30%, and $VPD_{max}$ exceeded 3.5 kPa. Thus, the seedlings and saplings surviving the fire suffered high temperature and low humidity stress as they usually grow in relatively cool and moist conditions (Fig. 2.8a). On 4 September 1998, six months after the fire, $T_{max}$ at the forest floor was still higher than before forest fire (Fig. 2.8c). This means that the forest structure, especially canopy structure had not recovered to that before the fire. High light intensity and high temperature due to canopy opening give physiological stress to the seedlings growing on such a site. The interspecific differences under such hot and bright conditions are discussed by Ishida et al. (Chapter 24, this volume) and are important for the composition of regenerating tree species.

## 2.5 Conclusion

Meteorological characteristics recorded in the BSEF from 1988 to 1998 were characterized by a tropical rainforest climate with occasional droughts. Wild fires occurred during occasional droughts but the fires rarely invaded well developed forests. Wild fires spread into forests and large-scale forest fires occurred only during the occasion of severe drought from January to April in 1998. An unusual increase in air temperature and an unusual decrease in air humidity were recorded during this severe drought period, the second year of the pronounced ENSO event in 1997–98. The burned area during this period in East Kalimantan was nearly the same as the burned area during the pronounced ENSO event in 1982–83 (Chapter 3, this volume). Study on climatic characteristics will contribute to preventing another large scale forest fire in East Kalimantan. Why the unusual increase in air temperature and the unusual decrease in air humidity occurred in the second year of the ENSO event in 1997–98 is need to be analyzed.

The drought and fires in 1998 induced leaf shedding and the death of individual trees in the BSEF. The changes in canopy and stand structure of the forest resulted in changes of microclimate, especially at the forest floor. The vertical and temporal variations in microclimatic characteristics will change with time, because of the changes in the forest structure with the recovery process. Monitoring of climatic changes during the recovery processes in the burned forests will contribute to understanding the interactive relationships between trees and microclimate in forests. Since there is little information about long-term changes in forest microclimate, performing a long-term microclimateological study will provide basic understanding of the forest ecosystems in the tropics.

# References

Brünig EF (1969) On the seasonality of drought in the lowlands of Sarawak (Borneo). Erdkunde 23:127–133

Goldammer JG, Seibert B, Schindele W (1996) Fire in dipterocarp forests. In: Schlte A, Schöne D (eds) Dipterocarp forest ecosystems: Towards sustainable management. World Scientific Publishing, Singapore, pp 155–185

Hastaniah (1993) Meteorological observations at Bukit Soeharto Research Station in 1992. Annual Report of PUSREHUT 3:167–181

Hastaniah, Morisada K (1992) Meteorological observations at Bukit Soeharto Research Station in 1991. Annual Report of PUSREHUT 2:61–76

Ishida A, Toma T, Marjenah (1999a) Leaf gas exchange and chlorophyll fluorescence in relation to leaf angle, azimuth, and canopy position in the tropical pioneer tree, *Macaranga conifera*. Tree Physiol 19:117–124

Ishida A, Toma T, Marjenah (1999b) Limitation of leaf carbon gain by stomatal and photochemical process at the top canopy of *Macaranga conifera*., a tropical pioneer tree. Tree Physiol 19:467–473

Ishida A, Toma T, Matsumoto Y, Yap SK, Maruyama Y (1996) Diurnal changes in leaf gas exchange characteristics in the uppermost canopy of a rainforest tree, *Dryobalanops aromatica* Gaertn. F. Tree Physiol 16:779–785

Kira T, Yoda K (1989) Vertical stratification in microclimate. In: Leigh H, Werger MJA (eds) Tropical rain forest ecosystems. Elsevier, Amsterdam, pp 55–71

Kiyono Y, Hastaniah (1996) Slash-and-burn agriculture and succeeding vegetation in East Kalimantan. PUSREHUT Special publication No. 6, PUSREHUT, Samarinda, East Kalimantan

Leighton M, Wirawan N (1986) Catastrophic drought and fire in Borneo rainforests associated with the 1982–3 El Niño southern oscillation event. In: Prance GT (ed) Tropical rain forests and the world atmosphere, Westview, Boulder, Colorado, pp 75–102

Longman KA, Jeník J (1987) Tropical forest and its environment, 2nd edn. Longman, Essex, England

Mackinnon K, Hatta G, Halim H, Mangalic A (1996) The ecology of Kalimantan. Periplus Editions, Singapore

Mohr ECJ (1933) Debodem der tropen in het algemeen en die van Ned.-Indie in het bijzonder. Meded Kom Ver Kol Inst Afd Handelsmuseum 31 Di.I, le stuk

Noguchi S, Nik AR, Sammori T, Tani M, Tsuboyama Y (1996) Rainfall characteristics of tropical rain forest and temperate forest: comparison between Bukit Tarek in Peninsular Malaysia and Hitachi Ohta in Japan. J Trop For Sci 9:206–220

Okimori Y (1991) Final report (II). Submitted to The tropical Rain Forest Research Project, Samarinda, East Kalimantan (unpublished)

Tang Y, Kachi N, Furukawa A, Muhamad A (1996) Light reduction by regional haze and its effect on simulated leaf photosynthesis in a tropical forest of Malaysia. For Ecol Manag 89:205–211

Walsh RPD (1996a) Drought frequency changes in Sabah and adjacent parts of northern Borneo since the late nineteenth century and possible implications for tropical rain forest dynamics. J Trop Ecol 12:385–407

Walsh RPD (1996b) Climate. In: Richards PW with Walsh RPD, Baillie IC, The tropical rain forest, 2nd edn. Cambridge University Press, Cambridge, pp 159–205

Walsh RPD (1996c) Microclimate and hydrology. In: Richards PW, with Walsh RPD, Baillie IC, The tropical rain forest, 2nd edn. Cambridge University Press, Cambridge, pp 206–237
Whitmore TC (1984) Tropical rainforests of the Far East, 2nd edn. Clarendon, Oxford

# 3 Effects of Droughts and Forest Fires on Dipterocarp Forest in East Kalimantan

Tokunori Mori

## 3.1 Introduction

Within two decades, a huge land area of Indonesia was burned twice by the largest forest fires in the world, in 1982–83 and 1997–98, after prolonged dry spells caused by strong El Niño Southern Oscillation (ENSO) events. The changes of monthly rainfall were very similar in both 1982–83 and 1997–98. Much lower than normal rainfall continued for about 11 months from June to April, except for a 2-month interruption in November and December when the rainfall was average (Fig. 2.1 in Chapter 2, this volume). The damages from both fires were especially serious in Borneo. It is believed the land area affected by the 1982–83 fires reached 5 million ha in Borneo and 3.6 million ha in the state of East Kalimantan alone (Lennertz and Panzer 1984, Goldammer and Seibert 1990). The extent and the visible impacts of these fires have been described by many studies (Wirawan 1984, Lennertz and Panzer 1984, Malingreau et al. 1985, Tagawa et al. 1988).

In the fires of 1997, the area of burned forest land reported to the Ministry of Forestry and Estate Crops, Indonesia, was 260 000 ha over all of Indonesia and 26 000 ha in East Kalimantan, respectively. On the other hand, from January to April 1998, the fires occurred mainly in eastern Borneo and the burned forest area reported to the Ministry reached about 500 000 ha in East Kalimantan (Table 3.1). However, judging from the distribution area of hot spots seen from NOAA satellites between February and April 1998 (FFPMP-JICA, 1999; see color plates, this volume), we estimated that the land area affected by the 1998 fires in East Kalimantan was similar to that (3.6 million ha) in the 1982–83 fires reported by Lennertz and Panzer (1984) and Richards (1996) (Fig. 3.1).

It has been believed that tropical rainforests are rarely struck by forest fires because of their humid environment (Mutch 1970). However, the great fires in Kalimantan in 1982–83 have focused the attention of many scientists on fire in tropical rainforests (Leighton 1984, Goldammer and Seibert 1990, Walsh 1996, Richards 1996). Within only 15 years, two forest fires of similar scale occurred in the tropical rainforests of East Kalimantan. It is of interest to analyze the factors that induced such extraordinary events in a short period. Such an analysis of these events will contribute to the conservation of tropical forests and the prevention of future fires.

Ecological Studies, Vol. 140
Guhardja et al. (eds.) Rainforest Ecosystems of East Kalimantan: El Niño, Drought, Fire and Human Impacts

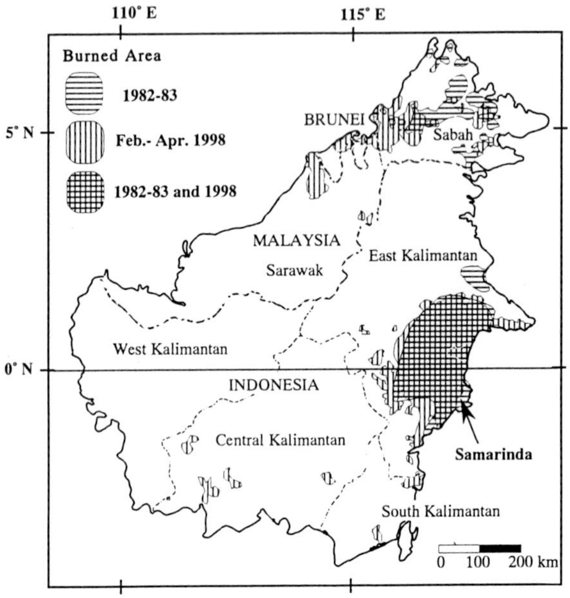


**Fig. 3.1.** Area affected by the fires in 1982-83 and 1998. Affected area in the 1982-83 fire was estimated by Richards (1996) based on the report of Leighton and Wirawan (1986) and that of the 1998 fire was estimated by the author based on the hot spot maps from FFPMP-JICA (1999)

**Table 3.1.** Area of forest burned in Indonesia over recent 14-year period (1000 ha)

| Year | 1985 | 1986 | 1987 | 1988 | 1989 | 1990 | 1991 |
|---|---|---|---|---|---|---|---|
| Burned area | 42.6 | 22 | 49.3 | 17.7 | 15.9 | 25.5 | 118.9 |
| Year | 1992 | 1993 | 1994 | 1995 | 1996 | 1997 | 1998[a] |
| Burned area | 14.5 | 40.9 | 161.8 | 6.7 | 10.3 | 264 | 533.3 |

[a]East Kalimantan until July 1998 (Forestry Department of East Kalimantan).
Source: Ministry of Forestry and Estate Crops

Our main observation site, Bukit Soeharto Education Forest (BSEF) of Mulawarman University, Samarinda (Fig. 3.1), which is located on a low hill on the east coast between Balikpapan and Samarinda, was burned in both 1982–83 and 1998. Many researchers, including several authors of this volume, have been studying these forests since 1985 and have been collecting many kinds of data. In addition to these data, detailed observations of the forest fires from February to April 1998 have clarified the characteristics of the fires in tropical lowland rainforests. This chapter will attempt to synthesize the results of many studies related to the effects of wildfire on the typical tropical lowland flora and fauna. It will also attempt to clarify the role of drought and wildfire in the dynamics of dipterocarp forests in lowland East Kalimantan. These considerations are based on observations made up to October 1998, 6 months after the end of the latest forest fires.

## 3.2 Cause of Forest Fires in 1998

As described in detail by Toma et al. in Chapter 2 of this volume, East Kalimantan was visited by extraordinary drought in 1997 and 1998. The drought was especially severe from January to mid April 1998, when there was no recorded rainfall at our observation forest at BSEF. The hot spots as seen from the NOAA satellites from January to April 1998 were concentrated in the lowland area less than 500 m above sea level, mostly in Kutai regency, East Kalimantan (Fig. 3.1).

The droughts of both years must have been brought about by a covering of high pressure air mass for an unusually long period connected to the strong ENSO event. The lowlands of Kutai had relatively strong winds from the north-north-east during January to April 1998. On the other hand, during the drought in 1997, weak winds blew from the south-south-east or from the sea. Furthermore, comparison of air temperature and humidity showed that these were much different between the drought periods in 1997 and 1998: lower air humidity and higher air temperature were observed in BSEF during the drought period in 1998 (Fig. 2.5 b,c and 2.7 in Chapter 2, this volume). This high air temperature and low air humidity during the period in 1998 may have been caused by the foehn phenomenon because the winds during the period came over the Sambaliung mountains (see Fig. 1.1 in Chapter 1, this volume) at the border between Kutai and Berau regency, which probably resulted in a foehn phenomenon in the lowlands of Kutai. This large difference may also explain the large difference in the burned area in East Kalimantan between years, i.e., the burned area was about 20 times larger in 1998 than in 1997, as mentioned above.

There are three main types of forest fires: underground fires (generally in peat forests), surface fires, and crown fires. The lowland dipterocarp forests mostly suffered from surface fire, so that the description in this text is restricted to surface fire. In order to describe how forest surface fire occurred in East Kalimantan in 1998, the fire will be analyzed with respect to three subjects: combustibles (floor debris, weeds, shrubs, etc.), their flammability, and fire sources.

In surface fires, combustibles are mostly fallen litter, grasses, and shrubs. Fallen litter generally decomposes quickly in tropical rainforests (Whitmore 1990), but when there is a prolonged drought, fallen litter does not decompose easily because of its low water content. In addition, many trees shed a large amount of leaves as a result of their water deficit during the drought (Richards 1996). According to the data of N. Tanaka (Unpublished 1998), the amount of fallen litter was 2 to 6 times greater than usual (0.2–0.4 kg $m^{-2}$ per 2 weeks) after the dry spells had continued for 30 to 40 days. This large and simultaneous leaf fall occurred just before the fire broke out in BSEF. The fact that it occurred during the dry period is consistent with the finding that the leaf area index of a *Macaranga conifera* forest dropped from 5.1 in the normal year of 1996 to 2.3 in February 1998 (Ishida et al. 1999).

The reduction of crown leaves also brought about high sunlight penetration to the forest floor, which stimulated desiccation of floor debris. The grasses also died as the drought advanced. Furthermore, just after the fire, the leaves of many small trees killed by the fire fell again. The leaf fall was about 10 times that of an average month. This large amount of fallen leaves resulted in a second fire at many places in natural and plantation forests. The abundance of combustibles such as shrubs and vines in secondary forests induced more serious fire damage than which occurred in primary forests, as many researchers have reported (Wirawan 1984, Lennertze and Panzer 1984, Tagawa et al. 1988).

Flammability is directly related to low water content of combustibles, which is generally brought about by low air humidity. In the temperate zone, when the water content of fallen leaves exceeds 26%, flammability becomes very low, but below the 20% level, precaution should be taken when using a match or a lighter, and below 10%, strict caution is required for every fire source (Kohei and Nakaoki 1971). When air humidity is below 50%, the water content of fallen leaves from temperate deciduous trees drops to about 14%. Within tropical rainforests, the average humidity is usually very high. Indeed, even during the extraordinary drought between June and October 1997, the monthly average of the daily minimum relative air humidity on the top of the forest canopy in BSEF always remained above 60% (Fig. 2.5c in Chapter 2, this volume). In March 1998, on the contrary, it dropped to 40%. Furthermore, if humidity is similar, the rates of desiccation of forest floor debris are greater than rates in the temperate zone because air temperatures in tropical areas are higher. Therefore, the forest floor debris must have dried very quickly after initiation of the drought in January 1998.

These speculations are supported by the fact that in East Kalimantan, fires in the primary forests were very rare in 1997, but they were common in 1998. Also, in Kutai National Park, primary forests located at the center of the park did not burn

**Fig. 3.2.** Extinguishing work on burning coal seam (*rear center*) by making a small dam. Burning coal seams are not extinguished by rain or small amounts of water, but submergence of the area with water is effective. (Photograph July 1998)

during the drought period in 1997, even though the surrounding buffer zone near the east coast burned in that period. However, almost the entire area of the park, 200 000 ha, was burned in March and April 1998.

One of the characteristics of the forest fires in Indonesia is that the fire sources were distributed over a wide area simultaneously. In this country, fire does not usually spread from one point to another, but fires occur numerously and simultaneously as can be seen in the hot spot data. It is well known there are many kinds of fire sources, e.g. fires used for land preparation at commercial plantations and at farm lands for slash-and-burn agriculture or fires due to illegal logging, hunting, carelessness, mischief, etc. (Ministry of Environment and UNDP 1998). These sources act independently or jointly at random times and locations. No one has been able to identify one or two definite fire sources as the main cause of the great fires. However, it is safe to say that the existence of numerous fire sources and their wide distribution are deeply connected to government policies, for which economic development is a priority, and the customs of the local people.

Also, there is a natural fire source in this district, which are the long-lasting fires of coal (lignite) seams (Goldammer and Seibert 1990). The coal seams extend to near the ground surface in various rainforests on the low hills along the east coast of Kutai regency. About 10 burning coal seam sites were present in the BSEF before the 1997 fires, which were ignited by the former great fires in 1982–83. Some of them acted as natural fire sources during the dry period of 1997–98. After the 1997–98 fires, the number of burning coal seams in BSEF increased to more than 60. About 50 of these fires have been extinguished by the Japan International Cooperation Agency (JICA) team (Fig. 3.2), but fires that are still burning at depths of several meters are very difficult to extinguish and may continue to burn until the coal is burned out. In the 1997–98 fires, however, the number of forest fires ignited by the burning coal seams was much less than the number of the fires due to human activities.

The great forest fires of both 1982–83 and 1997–98 occurred under the same conditions, i.e., they occurred as a result of the interaction between the unusual high temperature and low humidity during the prolonged droughts due to strong ENSO events and the presence of numerous fire sources scattered throughout lowland East Kalimantan. As many researchers have pointed out, droughts caused by ENSO events have often occurred in Indonesia, and especially in eastern Borneo from ancient eras (Leighton 1984, Goldammer and Seibert 1990, Walsh 1996). Leighton (1984) estimated that severe drought years have occurred 6 times since the late nineteen century: 1887–88, 1903–04, 1914–15, 1941–42, 1972, and 1982–83. However, there are no reliable reports of large fires in Indonesia before the 1982–83 fires. On the other hand, in the more recent period, medium scale droughts have been observed in 1987, 1991, 1994, and they have been accompanied by forest fires of increasing size (Table 3.1). This means that the recent increase in the burned forest areas may largely depend on the expansion of exploitation.

# 3.3 What the Fires Did to the Forests

## 3.3.1 Monitoring Forests at BSEF

Most primary forests of the lowlands and hills in East Kalimantan are dominated by Dipterocarpaceae, especially the genera *Shorea*, *Dipterocarpus*, and *Dryobalanops* (Kartawinata et al. 1981a, MacKinnon et al. 1996). Many of these forests have an abundance of ulin (*Eusideroxylon zwageri*) at the middle layer of their canopy. The undisturbed primary forests of this area have high species diversity, tall emergent trees and a large biomass (Kartawinata et al. 1981a,b, Proctor et al. 1983, Sukardjo et al. 1990, Sist and Saridan 1998) (Table 3.2). The tallest tree is 60 to 70 m in height, the basal area exceeds 30 m$^2$ ha$^{-1}$, and stem biomass reaches about 400 ton ha$^{-1}$ (Yamakura et al. 1986). Until the mid 1960s, typical primary forests in this area had this floristic composition and stand structure, and our main study forests in BSEF were originally covered by such primary forests of the dipterocarp-ulin type.

**Table 3.2.** Floristic composition and structure of the monitoring and undisturbed primary forests

| Forest types | Monitoring forest | | | Undisturbed primary forests | | | | |
|---|---|---|---|---|---|---|---|---|
| Study sites | LDF in BSEF | | | Wanariset[a] | Lempake[a] | Sebulu[a] | Berau[a] | Brunei |
| Survey area (ha) | 1 | 1 | 1 | 1.6 | 1.6 | 1 | 3×4[b] | 0.95 |
| Survey year | 1988 | 1997 | 1998 | | | | | |
| No. species ha$^{-1}$ | 130 | 151 | 108 | 175[c] | 153[c] | 276 | 182 | 214 |
| No. genera plot$^{-1}$ | 72 | 86 | 68 | 122 | 125 | 127 | - | - |
| No. families plot$^{-1}$ | 32 | 36 | 32 | 45 | 44 | 47 | - | - |
| No. individuals ha$^{-1}$ | 310 | 356 | 202 | 541 | 445 | 592 | 530 | 778 |
| No. ind./No. sp. | 2.4 | 2.4 | 1.9 | 3.1 | 2.9 | 2.1 | 2.9 | 3.6 |
| Basal area (m$^2$ ha$^{-1}$) | 25.5 | 26.7 | 21.4 | 29.7 | 36 | 33.9 | 31.1 | 57 |
| Dipterocarps | | | | | | | | |
| No. species ha$^{-1}$ | 13 | 14 | 12 | 14 | 12 | 23 | 29 | - |
| No. individuals(A) | 37 | 37 | 26 | 90 | 43 | 95 | 131 | 120 |
| % ocupancy of A | 11.9 | 10.4 | 12.9 | 17.5 | 9.7 | 16 | 25.1 | 15.4 |
| Basal area (B) (m$^2$ ha$^{-1}$) | 11.1 | 10.6 | 8.7 | 16.38 | 21.08 | 14.82 | 15.58 | 24.6 |
| % ocupancy of B | 43.5 | 39.7 | 40.7 | 55.2 | 56.2 | 43.7 | 50.1 | 43.2 |
| Biomass | | | | | | | | |
| Stem: m$^3$ ha$^{-1}$ (≥10cm) | 370.7 | | | | | | 402 | |
| AG:ton ha$^{-1}$ (≥10cm) | 265[d] | 315 | 266 | | | 509 | | 650 |
| Authors | 1 | 2 | 3 | 4 | 5 | 6 | 7 | 8 |

No.: Number, ind.: individuals, sp.: species, AG: Above ground.
Authors 1: Matius and Okimori (1991) and Kiyono and Hastaniah (1993), 2: Toma et al. (1997), 3: Toma et al. (Chapter 10, this volume), 4: Kartawinata et al. (1981b), 5: Kartawinata et al. (1981a), 6: Sukarjo et al. (1990) and Yamakura et al. (1986), 7: Sist and Saridan (1998), 8: Proctor et al. (1983).
[a]Locations of these study sites shown in Fig. 1.2 of Chapter 1 in this volume, [b]3 plot×4ha, [c]No. ha$^{-1}$ estimated from species area curve, [d]Stem.

However, these forests are presently impossible to find in the lowlands or near roads in East Kalimantan. This is because large-scale forest exploitation of East Kalimantan started after the establishment of new policies in 1967 when all Indonesian forests were declared to be property of the state and foreign investment was heavily promoted (Kartawinata et al. 1981a, MacKinnon et al. 1996, Chapter 1, this volume). The large-scale and organized transmigration to Kalimantan was accelerated at the same time. Consequently, by the late 1970s, undisturbed primary forests were already rare in the area around Balikpapan and Samarinda (Kawana 1984). The BSEF and its surrounding area (Total 77 000 ha) were designated as protected forests in 1978 and then as national forest park in 1982.

The natural forests of BSEF mainly consist of three types. There are dipterocarp-ulin forests, forests dominated by *Macaranga* spp., such as *Macaranga gigantea* or *M. triloba*, that were formed after the 1982–83 fires, and a third type which is intermediate between the first and the second types. We set up one monitoring site in each type of forest. i.e., Plot 1 was set up in a dipterocarp-ulin forest, which was designated as a lightly disturbed forest (LDF), plot 2 was set up in an intermediate forest, which was designated as a moderately disturbed forest (MDF), and plot 3 was set up in a *Macaranga* forest, which was designated as a heavily disturbed forest (HDF).

Most forests in BSEF had been subjected at least once to selective cutting of ulin and Kapur (*Dryobalanops* spp.) by the local people and partly by commercial logging companies. These forests were burned by the 1983 fires (Okimori 1991, Matius and Okimori 1991). Although LDF is presently one of the best forests in BSEF, it must have been degraded (Kiyono and Hastaniah 1997), especially in species diversity and in tree size as compared with the primary forests that had been studied previously (Table 3.2). For example, the low number of tree species in LDF may be the result of the fire damage in 1983, because forest fires generally kill many small and pole-sized trees. As to tree size, in the primary forests of this area before the 1970s, more than 70 trees per ha had stem diameters above 50 cm (Kawana 1984), but in the LDF in 1997, only 32 trees per ha had this size. This number of large trees per ha is almost the same as that (40 trees $ha^{-1}$) in the forests logged selectively at Sebulu in the late 1970s. The small number of large trees resulted from selective cutting in the 1970s. As would be expected, LDF was much higher in species diversity and in tree volumes than was a 30-year-old secondary forest reported by Kartawinata et al. (1981a).

## 3.3.2 Impacts of the 1998 Fire on the Forests

Most areas of the BSEF were burned between 25 February and 3 March, and some areas were wiped out by the second fire from late March to early April 1998. The fires were extinguished by rain that started on 16 April 1998. The damage caused by these fires was very heavy and the forest destruction was worse than that caused by the 1982–83 fires (Table 3.2, 3.3). Namely, in LDF, about 60% of the trees above 10 cm in DBH were killed and the mortality increased as the diameter of the trees decreased. Most of the small trees having DBHs less than 10 cm disappeared after the fires. Many canopy trees also died back as a result of the effects of the drought and fires (Fig. 3.3). At the time of writing, six months after the end of the fires, the amount of leaves in the canopy has not yet recovered.

The effects of the fires were more serious in the secondary forests dominated by *Macaranga* spp., for which mortality was almost 100% (Table 3.3), because the

**Table 3.3.** Changes of composition and structure of monitoring plots after the fire in 1998

| Monitoring site | LDF(1ha) | | MDF(0.72ha) | | HDF(0.36ha) | |
|---|---|---|---|---|---|---|
| Year | 1997 | 1998 | 1997 | 1998 | 1997 | 1998 |
| No. of species | 151 | 108 (72) | 110 | 60 (55) | 42 | 15 (36) |
| No. of individuals (ha$^{-1}$) | 356 | 202 (57) | 500 | 132 (26) | 553 | 50 ( 9) |
| Basal area (m$^2$ ha$^{-1}$) | 26.7 | 21.4 (80) | 29 | 14.9 (51) | 20.6 | 4.2 (20) |

For trees more than 10 cm in DBH. Figures in parentheses represent ratio(%) of 1998 (after the fire) to 1997 (before the fire). Source: modified from chapter 10, Toma et al., this volume.

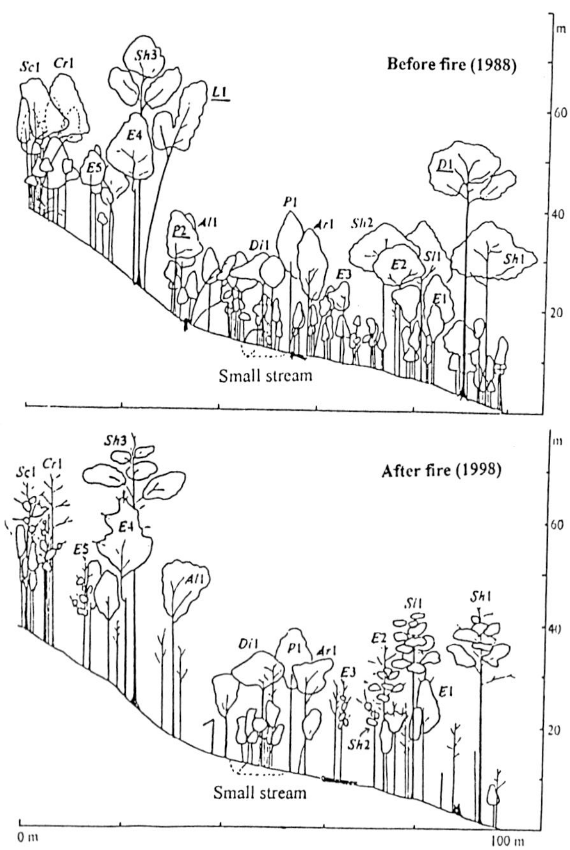


**Fig. 3.3.** Profile of lightly disturbed forest (LDF) monitoring site in 1988 (before 1998 fire) and 1998 (after the fire). *Underlined* genus symbols show trees that died before 1997 (before the fire) *Al*: *Alseodaphne*, *Ar*: *Artocarpus*, *Cr*: *Cratoxylum* (*Cr*1 died in the fire), *D*: *Dryobalanops*, *Di*: *Dipterocarpus*, *E*: *Eusideroxylon*, *L*: *Litsea*, *P*: *Palaquim*, *Sc*: *Scaphium*, *Sh*: *Shorea*, *Si*: *Sindora*. The profile before the fire (*upper*) was drawn by Okimori (1991) in 1988

secondary or degraded forests burned briskly as mentioned above. Within the same forest type, damage was light near streams even though the streams were dry during the drought period. On the contrary, many large dead trees were found on the ridge where the effects of drought were probably more severe than they were near the streams. For example, the large dipterocarp trees used for phenological observations (Chapter 11, this volume), which have 25 to 100 cm DBH and mostly stand near roads or ridges, showed very high (about 57%) mortality during the 1997–98 drought period, whereas this is less than 1% in an ordinary year. These visible impacts of the severe drought and fires are basically similar to the observations by many researchers after the great fires in 1982–83 (Wirawan 1984, Lennertze and Panzer 1984, Tagawa et al. 1988, Goldammer and Seibert 1990).

The drought tolerance of dipterocarp species varied considerably. Table 3.4 shows the damage to trees having DBHs between 5 to 70 cm (average 22 cm), which had been planted in an unburned area. *Cotylelobium melanoxylon* was very resistant to drought. In contrast, *Shorea smitiana*, *S. seminis*, *Dryobalanops beccarii*, etc. were very sensitive to water deficit. Similar results are described in Chapter 11 by Kiyono and Hastaniah in this volume. In general, many dipterocarp species are rather sensitive to drought and fire. Indeed, the various dipterocarp trees that were planted in BSEF over the last 10 years were completely destroyed by the 1998 fires. Some of the commercial tree species planted widely in this district, *Peronema canescens*, *Gmelina arborea*, *Pterocarpus* sp. and *Tectona grandis* showed strong tolerance to the drought and fires. In plantations of several-year-old *Acacia mangium* trees, the trees above 20 cm in DBH were able to survive at high rates but those below 15 cm could not. On the other hand, plantations with similar trees that suffered twice from the fires from February to April were completely destroyed.

In the burned forests, some surviving trees were found even among those with diameters less than 10 cm. Table 3.5 shows the numbers of surviving and sprouted tree species in LDF and HDF after the fires. The low number of surviving trees in HDF appears to reflect the abundance of undergrowth which induces fast burning.

**Table 3.4.** Damages to dipterocarp trees at the arboretum of PUSREHUT in 1997–98

| Species | Number of trees | Survival (%) | Dieback (%) | Sprout (%) | Death (%) |
|---|---|---|---|---|---|
| *Shorea smithiana* | 27 | 7.4 | 0.0 | 0.0 | 92.6 |
| *Shorea seminis* | 55 | 3.6 | 1.8 | 5.5 | 89.1 |
| *Shorea ovalis* | 16 | 18.8 | 0.0 | 12.5 | 68.7 |
| *Dryobalanops beccarii* | 35 | 20 | 0.0 | 11.4 | 68.6 |
| *Hopea mengerawan* | 73 | 16.4 | 15.1 | 8.2 | 60.3 |
| *Dipterocarpus gracilis* | 13 | 38.5 | 7.7 | 15.3 | 38.5 |
| *Anisoptera costata* | 16 | 56.3 | 12.5 | 6.2 | 25.0 |
| *Cotylelobium melanoxylon* | 23 | 91.3 | 0.0 | 8.7 | 0.0 |

Dieback: Trees defoliated for more than half top of stem.

On the other hand, the high number of sprouted trees in HDF was probably due to the high occupancy rates of the secondary forest species. The large or medium sized trees that were severely damaged often had sprouts on their trunk and boughs, but they mostly dried out, eventually leading to the death of the whole tree. The large trees that had a few leaves attached as a result of dieback appeared to be gradually weakened. This is consistent with the observations in LDF in which the death of large dipterocarp trees continued until 1988, 5 years after the 1982–83 fires (Okimori 1991).

The animals in the forests also suffered serious effects from the prolonged drought and fires. For example, in most species of the family Cerambycidae (longicorn beetles), both the numbers of species and the number of individuals were reduced after the start of the drought in January 1998, and then many of the beetles species disappeared just after the fires wiped out the forest floors (see Chapter 14, this volume). After the rain started in mid April 1998, the numbers of species and individuals of the Bostrychidae and Cerambycidae gradually increased but those of the Carabidae, Platypodidae and Scolytidae did not. In October 1998, the number of individuals of Cerambycidae jumped to more than 10 times above the level before the fires. This group could recover quickly because their larva or eggs are deposited in trunks or crowns of tall trees where the heat of the fires did not reach, and also because there are abundant food logs and food trees weakened by the drought and fires. Furthermore, the number of natural enemies might have been reduced during the post-fire period.

Insects that inhabit the forest floor, such as ground beetles and butterflies, dramatically declined in numbers after the fires, and some of them may have become extinct in the burned area. The tall and dense canopy layer of tropical forests may play an important role as a temporal escape refuge when disaster occurs on the floor. Mammals such as deer, monkeys, and bush-pigs returned soon after the end of the fires, although their numbers were lower than they were a few years before the fires. On the other hand, the recovery of carnivores, such as raptors and vespids, has been slow. In the BSEF, raptors were found only once at 5 months after the end of the fires and vespids have not been seen as of October 1998 (H. Makihara personal communication, 1999). The less mobile invertebrates, such as mountain leeches, completely disappeared and may not recover for several years.

Within 4 months after the rain recommenced, flowering and fruiting were observed in secondary forest tree species such as the genera *Macaranga*, *Peronema*, *Schima*, *Trema*, *Vitex*, *Vernonia*, *Acacia*, etc. but the primary forest tree species such as dipterocarps still have not flowered six months after the end of the fires. As food resources are scanty for herbivores, especially for fruit or seed eaters, the unbalanced species composition and population of animal communities in the burned forests have caused abnormal infestation of a few species. Also this infestation has probably been stimulated by a lack of natural enemies, as described by MacKinnon et al. (1996).

**Table 3.5.** Number of surviving and sprouting tree individuals less than 10 cm in DBH after the 1998 fire

| LDF (0.36 ha, upper part of LDF) | | | | HDF (0.36 ha) | | | |
|---|---|---|---|---|---|---|---|
| Survivals | No. | Sprouts | No. | Survivals | No. | Sprouts | No. |
| *Millettia sericea* | 10 | *Millettia sericea* | 155 | *Dillenia indica* | 2 | *Millettia sericea* | 134 |
| *Dipterocarpus humeratus* | 3 | *Urophyllum* sp. | 10 | *Syzygium* sp. | 2 | *Pternandra galeata* | 60 |
| *Pternandra rostrata* | 3 | *Aporusa* sp. | 8 | *Gironniera nervosa* | 1 | *Crypteronia paniculata* | 13 |
| *Aporusa elmeri* | 2 | *Eurycoma longifolia* | 6 | *Knema intermedia* | 1 | *Litsea* sp. | 9 |
| *Aglaia* sp. | 2 | *Macaranga hypoleuca* | 4 | *Millettia sericea* | 1 | *Vernonia arborea* | 9 |
| *Dillenia indica* | 2 | *Castanopsis megacarpa* | 3 | *Nauclea subdita* | 1 | *Syzygium* sp. | 8 |
| *Dillenia excelsa* | 2 | *Eugenia* sp. | 3 | *Pternandra galeata* | 1 | *Barringtonia racemosa* | 7 |
| *Litsea* sp. | 2 | *Eusideroxylon zwageri* | 3 | *Rhodamnia cinerea* | 1 | *Nauclea subdita* | 7 |
| *Polyalthia lateriflora* | 2 | *Ficus uncinata* | 3 | | | *Ardisia villosa* | 6 |
| *Pternandra galenta* | 2 | *Gironniera nervosa* | 3 | | | *Eusideroxylon zwageri* | 6 |
| | | *Glochidion obscurum* | 3 | | | *Ficus hispida* | 6 |
| | | *Macaranga lowii* | 3 | Minor sprouting sp. | | *Urophyllum* sp. | 6 |
| Minor species | | Minor species | | 1 individual : 28 species | | *Pternandora rostrata* | 5 |
| 1 individual : 25 species | | 1 individual : 20 species | | 2 individuals : 9 species | | *Gironniera nervosa* | 4 |
| | | 2 individuals : 8 species | | 3 individuals : 7 species | | *Symplocos* sp. | 4 |

Observation was in June 1998.

## 3.4 Recovery of the Forest Vegetation

Many shrub trees initiated sprouting after normal rainfall returned to the BSEF starting 16 April 1998 (Table 3.5). *Millettia sericea*, small legume trees, sprouted even before the rainfall recommenced and sprouted in extremely large numbers in the BSEF in June 1998, indicating its high tolerance for the drought and fires. Other species with a high capability for sprouting were *Pternandra rostrata*, *Crypteronia paniculata*, *Vernonia arborea*, *Litsea*, *Urophyllum*, etc. Ulin had a strong sprouting ability even in large trees (Fig. 3.3). Among dipterocarps, sprouts were observed in pole-sized trees of *Cotylelobium lanceolatum* and young *Shorea roxburghii*. The later species is naturally distributed in monsoon forests, and it is known as a fire-tolerant species (Symington 1943). The sprouts of undergrowth species may have an important role in the early covering of burned bare ground and also as food for herbivores.

One month after normal rainfall recommenced, many germinated seedlings were observed in the burned forest floors. These were *M. gigantea*, *M. hypoleuca*, *M. trichocarpa*, *M. triloba*, *Trema cannabina*, *T. orientalis*, *Homalanthus populneus*, *Leea indica*, *Vitex pubescens*, *Glochidion obscurum*, *Mallotus paniculatus*, *A. mangium,* etc. All of these pioneer tree species germinated from seed banks in the soil. For many of these species, the seeds can germinate readily and simultaneously after a forest fire because germination is stimulated by heat. Germination from the seed banks was completed within three months after recommencement of rain, in agreement with the observation by Kartawinata et al. (1981a). About 6 months later, the largest seedlings grew more than 100cm in height. Tree species with low height such as *M. trichocarpa*, and *T. cannabina* increased considerably, in comparison with pioneer species that were present before the 1998 fires.

Herbs and grasses also vigorously invaded the open burned area. *Imperata cylindrica* is well known for a habit that is adapted to fire and farming (Kiyono and Hastaniah 1997). This grass starts sprouting within one week after being burned and flowers 40 to 50 days later. Soon after, fruiting occurs and the seeds are scattered. Three months are enough time for its invasion of bare lands. Herb and grass species, such as *Solanum jamaicense* and *Scleria pupurascens* (at humid sites), recovered quickly, and some Pteridaceae newly and widely invaded the burned areas. The rhizome plants of the Marantaceae and Zingiberaceae, which were very common on the forest floor before the fires, have not yet been found six months after the end of the fires.

Within 3 months after the rainfall recommenced, the many canopy gaps that were opened up by the fires and the sites where dead trees were standing were mostly covered by a green carpet of the above grasses and young tree seedlings. In view of the normal species richness in humid tropical forests, the number of species represented by these pioneers and undergrowth was rather limited. Nevertheless, they will help to stabilize the soils. For the short period between the recommencement of rain and the recovery of vegetation, the black ash on the soil of the burned forest floors was washed away, resulting in widespread and heavy soil erosion.

In the dipterocarp forests, where there are few canopy gaps like those in LDF, new vegetation is rather scanty. Only the few pioneer tree species mentioned above have germinated in the gaps newly created by the death of pre-existing secondary forest tree species. The primary tree species generally do not make seed banks, so that their germination does not occur until the next fruiting time. Various dipterocarp species had fruited abundantly during a short wet period from November to December after the first dry period from June to October 1997. However, these fruits and seeds were killed as a result of the drought and fires from January to April 1998. After that, there was no mass fruiting of primary forest tree species. As many dipterocarp trees were damaged by the drought and fires, it is unclear when the next mass flowering and fruiting will occur.

In the case of the 1982–83 fires, the fruiting of dipterocarps was observed in the year following the fires (Goldammer and Seibert 1990). In Kutai National Park, on the other hand, a mass fruiting of dipterocarps occurred in 1986 and their seeds germinated and grew up to pole size. However, they were completely killed by the 1998 fires (N. P. Oka, personal communication). In addition to the difficulty of seed production of dipterocarp trees is the problem of low diversity of mycorrhizal fungi in burned forest. According to preliminary observations by Akema (1999), only one or two fungi were detected in the lightly burned forests as compared to 6 to 7 in a unburned forest. This is very important information for the regeneration of burned dipterocarp forest, because most dipterocarp species have ectomycorrhizal fungi and their seedlings are largely suppressed without the symbiosis of mycorrhizal fungi and often cannot survive (see Chapters 21 and 22, this volume).

According to the records of our monitoring sites (LDF, MDF, and HDF), in 1988, 5 years after the 1982–83 fires, HDF was already covered by *M. gigantea* which

**Fig. 3.4.** Colony of *Macaranga triloba* seedlings germinated after the 1998 fire (Photograph December 1998)

accounted for 80% of the number of individuals and their average DBH was 17.2 cm (Matius and Okimori 1991). Self thinning of *M. gigantea* started in 1991 and the basal area of these trees became saturated after reaching their maximum level in 1994. Therefore, *M. gigantea* is probably close to its age of alternation with the next succeeding tree species, in accordance with plant succession theory. In HDF after the 1998 fires, new seedlings of several pioneer species are presently growing. However, it is predicted that the nearly pure *M. gigantea* forest as mentioned above cannot recover in the near future because the number of its succeeding seedlings is probably not enough to make a pure stand. There are also many sites in BSEF that have been colonized by a single pioneer species, such as *M. gigantea*, *M. triloba*, *T. orientalis*, *A. mangium*, etc. (Fig.3.4) but generally these places were not as degraded as HDF before the fires.

In the dipterocarp forests, LDF in 1988 was still suffering from the after-effects of the 1982–83 fires. Namely the growth rate from 1988 to 1989 was negative, which was due to the death of the weakened emergent trees. The basal area continued to decrease until 1991 and it began to gradually increase from 1992 about 10 years after the fires, as a result of the strong recruitment of Euphorbiaceae species (Matius and Okimori 1991, Kiyono and Hastaniah 1993, Toma et al. 1997). Dipterocarpaceae, which was the dominant family in 1988, has still not recovered its occupancy rates in species number and basal area (Table 3.2). Instead of the Dipterocarpaceae, Euphorbiaceae became the dominant family in 1997.

Therefore, these results indicate that the recovery of dipterocarp forests takes a long time, i.e., more than several decades. Kiyono and Hastaniah (1997) have estimated that the recovery of the above ground biomass of LDF to the level of the primary forest studied by Yamakura et al. (1986) (Table 3.2) would take about 60 years.

## 3.5 Conclusion

From the late nineteenth century to the present, severe drought years occurred several times in Borneo, especially in eastern Borneo (Leighton 1984, Goldammer and Seibert 1990, Richards 1996). Although Walsh (1996) reported that the drought in Sabah was not perfectly associated with ENSO episodes, it is generally believed that the severe droughts in this district are connected to strong ENSO events (Leighton 1984, MacKinnon et al. 1996, Richards 1996). If this is so, it is likely that severe droughts have occurred periodically in eastern Borneo for a long time, probably for at least thousands of years (Goldammer and Seibert 1990). Wildfires would have occurred in East Kalimantan at times of severe drought as described by Goldammer and Seibert (1990), but the burned forest areas might not have exceeded several thousand or ten thousand ha during those drought times. Large forest fires, such as those having burned areas of more than a million ha, appear to be rather recent events, i.e., they appear to have occurred only in the last 30 years. This is because there are no reliable records of great forest fires even in the recent severe drought years in Indonesia (1887–88, 1903–04, 1914–15, 1941–42, and 1972), which are considered as severe as the drought period from 1982 to 1983 (Leighton 1984).

The great fires of both 1982–83 and 1997–98 resulted from the interaction between severe droughts and the creation of numerous fire sources connected closely with human activities. As mentioned above, neither of the great fires could have occurred without the existence of these fire sources. Before large-scale exploitation of Kalimantan started in the late 1960s, fire sources due to human activities must have been rare or in restricted places.

There is another observation that supports the above speculation that the great forest fires are recent events. That is the existence of dipterocarp forests with trees of huge height and diameter and with high species diversity. These giant trees are estimated to be 300 to 400 years old (Riswan et al. 1985). More than 85% of the lowlands and hills in East Kalimantan had been covered by such dense primary forests dominated by dipterocarp trees before the large-scale exploitation era (Kartawinata et al. 1981a, Kawana 1984). On the other hand, forest fires strongly degrade the forest constitution and structure as mentioned above. The occurrence of several great wildfires in a century would not have left such primary forests at almost all lowlands of the state. Therefore, the frequency that one forest site would be burned by natural fire sources might be once in several centuries or millennia if wildfires did occur.

It is also true that severe drought itself weakens large dipterocarp trees and often kills them. As a result, this helps the formation of canopy gaps, which is believed to be the driving force for natural regeneration of tropical rainforests (Whitmore 1990). On the other hand, drought alone rarely kills medium and small sized trees or succeeding trees. Furthermore, observations in the BSEF for 10 years indicated that mass flowering and fruiting of many dipterocarps occurred just after the end of each drought period (see Chapter 11, this volume) Therefore, the author presumes that in this district, the periodic occurrence of drought provides a chance for the regeneration of dipterocarps and has an important role in maintaining their dominance in the forests.

A period of 15 years (from 1983 to 1998) was too short for the burned forest

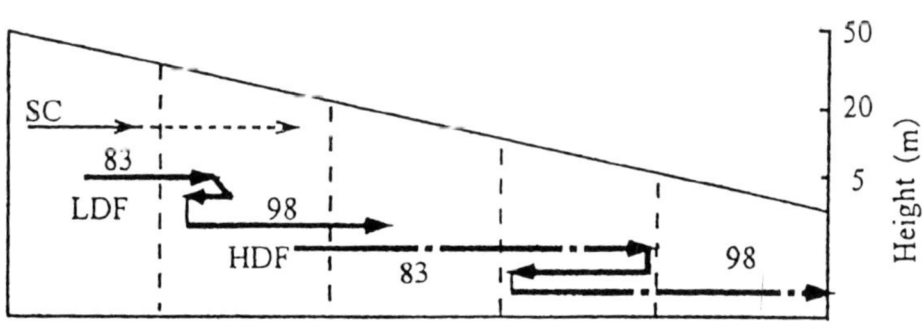


**Fig. 3.5.** Schematic model of forest degradation and recovery in the monitored forests. *Right-facing arrow,* degradation by forest fires or logging; *left-facing arrow,* forest recovery over 15 years from 1983 to 1997. Forest types: *A*, Dipterocarp forest; *B*, Dipterocarps dominate in basal area but not in number of individuals such as LDF; *C*, Non-dipterocarp primary species dominate such as moderately disturbed forest (MDF); *D*, Pioneer species dominate (>50% in number) such as heavily disturbed forest (HDF) or fallowlands after slash-and-burned agriculture; *E*, Bush forest with fire-tolerant species. Numbers 83 and 98 indicate the 1983 and 1998 fires, respectively. *SC*, Selective cutting (*dotted line* shows range of degradation by different logging grades)

vegetation to recover fully, and thus all the forests burned by the 1998 fires became much more degraded than did those that were burned by the 1983 fires (Fig. 3.5). Forest degradation by fire can occur in a moment, but recovery of burned forest takes time, especially in the vegetation phases shown on the left side of Fig. 3.5. A tentative conclusion that can be drawn from these observations is that the frequency of forest fires decides the vegetation types in this district: places that are burned every 1 to 2 years become *I. cylindrica* grass land. Places that are burned every 4 to 5 years become bush land near agriculturally exploitated areas. These areas are usually dominated by fire-tolerant species such as *Nauclea orientalis*, *Piper aduncum*, *Macaranga tanarius*, etc. Places that are burned every 10 to 20 years become *M. gigantea* and *M. triloba* forests or fallow lands associated with slash-and-burn agriculture (see Chapters 16 and 17, this volume). Places that are burned every several decades become intermediate phases of the rainforests such as MDF. And places that are burned every several hundred years remain primary forests of dipterocarps. Accordingly, in order to provide succeeding generations with the diverse species and forest resources of tropical rainforests, the occurrence of forest fires must be reduced as soon as possible. When air humidity dropped below 50%, even primary forest was left in a dangerous situation where it can catch fire very easily. This finding can be used as one of the indicators for a forest fire warning system which is acutely needed and which can be executed immediately in this country.

## References

Akema T (1999) Effects of forest fire on density and diversity of mycorrhiza in dipterocarp forests. Trop Rain For Res Project (III) JICA Expert Report 1997– No.3 (Unpublished)

FFPMP-JICA (Forest Fires Prevention Management Project-JICA) (1999) Annual report of the FFPMP. Bogor, Indonesia (Unpublished)

Goldammer JG, Seibert B (1990) The impact of droughts and forest fires on tropical lowland rain forest of East Kalimantan. In: Goldammer JG (ed) Fire in the tropical biota. Ecological Studies 84. Springer, Berlin, pp11–31

Ishida A, Toma T, Marjenah (1999) Limitation of leaf carbon gain by stomatal and photochemical processes in the top canopy of *Macaranga conifera*, a tropical pioneer tree. Tree Physiol 19:467–473

Kartawinata K, Adisoemarto S, Riswan S, Vayda A (1981a) The impact of man on a tropical forest in Indonesia. AMBIO, 10:115–119

Kartawinata K, Abdulhadi R, Partomihardjo T (1981b) Composition and structure of a lowland dipterocarp forest at Wanariset, East Karimantan. Malay For 44:397–406

Kawana A (ed) (1984) Relationship between the tropical rain forest and the people (Final report of joint research). JICA (Unpublished)

Kiyono Y, Hastaniah (1993) Growth of dipterocarps forest and trees at Bukit Soeharto, East Kalimantan, Indonesia. Ann Rep PUSREHUT. Mulawarman Univ, Samarinda

Kiyono Y, Hastaniah (1997) Slash and burn agriculture and succeeding vegetation in East Kalimantan. PUSREHUT special Pub. 6. Mulawarman Univ, Samarinda

Kohei H, Nakaoki Y (eds.) (1971) Forest fire handbook (in Japanese). Kohrin Times, Tokyo

Leighton M (1984) The El Niño-Southern Oscillation event in Southeast Asia: Effect of drought and fire in tropical forest in eastern Borneo. WWF Report, Indonesia
Lennertz R, Panzer KF (1984) Preliminary assessment of the drought and forest fire damage in Kalimantan Timur. Report by DFS German For Inven Serv Ltd for GTZ, Samarinda
MacKinnon K, Hatta G, Halim H, Mangalik A (1996) The ecology of Kalimantan. Periplus Editions (HK), Singapore
Malingreau JT, Stephens G, Fellows L (1985) Remote sensing of forest fires: Kalimantan and North Borneo in 1982–83. AMBO 14(6):314–321
Matius P, Okimori Y (1991) Floristic composition and dynamics of lowland dipterocarp forest largely burned in Bukit Soeharto protection forest, East Kalimantan. BIOTROP Special Pub, No 41:257-271
Ministry of Environment, UNDP (1998) Forest and land fires in Indonesia. Vol. 1 Impacts, Factors, and Evaluation. Ministry of Environment, Indonesia, Jakarta
Mutch RW (1970) Wildland fires and ecosystem — a hypothesis. Ecology 51:1046–1051
Okimori Y (1991) Research of regeneration process in secondary forest of tropical rain forest, East Kalimantan. Final Report. PUSREHUT Mulawarman Univ, Samarinda (Unpublished)
Proctor J, Anderson JM, Chai P, Vallack HW (1983) Ecological studies in four contrasting lowland rain forests in Gunung Mula National Park, Sarawak. I: Forest environment, structure, and floristic. J Ecol 71:237–260
Richards PW (1996) The tropical rain forest, 2nd edn. Cambridge, New York Melbourne
Riswan S, Kenworthy JB, Kartawinata K (1985) The estimation of temporal processes in the tropical rain forest: a study of primary mixed dipterocarp forest in Indonesia. J Trop Ecol 1:171–182
Sist P, Saridan A (1998) Description of the Primary lowland forest of Berau. In: Bertault J-G, Kadir K (eds) Silvicultural research in a lowland mixed dipterocarp forest of East Kalimantan. CIRAD-foret, FORDA, INHUTANI, Indonesia, pp51–73
Sukardjo S, Hagihara A, Yamakura T, Ogawa H (1990) Floristic composition of a tropical rain forest in Indonesian Borneo. Bull Nagoya Univ For 10:1–43
Symington CF (1943) Foresters' manual of dipterocarps. Malayan Forest Records No. 16. Penerbit Universiti Malaya, Kuala Lumpur
Tagawa H, Suzuki E, Wirawan N, Miyagi Y, Oka NP (1988) Change of vegetation in Kutai national park, East Kalimantan. In: Tagawa H, Wirawan N (eds) Res. Process Earlier Recov. Trop Rain For after Large Scale Fire Kal-Tim Indonesia, Res Cent South Pacific, Kagoshima Univ, Japan 12–50
Toma T, Matius P, Hastaniah, Kiyono Y, Watanabe R, Okimori Y (1997) Floristic composition and stand dynamics in logged-over and burned lowland dipterocarp forests in Bukit Soeharto, East Kalimantan. In: Suhartoyo H, Fatawi M (eds) Proc 2nd Intern Symp Asean Trop For Manage, PUSREHUT Univ Mulawarman, Samarinda, pp.198–206
Walsh RPD (1996) Drought frequency changes in Sabah and adjacent parts of northern Borneo since the late nineteenth century and possible implications for tropical rain forest dynamics. J Tropical Ecol 12:385–407
Whitmore TC (1990) Tropical rainforest of the far east, 2nd edn. Clarendon, Oxford
Wirawan N, Hadiyono (1983) Survey to the eastern part of the proposed Kutai National Park. June 17–27, 1983 (Mimeograph)
Wirawan N (1984) Kutai National Park and the great Kalimantan fire. WWF monthly Report, June 1984, 125–131
Yamakura T, Hagihara A, Sukardjo S, Ogawa H (1986) Aboveground biomass of tropical rain forest stands in Indonesian Borneo. Vegetatio 68:71–82

# Part 2

# Soil Characteristics and Human Impacts

# 4 Are Soils in Degraded Dipterocarp Forest Ecosystems Deteriorated? A Comparison of *Imperata* Grasslands, Degraded Secondary Forests, and Primary Forests

SEIICHI OHTA, KAZUHITO MORISADA, NAGAHARU TANAKA, YOSHIYUKI KIYONO, and SYARIF EFFENDI

## 4.1 Introduction

Many primary forests in the tropical regions of the world have been converted into degraded secondary forests and grasslands of species such as *Imperata cylindrica*. The main destructive agents responsible include illegal logging, slash-and-burn cultivation, extensive cattle grazing and natural forest fires (Fatawi and Mori, Chapter 1, this volume). Understanding the soil characteristics of these degraded forest ecosystems is important not only for the conversion of these areas into more productive land but also for evaluating the effects of forest degradation on global climatic change. However, information on the soils of degraded ecosystems is very scarce and many theories on the soil fertility of *Imperata* grasslands are still under debate. For instance, it has been reported that in the Philippines degraded grassland soils are severely eroded, extremely acidic, low in organic matter and lacking in key elements, particularly N, P and Mg (Ohta 1988, Dela Cruz 1986). Dela Cruz (1986) concluded that the poor chemical properties of these areas made forest regeneration in *Imperata* grasslands technically difficult, expensive and time consuming. However, Soepardi (1980) who investigated the chemical and physical properties of soils from *Imperata* grasslands in four locations on Sumatra, two on Java and one on Kalimantan concluded that *Imperata* soils are not infertile to such a great extent and that grasses can even improve the fertility and physical properties of surface soils. These contradictions suggest that there may be a large degree of variation in the fertility of *Imperata* grassland soils. In order to fully understand the soils of these degraded land areas, we need to know how soil characteristics vary and be able make generalisations about the relationship between soil fertility and other soil parameters.

According to Ohta et al., (Ohta and Effendi 1992ab, Ohta et al. 1992), there are relatively wide variations in the texture of soils generated from Tertiary deposits in East Kalimantan. Soil texture has been shown to be one of the most important factors governing soil properties in primary forests. Thus, we carried out a study of the soil characteristics of degraded land areas including degraded secondary forests and *Imperata* grasslands, and compared these with conditions in primary forests in the same region, with a particular emphasis on soil texture (Ohta and Effendi 1992ab, Ohta et al. 1992).

Ecological Studies, Vol. 140
Guhardja et al. (eds.) Rainforest Ecosystems of East Kalimantan: El Niño, Drought, Fire and Human Impacts

## 4.2 Study Area, Vegetation and Soils

The study area was located in a hilly region alongside the road between the cities of Balikpapan and Samarinda in East Kalimantan, Indonesia. The geology of the study area is characterised by Tertiary sedimentary rocks of various textures. The area was covered with thick primary forest in the early 1970's and began to degrade rapidly during the early 1980's due to a combination of slash-and-burn agriculture and frequent fire events (Kiyono and Hastaniah Chapter 17, this volume). As a result, the vegetation of the area now consists mostly of degraded secondary forest and *Imperata* grassland. The exception to this is the Bukit Soeharto Protection Forest area where the forest remains relatively intact. The secondary forests in the area were 3 to 11 years old and were further classified into 3 major types by Kiyono (1994) based on species composition. In this chapter, however, as the soil conditions of the various secondary forests were not found to be significantly different from each other in most of the parameters studied, the three different types of secondary forest were dealt with together irrespective of type.

One meter deep soil pits were dug at 28 of the secondary forest and 23 of the *Imperata* grassland locations. Soils with medium to fine textures were classified mainly as paleudults, with some classified as hapludults (Soil Survey Staff 1997), which are the major soil types associated with the Tertiary geology of the area (Ohta and Effendi 1992a). Some very sandy soils occurring in patches might fall under tropopsamments (Soil Survey Staff 1997). The soils of secondary forests were morphologically similar to those of the primary forests. However, *Imperata* soils were remarkably distinguished from those of the forest areas by the presence of various amounts of wormcasts (trace levels to some 50 ton $ha^{-1}$ on dry weight) on the soil surface, larger amounts of rhizomes, prominent rust and grey mottles in the Ah-E horizons, and larger amounts of fine roots in deeper horizons, especially of coarse textured soils.

The soils samples from each soil horizon were examined for bulk density (B.D.), total C and N contents (T-C and T-N), total P content (T-P) (Kitson and Mellon 1944), mechanical composition, available N content (Av-N), N availability expressed as the percentage of Av-N to T-N, available P (Av-P), P availability expressed as the percentage of Av-P to T-P, pH ($H_2O$, KCl), ΔpH, exchangeable Ca, Mg, K, Na, Al and H contents (Yuan's 1959), ECEC (sum of the exchangeable Ca, Mg, K, Na, Al and H contents), base and Al saturation in ECEC, and the amount of each element. The results obtained were compared with 20 soil profiles in the undisturbed primary forest of the Bukit Soeharto Protection Forest (Ohta and Effendi 1992ab; Ohta et al. 1992).

## 4.3 Chemical Properties

The soils of degraded ecosystems had diverse textures with most chemical parameters tending to be influenced by clay content, as well as the soils of the primary forests (Ohta and Effendi 1992ab; Ohta et al. 1992). As a result of this, soils associated with different vegetation types could be easily compared with specific reference to this one parameter. As most subsoil characteristics showed little variation in the areas studied with different vegetation, comparisons will be discussed only for the top Ah horizon and subsurface eluvial E horizon.

## 4.3.1 Total Carbon, Nitrogen and Phosphorus

In the Ah horizon T-C was weakly controlled by clay content whereas T-N was strongly controlled, and this was consistent amongst all the vegetation types studied (Fig. 4.1). The positive correlation of the levels of T-C and T-N with the clay contents suggests a greater magnitude of organic matter stabilisation in more clayey soils by formation of stable organo-mineral complexes. In contrast, in the E horizon T-C (which showed no correlation) and T-N (which showed relatively clear correlations with clay content) were both higher in the degraded ecosystems, particularly in the *Imperata* grassland possibly due to its higher production of underground biomass than in the primary forests.

T-P could also be correlated to clay content in each layer investigated. This correlation relates to the fact that a major proportion of T-P in an ecosystem is originally derived from parent materials, and finer sedimentary rocks contain larger amounts of P than coarser ones. For Ah horizons this tended to be lower in the degraded ecosystems than in the primary forests (Fig. 4.1), whereas E horizons showed the opposite pattern. Subsoil N and P levels showed no significant difference amongst the different vegetation types studied.

## 4.3.2 Available Nitrogen and Phosphorus

Low levels of Av-N in surface soils are one of the important features of degraded ecosystems. Av-N levels were lower in the degraded ecosystems compared with the primary forest in the Ah horizon, whereas in the E horizons there were no consistent differences among the different vegetation types (Fig 4.2). This might suggest a higher consumption of N by plant uptake and/or leaching loss in forms of inorganic nitrogen in the surface soils of the degraded ecosystems, which are possibly less efficient at conserving nutrients than in those of the primary forests. N availability was also significantly lower for the Ah and E horizons of the *Imperata* grasslands and the Ah horizon of secondary forests than in the primary forests. The production of Av-N is known to be generally associated with the C/N ratio, in fact the N availability tended to be higher in the Ah horizons with lower C/N ratios throughout each vegetation type. However, each group displayed similar patterns in the C/N ratio-N availability diagram (Fig. 4.2), and the C/N ratio could not explain the lower N availability in the degraded ecosystems. One possible explanation for this is that nitrogenous organic compounds were altered qualitatively in the processes of biomass burning and/or repeated biochemical transformation of soil organic matter under the harsher conditions of the soil surface of degraded ecosystems, and the nitrogen

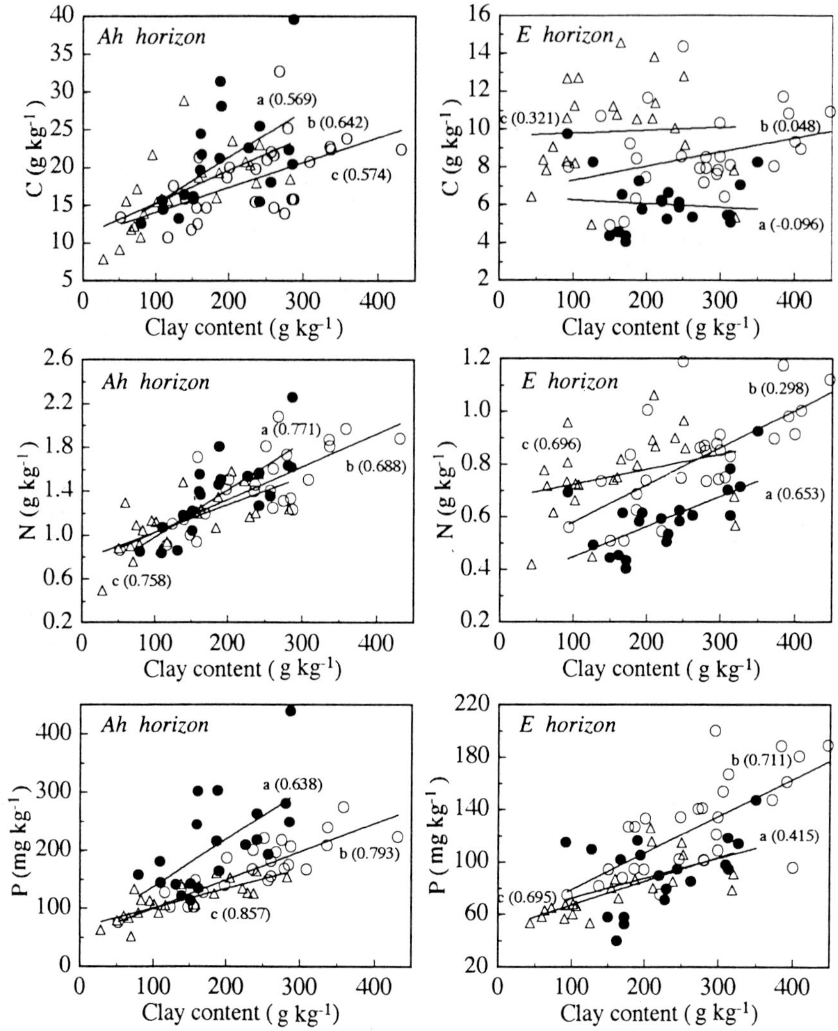


**Fig. 4.1.** Comparison of Ah and E horizons under primary forest (*solid circles*), secondary forest (*open circles*), and *Imperata* grassland (*open triangles*) in concentrations of total C, N, and P. Letters *a*, *b*, and *c* attached to regression lines in the figure indicate primary forest, secondary forest, and grassland, respectively, and are followed by correlation coefficients in *parentheses*

became incorporated into stable organic substances which are more resistant to decomposition. Obviously lower levels of Av-P of Ah horizon than in the primary forests were also one of the outstanding features of the surface soils in degraded ecosystems (Fig. 4.2). This distinct deterioration in Av-P may be associated with a possible decline in mycorrhizal activity in disturbed Dipterocarp forest ecosystems, which are known to solubilize less soluble P compounds. P availability was also lower in the degraded ecosystems compared with the primary forests for both the Ah

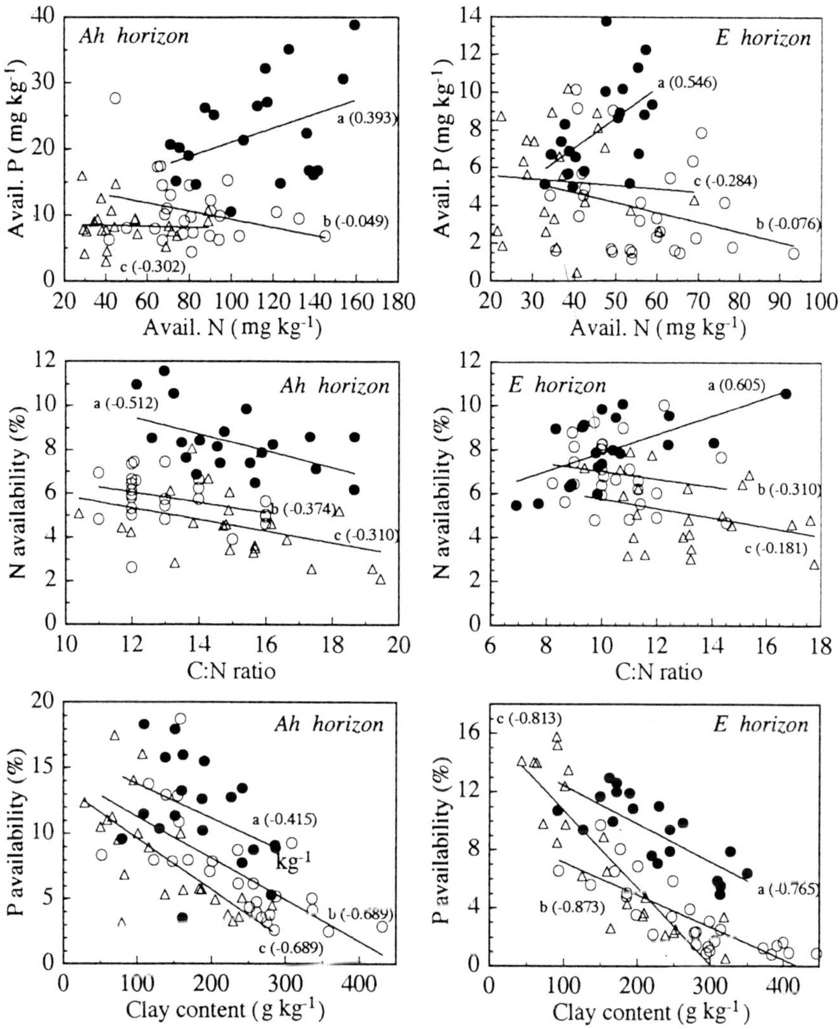


**Fig. 4.2.** Comparison of Ah and E horizons under primary forest (*solid circles*), secondary forest (*open circles*), and *Imperata* grassland (*open triangles*) in available forms of N and P. Letters and figures: see legend for Fig. 4.1

and E horizons, and was negatively correlated with clay content in each vegetation type and over all soil horizons (Fig. 4.2) due to a higher content of Al and Fe oxides and exchangeable Al in the more clayey soils that react with P to form relatively insoluble Al and Fe phosphate (Ohta et al. 1992).

Consequently, it can be concluded that depleted levels of Av-N and Av-P are the outstanding characterising features in soils of the degraded ecosystems in the area studied. Similar patterns were reported in the *Imperata* grasslands of the Philippines (Ohta 1988) and Sumatra (Ohta unpublished), and it is highly possible that it is a common phenomenon in the humid and monsoon tropics that forest degradation results in the depletion of Av-N and Av-P particularly in surface soils.

### 4.3.3 pH and Exchangeable Cation

Also, for pH and exchangeable cations there were clear differences between the vegetation types for the Ah and E horizons, though there was no consistent difference in the sub-horizon. The results for Ah horizon are shown in Fig. 4.3. The pH

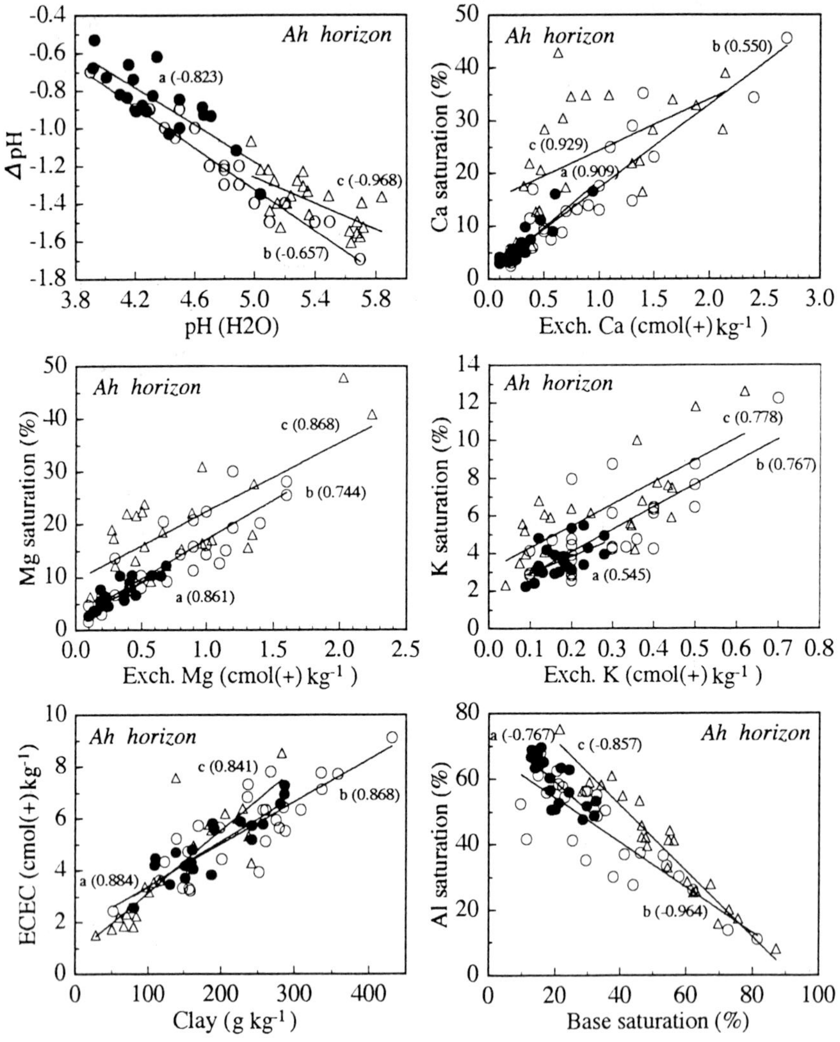


**Fig. 4.3.** Comparison of Ah horizon under primary forest (*solid circles*), secondary forest (*open circles*), and *Imperata* grassland (*open triangles*) in acidity and exchangeable cations. Letters and figures: see legend for Fig.4. 1

($H_2O$) was markedly higher in the degraded ecosystems, particularly the *Imperata* grasslands, than in the primary forests for both the Ah and E horizons. This pronouncedly lower soil acidity in the degraded ecosystems could be ascribed to the higher base status as mentioned below. Many of the Ah and E horizons of the degraded ecosystems were significantly higher in exchangeable Ca, Mg and K levels, and in each respective saturation percentage than those of the primary forests. This is definitely attributable to the land use history of the degraded ecosystems where vegetation burning supplies base elements to the soil surface in the form of ash, whereas ECEC did not differ consistently between the different vegetation types. As a result of the enrichment of base elements and the constant ECEC, base saturation percentages were distinguishably higher and Al saturation percentages lower in the top two horizons of degraded ecosystems than in those of primary forest. It is worth noting that in degraded ecosystems base cations are present in the soil system at higher concentrations compared with more intact vegetation areas where a larger portion of base elements are retained in the larger biomass. Thus soils under degraded ecosystems are not deteriorated in base status, but are however ameliorated in comparison with the primary forest soils. These findings indicate that ecosystem degradation does not always accompany deterioration of soils with respect to certain soil characteristics.

However, in the areas with a longer history of land degradation such as the Philippines (Ohta 1988) and Sumatra (Ohta unpublished), the base status of the surface soils of *Imperata* grasslands was not higher than primary forest soils. This discrepancy between the different ecosystems suggests that the deterioration of base status may be associated with the history of land degradation. The burning of primary vegetation is accompanied with a large input of ash to the soil surface, and drastically raises the concentrations of exchangeable base cations and the pH of the topsoil. The increased base status will be gradually deteriorated through leaching and the wash-off of ash during the course of successive destructive land use episodes accompanied with repeated burning. However, the raised status will be maintained for a period during the earlier stages of degradation. Therefore, the raised base level of the surface soils in the study area is presumed to be the result of the relatively short period (10–20 years) of the degradation history of the area.

## 4.4 Conclusion

The average stock of each element in the 0–20 cm and 0–100 cm layers is shown in Fig. 4 in order to allow comparison between soils. Stock levels of T-C did not decrease with forest degradation and in fact increased over every layer during the conversion of forests to grasslands; this is possibly the result of a high production of underground grass biomass and also of accumulation of stable organic matter such as partially carbonised materials supplied through burning. Stock of T-N was also slightly higher in the degraded ecosystems in the upper layers. The differences found in T-P over every layer were not consistent with vegetation change, but were judged to be the result of differences in the quantity of clay. Thus it is concluded that surface and sub-surface soils in the degraded ecosystems studied were not deteriorated in terms of T-C and T-N and were actually somewhat improved in the degraded ecosystems forests compared to primary forests.

Stock levels of exchangeable Ca, Mg, and K were always the greatest in *Imperata* grasslands for the upper soil layer. Exchangeable K stock level in the upper soil layers in the secondary forests were also significantly higher than the primary forest areas. This resulted in markedly higher pH of surface and subsurface soils in the degraded ecosystems. These findings demonstrate clearly that the status of exchangeable base cations in surface and subsurface soils under degraded ecosystems were obviously improved. At the same time however, it should be also noted that when taking account of the amounts of these elements in terms of total biomass, the levels of these elements existing within a whole ecosystem must be significantly smaller in the degraded ecosystems compared with the primary forests. The relatively short history (10–20 years) of forest degradation in the area studied is presumed to explain the increased levels of exchangeable bases in surface and subsurface soils originating from ash input through the burning of primary vegetation during the course of forest degradation. The widely held belief that the degradation of forest ecosystems and subsequent conversion to grasslands or secondary forests causes soil deteriora-

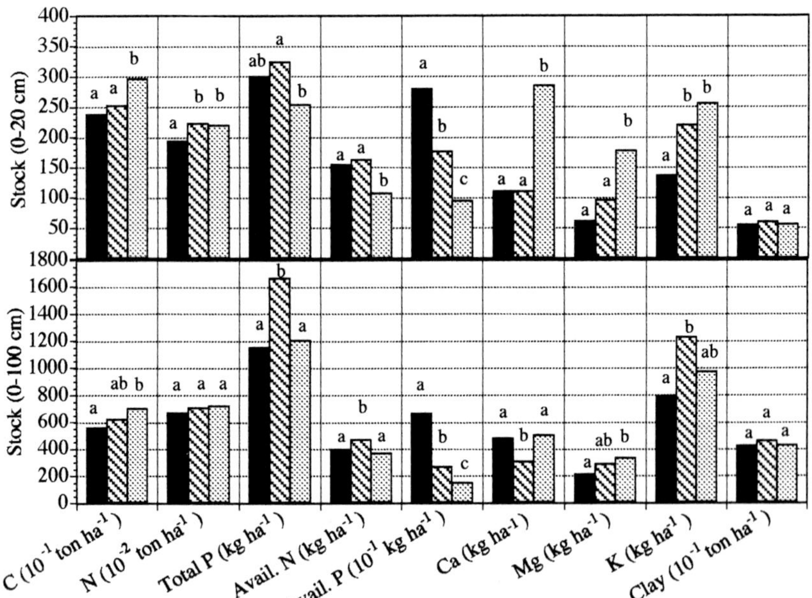


**Fig. 4.4.** Comparison of average amounts of each element in 0–20 and 0–100 cm layers for soils with clay levels between 2800 to 5700 ton/ha in a 0–100 cm layer among primary forest (*solid bars*), secondary forest (*hatched bars*), and grassland (*dotted bars*). The letters *a, b,* and *c* attached to the bars indicate that means with the same letter are not significantly different at the 5 % level

tion is not always true, and that the pattern of deterioration in base status may differ depending on the degradation history as mentioned above.

Meanwhile, it must also be emphasised that the surface and subsurface soils under degraded ecosystems were distinctly deteriorated with respect to the status of N and P. It was clearly shown that Av-N stock levels in the 0–20 cm soil layers were significantly lower in the *Imperata* grassland than in the other areas. Also stock level of Av-P decreased remarkably in the order of primary forests > secondary forests > grassland in both the 0–20 cm and 0–100 cm soil layers. Increased stabilisation of organic matter and a decreased population of mycorrhizal fungi may be associated with the qualitative deterioration of N and P in the soils of the degraded ecosystems, respectively. This deterioration in N and P is suspected to be a common phenomenon caused by forest degradation in the humid and monsoon tropics as recorded in other tropical areas. Qualitative soil N and P deterioration has the practical implication that in order to achieve successful future land utilisation these elements in available forms will have to be replenished, depending on the N and P requirements of crop species introduced.

Finally, it has to be stressed that degraded forests urgently need to be reforested so that nutrients are fixed in biomass and conserved more efficiently in restored forest ecosystems while soils still hold substantial amounts of nutrients. This needs to be achieved before continuous destructive land use leads to irreversible soil deterioration.

## References

Dela Cruz RE (1986) Constraints and strategies for the regeneration of *Imperata* grassland. In: Forest Regeneration in Southeast Asia, Biotrop special pub, Seameo-Biotrop, Bogor, Indonesia, 25:23–34

Kitson RE, Mellon MG (1944) Colorimetric determination of phosphorus as molybdivanadophosphoric acid. Inc Eng Chem Anal 16:379–383

Kiyono Y (1994) Degradation of tropical rainforests of East Kalimantan by man's impact—Forests in areas of swidden agriculture (in Japanese). The Tropical Forestry 29:46–54

Ohta S (1988) Influence of deforestation on the soils of the Pantabangan area, Central Luzon, the Philippines. Soil Sci Plant Nutr 36:561–573

Ohta S, Effendi S (1992a) Ultisols of "Lowland Dipterocarp Forest" in East Kalimantan, Indonesia, I. Morphology and physical properties. Soil Sci Plant Nutr 38:197–206

Ohta S, Effendi S (1992b) Ultisols of "Lowland Dipterocarp Forest" in East Kalimantan, Indonesia, II. Status of carbon, nitrogen and phosphorus. Soil Sci Plant Nutr 38:207–216

Ohta S, Effendi S, Tanaka N, Miura S (1992) Ultisols of "Lowland Dipterocarp Forest" in East Kalimantan, Indonesia, III. Clay minerals, free oxides and exchangeable cations. Soil Sci Plant Nutr 39:1–12

Soepardi G (1980) Alang-alang (*Imperata cylindrica* (L.) Beauv.) and soil fertility. In Proc. of Biotrop workshop on *Imperata*, Biotrop special pub No. 5, Biotrop, Bogor, Indonesia, pp 57–69

Soil Survey Staff (1997) Keys to Soil Taxonomy, 7th edn, USDA Soil Conservation Service, Pocahontas Press, Inc., Virginia

# 5 Vertical Transportation of Chemical Elements Through Water Movement in Different Vegetation

NAGAHARU TANAKA, DARUL AKSA, SYARIF EFFENDI, and SEIICHI OHTA

## 5.1 Introduction

Strongly weathered soils, such as Acrisols and Ferralsols, are found in the humid tropics. Due to the high rate of weathering under the constantly warm, humid climate of this region, these soils consist mainly of kaolinite, a highly weathered clay mineral that is composed mostly of silica, aluminum, hydrogen, and oxygen (Jordan 1985). As a result, these soils have a low potential for holding nutrients or supplying mineral nutrients through mineral weathering, and these characteristics are recognized as one of the most outstanding features of tropical soils. In East Kalimantan, the major soils are Acrisols derived from Neogene sedimentary rocks (Ohta and Effendi 1992).

Tropical rainforests grow on these highly weathered nutrient-poor substrates, and they do not receive an adequate supply of nutrients from the weathering of soils and rocks. Therefore, nutrient input via precipitation is very important in tropical rainforest ecosystems. The magnitude of the nutrient input through precipitation is often less than the amounts of nutrients cycled within the ecosystem, and much less than the amounts stored in the biomass of mature forests. However, the input via precipitation over tens to hundreds of years can equal the amount stored in the biomass. Golley (1983) suggested that it is not unreasonable to conclude that rain input alone could supply the nutrients to the forest, since the ecological succession in a tropical forest where individual trees may be hundreds of years old, may require hundreds of years. Therefore, to understand nutrient cycling in the tropics it is quite important to know how large the supply of nutrients by precipitation is.

Although few nutrients are contained in rainwater, considerable quantities of nutrients are picked up as the rain washes through the vegetation to reach the soil (Sanchez 1976). The types and concentrations of solutes in rainwater change depending on the vegetation it flows through in the forest canopy, on the trunks of trees, and in the litter layer on the soil surface. Different vegetation might contribute differently to the types and concentrations of nutrients that flow into mineral soils, and this might be closely associated with the sustainability of the vegetation in terms of nutrients.

Ecological Studies, Vol. 140
Guhardja et al. (eds.) Rainforest Ecosystems of East Kalimantan: El Niño, Drought, Fire and Human Impacts

In this chapter, the vertical transportation of chemical elements, mainly nutrients, via water movement is compared in different types of vegetation, including natural forest, plantations of fast-growing leguminous tree species, and devastated grassland. The characteristics of these vegetation types in element cycling are discussed. Nutrients carried by stemflow were not taken into consideration, because observations in the field confirmed that the amount of stemflow is negligible. It is also reported that the volume of stemflow is less than 1 % of the volume of precipi-

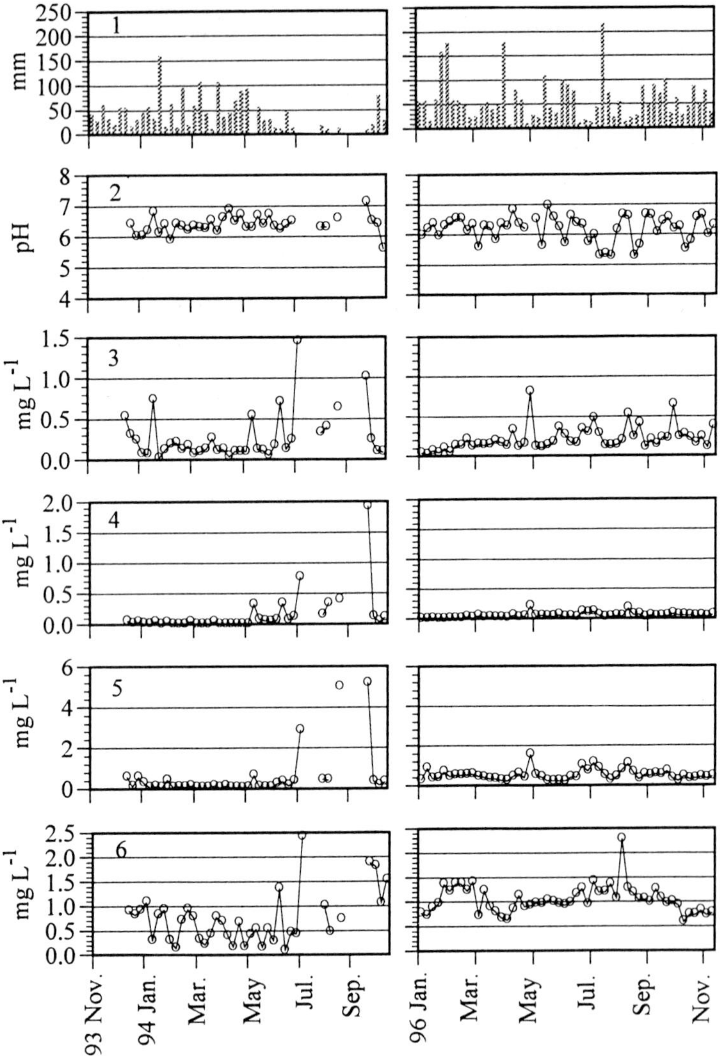


**Fig. 5.1.** Temporal fluctuations of pH and volume-weighted concentration (mg $L^{-1}$) of elements dissolved in precipitation. The numbers *1, 2, 3, 4, 5,* and *6* indicate the weekly precipitation, pH, Ca, Mg, K, and inorganic-N, respectively

tation (Strigel et al. 1994). We also observed temporal fluctuations in the amounts of dissolved elements in precipitation, comparing a year that had no distinct dry period with one that had a dry period.

# 5.2 Sites

Three study sites were established in Bukit Soeharto Experimental Forest, in a natural forest, an *Acacia* plantation, and *Imperata* grassland. All the plots were located on the upper middle part of slopes. The vegetation in the natural forest is a lowland Dipterocarp forest, which has been logged to an extent, but is still well conserved. The dominant tree species was *Shorea laevis*. The *Acacia mangium* plantation was 12 years old, and the tree density was 773 $ha^{-1}$ with an average DBH of 16.3 cm according to measurements made in 1996. The vegetation in the grassland in 1996 was dominated by *Imperata cylindrica* (alang-alang). According to S. Ohta (unpublished work), the soils in these areas were ferric Acrisols.

In each plot, bulk precipitation, throughfall, Ao-percolated-water, soil water, and stream water were collected weekly from November 1993 until October 1994 and from December 1995 until January 1997. The bulk precipitation was collected at two different locations in the area. The throughfall was captured at 8 different points in each plot. The Ao-percolated-water was collected using tension-free 0.019 $m^2$ lysimeters, developed by Arimitsu (1982), at 6 different points in each plot. The soil water was collected using a microlysimeter 0.2 m in diameter filled with undisturbed soil samples in each plot. Fine litter fall was collected monthly from January 1996 to December 1997, using 25 litter traps set up in 1 ha of the natural forest and 10 litter traps in 0.15 ha of the plantation. Water samples were also taken weekly from the stream at the base of the slope in each plot.

During the observation period, the summer of 1994 was dry, and no precipitation was recorded for more than 2 weeks in July and September, while there were no distinctly dry periods in 1996. The drought period in 1994 was a long and exceptionally severe (see Fig. 2.3, Chapter 2, this volume). The annual precipitation was 1842 and 2959 mm in 1994 and 1996, respectively.

# 5.3 Change in Water Chemistry Along the Pathways in Different Vegetation

## 5.3.1 Bulk Precipitation

The mean pH of precipitation during the study period was 6.4, and most of the values were obviously higher than 5.6, the ultimate pH of rainwater equilibrated with atmospheric carbon dioxide. A pH value lower than 5.6 was only recorded 5 times during the observation period (Fig.5.1). The Mg concentration was stable, and stayed around the average value of 0.03 mg $L^{-1}$. Likewise, the K concentration remained around 0.2 mg $L^{-1}$. Meanwhile, the concentrations of other elements fluctuated, and with decreasing rainfall the pH tended to drop while the concentration of elements tended to increase. In particular, in the dry season and immediately after it, the concentration of elements rose dramatically, possibly because during this period the rain dissolved large quantities of elements that were released into the atmosphere in the smoke produced by the slash-and-burn agriculture practiced in many places.

Although the volume-weighted concentrations of nutrients were lower in the wetter period than in the drier period, the nutrient fluxes were not lower in the wetter period because the nutrient concentrations did not drop radically, even in the rainy season. As a result, the annual flux of elements was larger in 1996 than in 1994, as was the amount of rainfall, and were in the order N > K > Ca > Mg every year. This result indicates that higher and constant precipitation is advantageous in terms of nutrient supply.

## 5.3.2 Chemistry of Throughfall, Ao-Percolated-Water, and Stream Water

As mentioned above, the chemical composition of bulk precipitation changes as it dissolves substances while passing through the canopy or Ao layers (Fig.5.2). In the natural forest, the pH value of the throughfall was slightly lower than that of the bulk precipitation, whereas the value rose in the *Acacia* plantation and grassland. The pH value rose after the water percolated through the Ao layers, and dropped remarkably in soil-percolated-water, irrespective of the vegetation. In the stream water, the pH value increased again, and was similar to the value in bulk precipitation or throughfall. The pH changes along the pathways were closely connected with the behavior of base cations.

The concentrations of $Ca^{2+}$, $K^{+}$, and $Mg^{2+}$ increased from the bulk precipitation to the throughfall and Ao-percolated-water because of element enrichment when passing through the canopy and Ao layers, where elements are leached from plant bodies and released through litter decomposition, respectively. Subsequently, the element concentrations decreased in the soil-percolated-water, and increased again in the stream water. The decrease in the concentration when the water passed through the mineral soil layer was attributed to base cation absorption by the soil. It is not known why the concentrations increased in stream water, especially that of $Mg^{2+}$, although it is possible that some elements are supplied from deeper soil, which would suggest that deeper soils have the potential to release nutrients to some extent, possibly through weathering.

The $NH_4^+$ concentration did not change significantly, except for slight increases in the throughfall in every vegetation type. On the other hand, the $NO_3^-$ concentration increased markedly in soil-percolated-water, especially in the *Acacia* plantation, although it did not change very much in other pathways. This significant increase in $NO_3^-$ in mineral soil water is explained by the translocation of considerable amounts of organic nitrogen from the Ao layer to the mineral soil horizons, where $NO_3^-$ is formed though microbial activity in the soil. Fig. 5.3 indicates that the charge is not balanced in most of the Ao-percolated-water and some of the throughfall; these only contain small amounts of inorganic anions, even though the bulk precipitation, soil water, and stream water are mostly balanced in charge. At the same time, the total carbon concentration was highest in the Ao-percolated-water and decreased

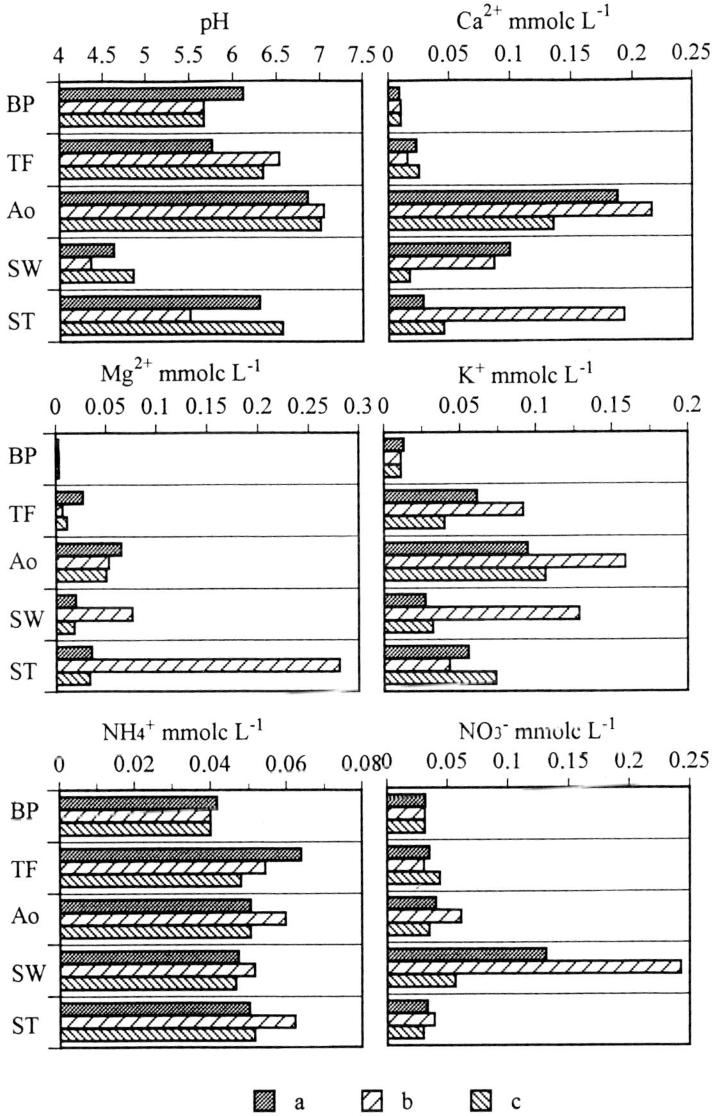


**Fig. 5.2.** Changes in mean pH value and the mean concentration of dissolved elements in bulk precipitation (BP), throughfall (TF), Ao-percolated-water (Ao), soil water (SW), and stream water (ST) for the natural forest (*a*), *Acacia mangium* plantation (*b*), and grassland (*c*), during the study period

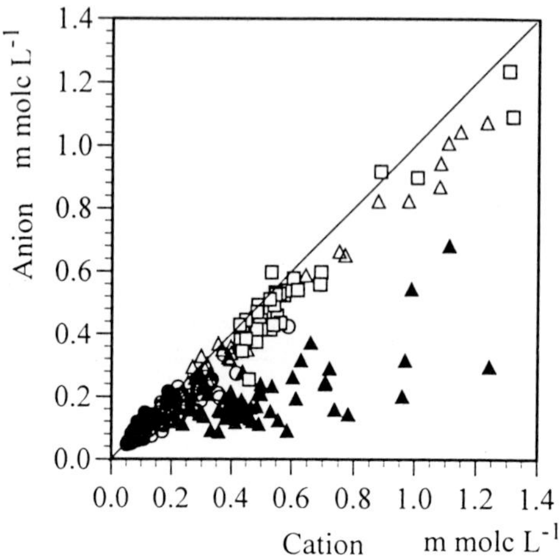


**Fig. 5.3.** Relationships between total inorganic cations and anions in bulk precipitation (*solid circles*), throughfall (*open circles*), Ao-percolated-water (*solid triangles*), soil water (*open triangles*), and stream water (*open squares*) for the *Acacia mangium* plantation

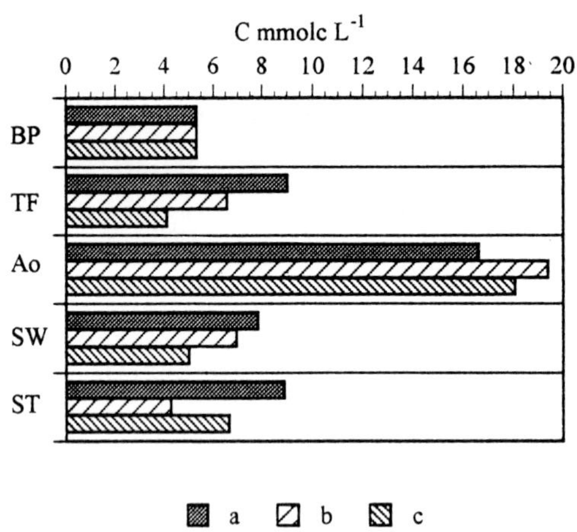


**Fig. 5.4.** The change in mean total carbon concentrations (mg $L^{-1}$) in bulk precipitation, throughfall, Ao-percolated-water, soil water, and stream water for the natural forest (*a*), *Acacia mangium* plantation (*b*), and grassland (*c*), during the study period. The abbreviations used are the same as in Fig.5.2

again in soil water, as shown in Fig. 5.4. This also suggests that substantial amounts of dissolved and particulate organic matter are supplied from the Ao layers to the mineral soil horizon, and they are subsequently lost through mineralization into $CO_2$. The input of dissolved and particulate organic matter from the Ao layer to the mineral layer might also explain the increases in $NO_3^-$ in the soil water. Therefore, these findings suggest that considerable amounts of anions, possibly of organic acids produced in the course of litter decomposition, are dissolved into Ao-percolated-water, and serve to balance the cations and supply substantial amounts of nutrients. The pale brown color of the Ao-percolate-water and the throughfall water also support this assumption.

## 5.4 Fine Litter Fall and Its Chemical Composition

The annual total amount of fine litter fall in the natural forest was 0.77 kg $m^{-2}$ (7.7 ton $ha^{-1}$) in 1996 and 1.00 kg $m^{-2}$ (10.0 ton $ha^{-1}$) in 1997, while in the *Acacia* plantation the amount was 1.12 kg $m^{-2}$ (11.2 ton $ha^{-1}$) in 1996 and 0.94 kg $m^{-2}$ (9.4 ton $ha^{-1}$) in 1997. In the grassland, the litter fall was not measured because the dead leaves of *Imperata* remained standing and did not fall down during the monitoring periods.

The mean concentrations of Ca, Mg, K, and N in the fine litter fall were 1.3, 0.6, 1.5, and 0.8 %, and 0.6, 0.5, 1.1, and 1.6 % for the natural forest and the *Acacia* plantation, respectively. There were striking differences in the Ca and N composition of the vegetation. The Ca content in the natural forest litter was twice that of the *Acacia*, while the *Acacia* litter contained twice as much N as that of the natural forest, probably reflecting the N fixing capability of leguminous species.

## 5.5 Annual Element Fluxes

In each vegetation type, the flux of every element, especially N and K, increased from the bulk precipitation to the throughfall, as elements contained in the plant bodies dissolved (Fig.5.5). The flux was larger in Ao-percolated-water than in throughfall; however, there were no significant differences between the vegetation types. Comparing the fluxes in litter fall and throughfall revealed that litter fall supplied 10 to 50 times as much Ca and Mg to the soil surface as throughfall in the natural forest and *Acacia* plantation, while it only supplied 2 to 4 times as much K and N, since these elements are found in higher proportions in bulk precipitation. An outstanding difference between the natural forest and the *Acacia* plantation was the amount of N input through litterfall. The amount in the *Acacia* plantation was three times larger than that in the natural forest, and this was attributed to the high N content in the foliage of *Acacia*, which is a N fixing leguminous tree species. Compared with other vegetation and locations, the trend is conspicuous. For example, in the low montane rainforest at Kengomna, New Guinea, the amount of N input through throughfall and litter fall was 30 and 91 kg $ha^{-1}$ $year^{-1}$, respectively (Whitmore 1984); in a deciduous broad-leaved forest in Japan, it was 4 and 80 kg $ha^{-1}$ $year^{-1}$, respectively; and in a *Chamaecyparis obtusa* plantation it was 20 and 43 kg $ha^{-1}$ $year^{-1}$, respectively (Katagiri 1996). However, there were no obvious differences in the N flux in Ao-percolated-water among the ecosystems, probably because organic N was not included in the N flux. The significant difference in the N flux in litterfall and further differences between the Ao-percolated-water in the vegetation types greatly influenced the element fluxes in the mineral soil layers. The N flux in the mineral surface soil of the *Acacia* plantation was 2 to 3 times larger than in the natural forest or grassland. The fluxes of Mg and K in the *Acacia* were also 4 to 5 times larger than in the other types of vegetation. In contrast, in the natural forest and grassland only the Ca flux changed drastically in soil. These findings suggest that the larger N flux accompanies the leaching of basic elements such as Mg and K in the soil under the *Acacia* plantation to a greater degree. Assuming that the supply of these basic ele-

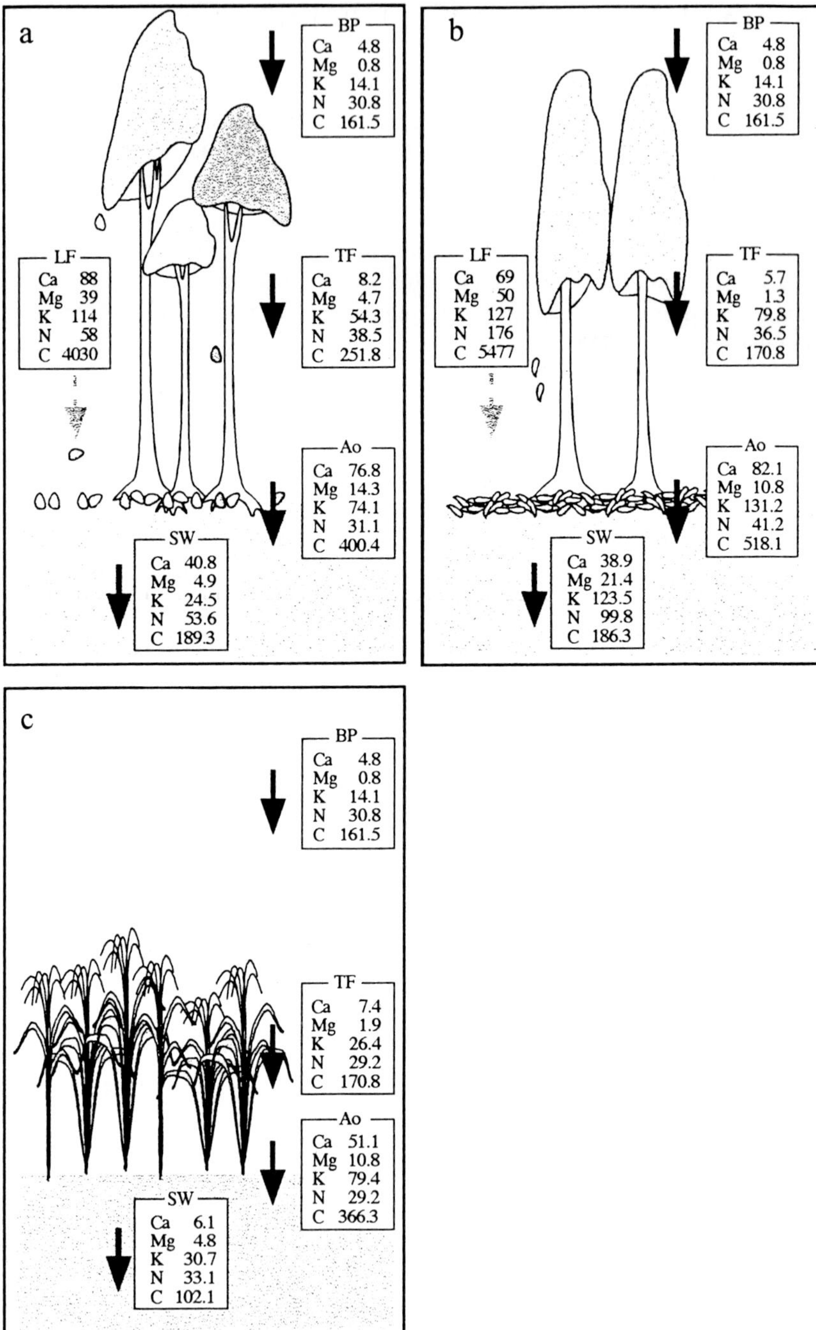


**Fig. 5.5.** The annual flux of each element in each pathway in different vegetation (in kg $ha^{-1}$ $year^{-1}$): natural forest (**a**), *Acacia mangium* plantation (**b**), and grassland (**c**). The abbreviations used are the same as in Fig. 5.2

ments to the soil is similar in all three types of vegetation, it is suspected that the soils under the *Acacia* plantation will gradually lose basic elements and become acidified. On the other hand, the fluxes of nutrient elements in soil water in the grassland were almost the same as in the natural forest, except for Ca, which was even smaller in the grassland than in the natural forest. Consequently, the *Imperata* grassland is assumed to be capable of conserving basic elements to a similar extent as natural forests, while the *Acacia* plantation tends to lose more Mg and K than the other vegetation.

## 5.6 Conclusion

In conclusion, in terms of nutrient conservation, the *Imperata* grassland system might have a well-developed mechanism that is comparable to that of natural forest vegetation. This might be one reason why the soils under *Imperata* grasslands are not deteriorated, at least in terms of basic cations (see Chapter 4 , this volume). On the other hand, the *Acacia* plantation is less capable of conserving nutrients than the natural forest or grassland, because of the larger flux of nitrogen as $NO_3^-$ in the system due to the N fixation capacity of the species. This could accelerate the loss of basic cations from the soil system through leaching, which might become a serious problem for the sustainability of *Acacia* plantations. Studies of the *Acacia* plantation system have to evaluate the influences of the characteristics of element cycling peculiar to the system, and not focus only on the impact of short-rotation harvesting. The human transformation of the natural vegetation in this region will lead to the destruction of balanced nutrient cycling and result in the degradation of soils that are originally poor in nutrients.

## References

Arimitsu K (1982) Studies of the dynamics of water in forest soils II. Movement of soil water and solute elements. Bull For & For Prod Inst 318:11–78 (in Japanese)

Golley FB (1983) Nutrient cycling and nutrient conservation. In: Golley FB (Ed) Tropical rainforest ecosystems, structure and function. (Ecosystems of the world; 14A). Elsevier, Amsterdam, pp 137–155

Jordan CF (1985) Nutrient cycling in tropical forest ecosystems. J Wiley, Chichester,

Katagiri S (1996) Production and nutrient cycling at different sites. In: Iwatsubo G (ed) Forest ecosystems (in Japanese). Buneidou, Tokyo, pp 24–258

Ohta S, Effendi S (1992) Ultisols of 'Lowland Dipterocarp Forest' in East Kalimantan, Indonesia, I. Morphology and physical properties. Soil Sci Plant Nutr 38:197–206

Sanchez PA (1976) Properties and Management of soil in the Tropics. J Wiley, New York, pp 349–359

Strigel G, Ruhiyat D, Prayitno D, Sarmina S (1994) Nutrient input by rainfall into secondary forests in East Kalimantan, Indonesia. J Trop Ecol 10:285–288

Whitmore TC (1984) Tropical rainforests of the Far East. Oxford Univerity Press, London, pp 123–134

# 6 Changes in Soil Nutrient Status After Abandonment of Swidden Agriculture at Benuaq Dayak Village

Kazuhito Morisada, Syarif Effendi, and Seiichi Ohta

## 6.1 Introduction

Swidden agriculture has a long history in Borneo, and there is evidence of major forest clearing 2500 B.P (before present). (Maloney 1985). Swidden agriculture is still the major source of livelihood for the indigenous people of inland Borneo (MacKinnon et al. 1996). Traditional swidden agriculture is characterized by long fallow periods between short periods of crop production (Whitten et al. 1987). The importance of the fallow period for maintaining the forest vegetation under shifting cultivation was well appreciated by Nye and Greenland (1960). Within the last few decades, however, increasing population pressure and the infiltration of a monetary economy have altered traditional land use, with the conversion of forests to permanent crop fields, shortening of fallow periods, and so on (e.g., Robison and McKean 1992; Chapter 15, this volume).

Swidden agriculture is based on the increase in the nutrients available in the soil when the forest, including the leaf litter, is burnt (Hirai 1998). It is a sustainable land use practice when the forest is allowed to grow back, because forest vegetation can store more nutrients in the aboveground biomass and surface litter than other vegetation communities such as grassland. The conversion of forest to permanent crop fields acidifies the soil and exhausts soil nutrients, such as available nitrogen and phosphorus (Funakawa et al. 1997), so that the fields require artificial fertilizers. The present controversy over the effect of swidden agriculture focuses on how altered land use, such as the shortening of the fallow period, may lead to degradation of the forest vegetation (e.g., Collins et al. 1991, MacKinnon et al. 1996). The current concern is whether the altered swidden agriculture is sustainable in terms of soil productivity for forest regeneration.

This chapter describes changes in soil nutrient status under swidden agriculture in the transitional stage (see Chapter 15, this volume), as practiced by the Benuaq people, one of the major Dayak tribes.

Ecological Studies, Vol. 140
Guhardja et al. (eds.) Rainforest Ecosystems of East Kalimantan: El Niño, Drought, Fire and Human Impacts

## 6.2 Fallow Forests in the Mencimai Village Area

The village of Mencimai, Barong Tongkok subdistrict, East Kalimantan province, is located in the middle course of the Mahakam basin (0°17'S, 115°40'E). The area ranges in altitude from 80 to 120 m above sea level. The geology is characterized by sandy Tertiary deposits, which were partially affected by Neogene volcanic eruptions. The climate is classified into Köppen's Af, and the mean annual precipitation is about 3000 mm (Voss 1982).

The Benuaq people established a settlement at Mencimai more than 300 years ago, and have supported themselves using traditional swidden agriculture for a long period. Their prolonged land use has changed the surrounding forest vegetation, and presently pyrophytic trees such as *Schima wallichii*, *Vitex pinnata*, *Peronema canescens*, and *Vernonia arborea* predominate near the village (see Chapter 17, this volume). Since Javanese people moved to the area and established a new settlement in the 1960s, the villagers' land use practices have changed rapidly. The Javanese people converted the forests into annual and perennial crop fields, and the Benuaq people also converted some fallow forests into rubber plantations. The vegetation presently surrounding the village includes primary and secondary forest, shrubs, and crop fields, including rubber plantations and fruit gardens (see Chapter16, this volume). Kiyono estimated the present fallow period in the village to be about 10.6 years (Chapter 17, this volume), which is remarkably short in comparison with the traditional 29 years (Inoue 1990).

The study area was about 2 km west of the village center. To relate soil nutrient status to fallow period, four plots with different fallow periods were set up following Okimori and Matius (1990) (Table 6.1). The four plots had been fallow for 1, 3, 15, and 30 years; shrubs predominated in the first two plots, while the last two were forest. The basal area of the 30-year-old stand was almost the same as that of the 15-year-old stand, but its canopy was higher. The aboveground biomass was smallest in the 1-year-old shrubs, and increased with the fallow period. Organic matter layers

**Table 6.1.** A brief description of the soil-sampling plots

| Plot name | Description [vegetation features, Maximum tree height[a], basal area[a] (dbh > 2 cm and height > 1.5 m), remarks] |
|---|---|
| 1-year-old shrubs | Abandoned upland rice field, shrub covered with herbaceous plants, signs of cassava (*Manihot esculenta*) cropping were noticed. |
| 3-year-old shrubs | Shrub dominated by *Macaranga* spp. and *Trema orientalis*, 6 m, 2.3 $m^2$ $ha^{-1}$, corresponds to "3-year-old shrubs" in Chapter 16, this volume. |
| 15-year-old stand | Secondary forest dominated by *Macaranga gigantia* and *M. triloba*, 20m, 13.8 $m^2$ $ha^{-1}$, the same plot as "15-year-old stand" in Chapter 16, this volume. |
| 30-year-old stand | Secondary forest dominated by *Pentace laxiflora* and *Shorea parvifolia*, 30 m, 13.3 $m^2$ $ha^{-1}$, corresponds to "30-year-old stand" in Chapter 16, this volume. |

[a] Compiled from Okimori and Matius (1990).

on the soil surface were rarely found or were very thin in the 1-, 3-, and 15-year-old stands. F layers 7 to 8 cm thick with many fine roots developed on the soil surface of the 30-year-old stand. In every plot, the soil color changed abruptly from dark brown in the surface horizons to yellowish brown in underlying horizons. Separate composite samples of surface (0–5 cm in depth), subsurface (10–15 cm), and sub (30–35 cm) soils were taken from the upper, middle, and lower slope in each plot. Each composite sample consisted of samples taken in triplicate using a 100mL soil-sampling cylinder (20 $cm^2$ × 5 cm).

## 6.3 Changes in the Physio-Chemical Properties of Soil with Fallow Period

An analysis of variance of the data indicated that the properties of samples from the same depth were significantly different ($p$<0.05) between the plots. On the other hand, except for pH($H_2O$), total carbon (T-C), and available phosphorus (Av-P), they did not differ significantly ($p$<0.05) between the different slope positions. Therefore, to examine the changes in soil nutrient status with fallow period, the data from the same depth at different slope positions were grouped together for each plot, and the mean values of each parameter were calculated.

In every plot, the texture of the 0–5 cm soil layer expressed as percentage clay was sandy (Table 6.2). The values of percentage clay for every soil were remarkably low in comparison with those of Ultisols, which are the major soils in the region (Ohta and Effendi 1992). The soil texture of the 30-year-old stand was only slightly coarser than that of the 1-, 3-, and 15-year-old stands, but the difference was significant ($p$<0.05).

The soil pH($H_2O$) was lowest in the 0–5 cm layer, and increased with depth in every plot. The pH($H_2O$) generally increased with the fallow period for every depth except the 0–5 cm layer of soil in the 30-year-old stand, which was significantly more acidic than the same layer in the other plots. The exceptionally low pH($H_2O$) in the 0–5 cm layer of the 30-year-old stand might be attributed to the accumulation of larger amounts of organic matter on the soil surface, which might have acidified the surface mineral soil. The effective cation exchange capacity (ECEC) was highest for the 0–5 cm layer and generally decreased with depth in every plot. The ECEC in the

**Table 6.2.** Clay content in the 0–5 cm soil layer

| Plot | Clay content (percentage clay in soil) |
|---|---|
| 1-year-old shrubs | 13.9 a |
| 3-year-old shrubs | 18.4 a |
| 15-year-old stand | 14.8 a |
| 30-year-old stand | 6.8 b |

Means ($n$=9) followed by the same letter are not significantly different ($p$<0.05).

0–5 cm layer was highest in the 1-year-old shrubs, and generally decreased with the fallow period.

There were no clear trends for parameters in the 10–15 cm and 30–35 cm soil layers with the fallow period.

The T-C, total nitrogen (T-N), Av-P, and the exchangeable calcium (ex-Ca), magnesium (ex-Mg), and potassium (ex-K) were generally higher in the 0–5 cm layer than in the underlying soil layers within a plot, and were generally highest in the 1-year-old shrubs. This implies that nutrients supplied by burning the forest slash still remained in the surface soil 1 year after the fields had been left fallow.

## 6.4 Changes in Nutrient Stock with Fallow Period

Guillemin (1956) illustrated the relationship between the length of the fallow period and soil productivity schematically as shown in Fig. 6.1. After a cultivated field is abandoned, the nutrients are transferred from the soil into the vegetation through absorption by plants and accumulate in the biomass with the progress of plant succession. A fraction of the nutrients absorbed into the plants is returned to the soil through litter production, and this amount increases as the forest matures. As a result, the nutrients stored in the litter layer and surface mineral soil build up gradually with growth of the vegetation. Under the traditional swidden system of agriculture, the fallow period is sufficiently long for the productivity of the surface soil to accumulate.

The nutrient stock in a soil is an important parameter for judging the productivity of the soil, and changes reflecting the length of the fallow period are believed to explain changes in soil productivity. In this section, the effects on soil productivity of changes in the fallow period are discussed in terms of nutrient stock. The stock of each nutrient in the soil 0–30cm deep was calculated, and the means for different fallow periods are listed in Fig. 6.2.

### 6.4.1 Carbon, Nitrogen, and Phosphorus

The T-C decreased significantly with the fallow period; the highest amount (8.7 kg/$m^2$) was found in the 1-year-old shrubs and the lowest (7.0 kg/$m^2$) in the 30-year-old stand. The vegetation in the first three plots remained at an early succession stage. In the earlier stages of succession, the production of aboveground biomass remains low, and the supply of organic matter to the soil system through litter is limited. Meanwhile, the loss of soil carbon through organic decomposition is believed to be maintained or even accelerated by increased soil temperatures under thinner forest canopies, and exceeds the carbon input through litter production. This imbalance in the soil carbon budget may be responsible for the decrease in the T-C with the fallow period. At the same time, however, the lower T-C in the 30-year-old stand might also be at least partially attributed to the lower clay content of the soil, because the soil T-C is generally controlled by the clay content in the Ultisols of the region (Ohta and

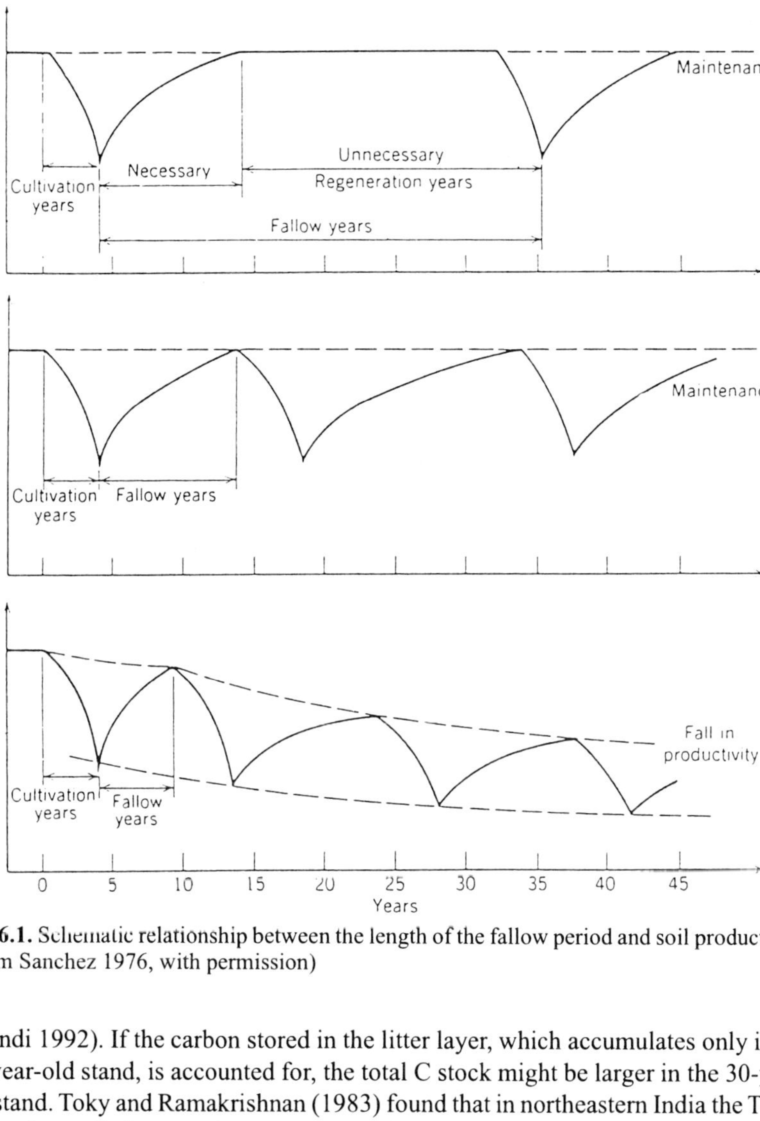


**Fig. 6.1.** Schematic relationship between the length of the fallow period and soil productivity. (From Sanchez 1976, with permission)

Effendi 1992). If the carbon stored in the litter layer, which accumulates only in the 30-year-old stand, is accounted for, the total C stock might be larger in the 30-year-old stand. Toky and Ramakrishnan (1983) found that in northeastern India the T-C in the surface soils increased over 50 years of fallow, while Sanchez (1976) mentioned that the stock usually decreased with a lengthening fallow period. Our results support Sanchez's opinion (1976), and we suspect that the stock of carbon in the soil requires a fallow period of at least 15 years to recover.

The level of T-N was roughly the same in the first three plots, about 0.5kg/m$^2$, but clearly decreased to 0.4kg/m$^2$ in the 30-year-old stand. The T-N in the biomass of the 30-year-old stand is probably far larger than that of the other plots, because of the larger biomass. In the 30-year-old stand, the progress of nitrogen transfer from the soil to the biomass and its accumulation in the surface litter layer through absorption and litter production by trees might lead to the low T-N in the mineral soil.

Scott (1978) found that in Gran Pajonal, Peru, nitrogen storage in the soil decreased with the progress of succession. However, the T-N in the soil of the 30-year-old stand might exceed that of the other plots if the nitrogen stored in the litter layer were included. The low T-N content in the 30-year-old stand might also be partially attributed to the lower clay content of the soil, since clay content also governs T-N (Ohta and Effendi 1992).

The Av-P was maintained for 15 years, and then increased in the 30-year-old stand. The Av-P in soil increases after burning because of the high P content of the ash. It decreases during cultivation, and this is usually one of the important reasons

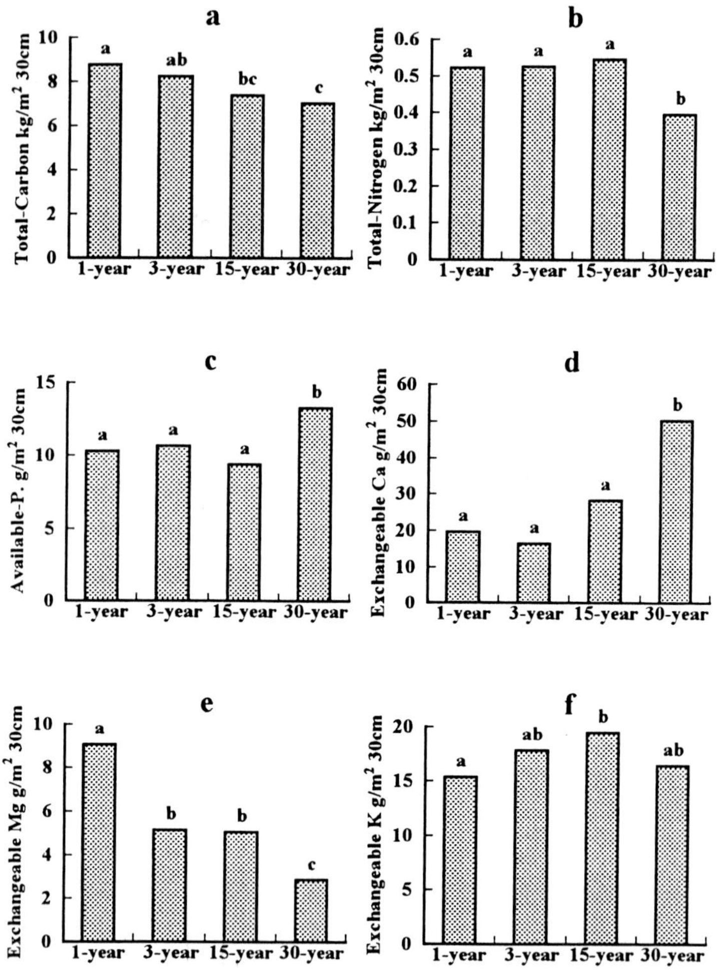


**Fig. 6.2.** Comparison of nutrient stock in the soil (0–30cm depth) with different fallow periods. **a** Total carbon, **b** total nitrogen, **c** available phosphorus, **d** exchangeable calcium, **e** exchangeable magnesium, and **f** exchangeable potassium. Means ($n$=9) followed by the same letter are not significantly different ($p$<0.05)

for abandoning the cultivation. The Av-P under the degraded vegetation was clearly lower than under the primary forest. It increased as the vegetation recovered (see Chapter 4, this volume), possibly in association with soil faunal development, particularly the mycorrhiza population, which is thought to play an important role in solubilizing insoluble phosphate. The Av-P in the 30-year-old stand was probably related to the amount of vegetation. In terms of the Av-P, the condition of the soil only improved after a fallow period of at least 15 years.

## 6.4.2 Exchangeable Cations

The ex-Ca in the soil in this study was initially 19.7 kg/m$^2$ in the 1-year-old shrubs. The value increased progressively to 16.5, 28.3, and 50.2 kg/m$^2$ in the 3-, 15-, and 30-year-old stands, respectively. This suggests that the transfer of Ca from the soil to the aboveground biomass and its return to the soil progressed successfully, resulting in the increased level seen in the 30-year-old stand. The ex-Mg was highest in the 1-year-old shrubs (9.1 kg/m$^2$), and decreased significantly with the fallow period: 5.1, 5.0, and 2.9 kg/m$^2$ for the 3-, 15-, and 30-year-old stands, respectively. The ex-Mg in the soil in this study was low compared with that in Ultisols in the region (Ohta et al. 1993). Therefore, the absorption of ex-Mg by the biomass markedly influenced the stock in the soil, and might cause the rapid decrease with fallow period. The ex-K level remained fairly constant with fallow period: 15.3, 17.8, 19.4, and 16.4 kg/m$^2$ for the 1-, 3-, 15-, and 30-year-old stands, respectively.

The cation stocks in soil increase greatly after burning, and decrease by various amounts during the cultivation period due to leaching and crop uptake. After abandonment, cations are translocated from the soil to the regenerated biomass and accumulate in the biomass and litter. In Gran Pajonal, Peru, Scott (1978) showed that the stock of cations (Ca, Mg, and K) in the vegetation, which included above and below ground biomass, tended to increase with vegetation succession, but the amount actually stored in the soil clearly declined. Zinke et al. (1978) also showed that the ex-Ca and ex-Mg stocks in soil decreased with the fallow period in the Lua forest of northern Thailand. In northeastern India, Toky and Ramakrishnan (1983) showed that regeneration of bamboo, a potassium accumulator, increased the stock of potassium in the soil. The changes in the ex-Mg stock found in this study concur with the results just mentioned, but the changes in the ex-Ca and ex-K do not. The change in the stock of bases in soil with fallow period could vary with the characteristic turnover patterns for base cations in the regenerating species. Masunaga et al. (1997, 1998) found that in the tropical rainforest of West Sumatra, Indonesia, specific tree species accumulated various forms of aluminum, potassium, and calcium, and the distribution of the accumulating species was closely related to the soil edaphic status. The different nutrient uptake characteristics of the regenerating species must be taken into consideration in any discussion of the build up of soil productivity during fallow.

## 6.5 Conclusion

Comparison of the soil after different fallow periods revealed that the nutrient stock changed with the fallow period. As a whole, the nutrient stock in the soil was not necessarily replenished with 30 years of fallow. In the soil, soil carbon and available phosphorus were not built up for at least 15 years, and soil productivity is thought to be at the midpoint after a 15-year fallow period. However, the soils in the study area are much sandier than the soils in the region, which might influence the nutrient stock, since clay particles play an important role in retaining nutrients in soil (Ohta and Effendi 1992, Ohta et al. 1993). Therefore, it is possible that the fallow period required for nutrient build up might be shorter in more clayey soil. The species composition of the regenerated vegetation is also thought to influence the build up of nutrient stock (Uhl and Jordan 1984).

These results indicate that current land use in the village with the present 10.6-year estimated average fallow period is not sustainable and may threaten soil productivity. A fallow period of more than 30 years that is needed to build up soil productivity will be difficult to provide with the pressure of increasing population and the infiltration of a market economy in the village. A land classification based on soil productivity should be considered to help solve this difficulty. Lal (1995) also stressed the importance of a land use capability assessment from the viewpoint of sustainable management of soil resources in the humid tropics. Infertile fallow sites, such as those examined in this study, should remain forest and be allowed to build up soil productivity with forest succession. On land that is suitable for crop production, new land use practices should be studied to satisfy people's demands.

## References

Collins, NM, Sayer, JF, Whitmore, TC (eds) (1991) The conservation atlas of tropical forests: Asia and the Pacific. Macmillan, London

Funakawa S, Tanaka S, Shinjyo H, Kaewkhongkha T, Hattori T, Yonebayashi K (1997) Ecological study of soil organic matter and its related properties in shifting cultivation systems of Northern Thailand. Soil Sci Plant Nutr 43:681–693

Guillemin R (1956) Evolution de l'agriculture autochtone dans les savannes de l'Obangui. Agron Tropicale 11:143–176 (cited from Sanchez (1976))

Hirai H (1998) Ecological systems under shifting cultivation and soils (in Japanese). Jpn J For Environ 40:49–54

Inoue M (1990) Changes in the swidden system of the Kenyah Dayak tribe (in Japanese with English abstract). Southeast Asian Stud 28:222–255

Lal R (1995) Sustainable management of soil resources in the humid Tropics. UN University Press, Tokyo

MacKinnon K, Hatta G, Halim H, Mangalik A (1996) The ecology of Kalimantan. Periplus Editions, Singapore

Maloney BK (1985) Man's impact on the rainforests of West Malaysia: the palynological record. J Biog 12:537–558

Masunaga T, Kubota D, Hotta M, Wakatsuki T (1997) Nutritional characteristics of mineral elements in tree species of tropical rainforest, West Sumatra, Indonesia. Soil Sci Plant Nutr 43:405–418

Masunaga T, Kubota D, William U, Hotta M, Shinmura Y, Wakatsuki T (1998) Spatial distribution pattern of trees in relation to soil edaphic status in tropical rainforest in West Sumatra, Indonesia I. Distribution of accumulating trees. Tropics 7:209–222

Nye PH, Greenland DJ (1960) The soil under shifting cultivation. Commonwealth Bureau of Soils Tech Comm 51, CAB, Bucks

Ohta S, Effendi S (1992) Ultisols of "lowland dipterocarp forest" in East Kalimantan, Indonesia II. Status of carbon, nitrogen, and phosphorus. Soil Sci Plant Nutr 38:207–216

Ohta S, Effendi S, Tanaka N, Miura S (1993) Ultisols of "lowland dipterocarp forest" in East Kalimantan, Indonesia III. Clay minerals, free oxides, and exchangeable cations. Soil Sci Plant Nutr 39:1–12

Okimori Y, Matius P (1990) Regeneration of secondary forest after swidden cultivation by the Dayak people in Indonesia (in Japanese). Jpn J For Environ 32(2):79–87

Robison DM, McKean SJ (1992) Shifting cultivation and alternatives, an annotated bibliography, 1972–1989. CAB, Wallingford

Sanchez PA (1976) Properties and management of soils in the tropics. J Wiley, New York

Scott GA (1978) Grassland development in the Gran Pajonal of eastern Peru. PhD dissertation, Department of Geography, University of Hawaii

Toky OP, Ramakrishnan PS (1983) Secondary succession following slash and burn agriculture in north-eastern India II. Nutrient cycling. J Ecol 71:747–757

Uhl C, Jordan CF (1984) Succession and nutrient dynamics following forest cutting and burning in Amazon. Ecology 65:1476–1490

Voss F (1982) Atlas of East Kalimantan, Indonesia. Transmigration Area Development Project. HWWA-Institut für Wirtschaftsforschung, Hamburg (Unpublished)

Whitten AJ, Mustafa M, Henderson GS (1987) The ecology of Sulawesi. Gadjah Mada University Press, Yogyakarta

Zinke PJ, Sabhastri S, Kunstadter P (1978) Soil fertility aspects of the Lua's forest fallow system of shifting cultivation. In: Kunstadter P, Chapman EC, Sabhasri S (eds) Farmers in the forest, University Press of Hawaii, Honolulu, pp 134–159

# 7 Serpentine Soils on Catena in the Southern Part of East Kalimantan, Indonesia

SYARIF EFFENDI, SATORU MIURA, NAGAHARU TANAKA, and SEIICHI OHTA

## 7.1 Introduction

Serpentinite is a well-known ultramafic rock, which is closely associated with tectonic plate boundaries throughout the world (Brooks 1987). Unique soils and stunted flora develop on serpentinite because of its peculiar chemical composition. Proctor and Woodell (1975) and Brooks (1987) have reviewed the relationships between the flora and specific properties of the soil, e.g., excessive magnesium, calcium deficiency, Ca/Mg imbalance, heavy metal toxicity, low nutrient levels, and so on. They pointed out that more species are generally found on serpentine soil than on non-serpentine soil, although the growth of trees on serpentine soil is lower than on non-serpentine soil. Many endemic species are also commonly found in serpentine vegetation. These features of serpentine ecosystems contribute to the biological diversity of a region. Nevertheless, serpentine soils tend to be exposed to the danger of erosion, because the cover consists of scattered stunted shrubs. Spence (1970), and Proctor and Woodell (1971) have reported examples of barrens or fellfields caused by erosion.

In the last three decades, many settlers have migrated into the central lowland of East Kalimantan. They have shifted into the interior as commercial deforestation has moved towards the high country (see Inoue, Chapter 15 and Kiyono and Hastaniah Chapter 17, this volume ). Some illegal farmers have exploited shifting cultivation, which has degraded the soil irreversibly (Ohta et al., Chapter 4, this volume ). This situation also occurs in the southern part of East Kalimantan, where ultramafic rocks including serpentinite occur in hilly to mountainous regions. Because of the vulnerability of serpentine soils, we are concerned about the destruction of this ecosystem with agricultural development. In this chapter, we discuss the physico-chemical properties of serpentine soils in the humid tropics, focusing on catena, and make suggestions for managing such delicate soils.

Ecological Studies, Vol. 140
Guhardja et al. (eds.) Rainforest Ecosystems of East Kalimantan: El Niño, Drought, Fire and Human Impacts

## 7.2 Study Area and Soils

The study area was located at 1°50' S and 116°00' E, near Kuaro town, in Pasir district, East Kalimantan. The mean annual rainfall in the area is about 2800 mm, and the mean annual temperature is about 28°C. The climate in the region is classified as Köppen's Aaw'f, or a tropical rainy isothermal climate with a hot summer, no dry season, and a single rainfall peak each year. The soil climate in the region falls into isohyperthermic soil temperature and udic soil moisture regimes. The geology of the area is, in general, older metamorphosed pre-tertiary rocks such as schist, and some intrusions of ultrabasic dunites, peridotites, and serpentinites are also found. The geomorphology of the area consists of high, rugged, strongly folded, and faulted mountains with large differences in elevation and very steep slopes in many places (Land Resource Development Centre, 1987). The elevations in the area studied range from ca. 65 to 365 m.

Fourteen pedons were examined along catena developed on the serpentinite, and their physico-chemical properties were determined. The subgroup names (Soil Survey Staff, 1997) and horizon designations are presented in Table 7.1. Pedons representing each soil group are illustrated in Fig. 7.1. Four groups of soils were recognized in the pedons according to drainage, geomorphology, and pedological properties.

The geomorphological situation and defining characteristics of the pedons are as follows. Group 1 was characterized primarily by poor drainage due to consolidated fresh serpentinite, and was located on gentle slopes on plateaus or steep seepage

**Table 7.1.** Subgroups in Soil Taxonomy (Soil Survey Staff 1997) and horizon designations of 14 pedons present on the serpentinite catena in Pasir district in East Kalimantan

| Profile | Inclination | Depth(cm) | Subgroup name | Horizon designation[a] |
|---|---|---|---|---|
| **Group 1** | | | | |
| KR35 | 9 | 48 | Lithic Hapludoll | A1-A2-AC-C |
| KR28 | 12 | 48 | Lithic Hapludoll | A1-A2-AC-C |
| KR29 | 3 | 80 | Typic Hapludoll | A1-A2-CA-C |
| KR18 | 28 | 80 | Typic Hapludoll | A1-A2-C |
| KR19 | 27 | 89 | Typic Hapludoll | A1-A2-CA-C |
| **Group 2** | | | | |
| KR6 | 34 | 110 | Typic Eutropept | A-BA-Bo-BC |
| KR5 | 10 | 58 | Typic Eutropept | A1-A2-BA-BC |
| **Group 3** | | | | |
| KR34 | 19 | 225 | Typic Dystropept | A-AB-Bo1-Bo2-Bo3-BC-C |
| KR4 | 11 | 195 | Typic Dystropept | A-BA-Bo1-Bo2-Bo3-BCo-C |
| KR1 | 12 | 110 | Lithic Kandiperox | A-BA-Bo1-Bo2-BCoc |
| KR36 | 6 | 134 | Rhodic Haploperox | A-Bo1-Bo2-BC-C |
| **Group 4** | | | | |
| KR3 | 5 | 220 | Rhodic Kandiperox | A1-A2-BA-Bo-Boc1-Boc2-Btoc |
| KR20 | 5 | 215 | Anionic Acroperox | A-AB-Boc1-Boc2-Boc3-Boc4-Boc5-Boc6 |
| KR31 | 3 | 220 | Anionic Acroperox | A-AB-Boc1-Boc2-Boc3-Boc4 |

[a]c, nodules (iron); o, residual accumulation of sesquioxides; t, accumulation of silicate clay

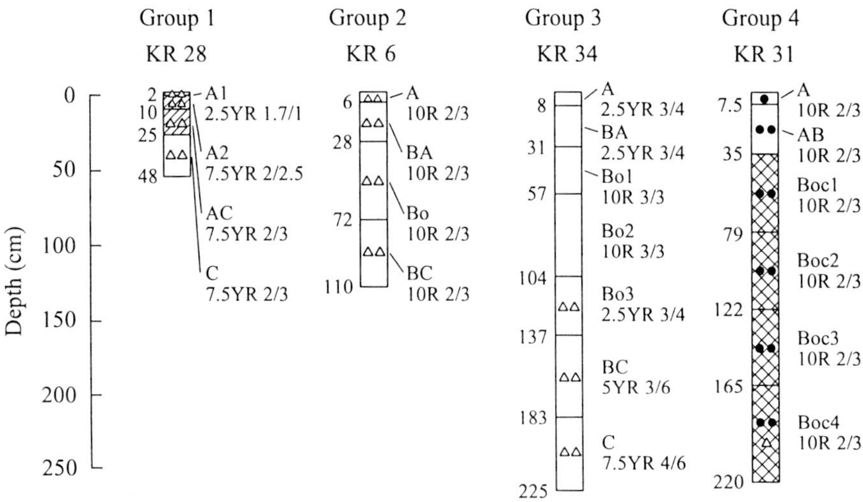


**Fig. 7.1.** Representative soil profiles of the four groups occurring on the serpentine catena. *Diagonal lines*, mollic epipedon; *cross-hatched lines*, oxic horizon; *triangles*, gravel and boulders; *dots*, iron oxide nodules

slopes. Outcrops of serpentinite rock were common, and gravel was found very frequently throughout the profile. The profiles were restricted by lithic contact at depths of 50 to 90 cm. They had brownish black mollic A horizons (7.5YR3/2), with a high organic content. The pedons in Group 1 included A and C horizons, but lacked B horizons, and were classified as Lithic Hapludolls or Typic Hapludolls.

The pedons in Groups 2 to 4 were located on well-drained slopes. The soils of these groups had very dark reddish-brown (10R or 7.5R) A and B horizons. These three groups had different slope stability and profile structures. The pedons in Group 2 were found on very steep unstable slopes or narrow ridges, and frequently contained weathered serpentinite gravel in almost every horizon and boulders at the bottom of the profile. Since the A horizons of Group 2 were shallower than 18 cm, which does not satisfy the requirement for a mollic epipedon, they were classified as Typic Eutropepts. The pedons in Group 3 occurred on moderately stable slopes, and had thick solums over 100 to 150 cm. Gravel and boulders of weathered or unweathered serpentinite were frequently found in the lower parts of the B and C horizons. Some of the pedons had bright C horizons 5YR to 7.5YR in hue. This group included Typic Dystropepts and Rhodic Haploperoxs. The pedons in Group 4 were found on very stable plateaus or lower toe slopes. They were characterized by a very thick solum and there were no signs of saprolite or fresh bedrock at 200 cm depths. Little or no gravel was found in the horizons, but spherical iron oxide nodules 0.5 to 2 mm in diameter were commonly found in the Bc horizons. These soils were Rhodic Haploperoxs or Anionic Haploperoxs.

The forest conditions for each group were as follows. Stunted scrubs grew at

Group 1 to 3 sites, and there were no trees larger than 30cm in diameter. On the other hand, forests similar in appearance to mature Dipterocarp forest formed at the Group 4 sites, with a canopy layer and a sub-canopy layer and trees larger than 50cm in diameter.

## 7.3 Chemical Properties

The pedogenic process in serpentine soils is mainly characterized by laterization, which involves the removal of silica and bases, and the residual accumulation of sesquioxides. This might be accelerated in a warm humid tropical climate (Brooks 1987). In the study area, both thick lateritic Oxisols (Group 4) and less laterized Molisols (Group 1) were found. The clear differences in the chemical properties of the soils were considered to be evidence of the existence of a variety of pedogenic stages in the soils studied. This was interpreted to be a consequence of different weathering, affected by drainage and geomorphology.

### 7.3.1 Bases and pH

The characteristic high exchangeable magnesium in serpentine soils is ascribed to the serpentinite parent material, which has a high magnesium content, (Wildman et al. 1968, Brooks 1987, Rabenhorst et al. 1982, Bonifacio et al. 1997). The small amounts of exchangeable calcium and potassium in the soils have predominantly biogenic origins (Kanno et al. 1965, Proctor and Woodell 1975).

The change in the exchangeable base contents with soil depth is presented in Fig. 7.2. The exchangeable magnesium level in the Groups 1 and 2 soils was remarkably

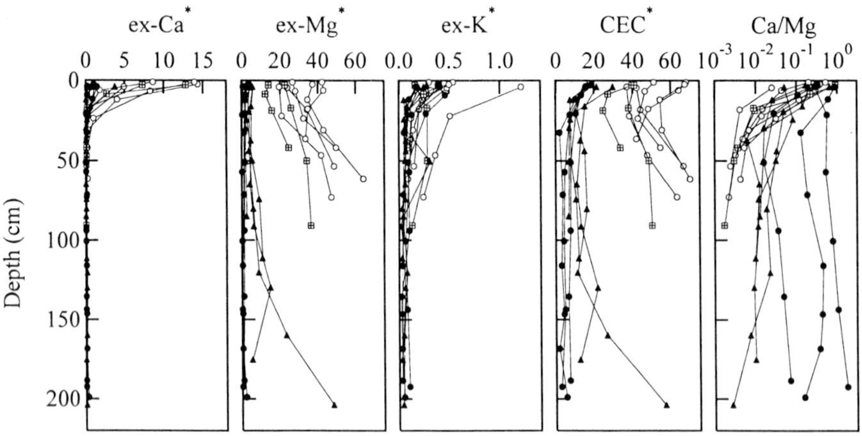


**Fig. 7.2.** Changes in the exchangeable bases, cation exchange capacity (CEC), and Ca/Mg ratio of the soils in the four groups with depth. *Open circles*, Group 1; *hatched squares*, Group 2; *solid triangles*, Group 3; *solid circles*, Group 4. *cmol+ $kg^{-1}$

high, while that of Group 4 was lower than 1.0 cmol+ $kg^{-1}$. The high values around 40 cmol+ $kg^{-1}$ for Group 1 are comparable with those found in temperate regions (Rabenhorst et al. 1982, Bonifacio et al. 1997). The difference in the exchangeable magnesium status of the four groups reflects the extent of leaching. The exchangeable calcium and potassium levels in these soils are low, and there is little difference between groups. The remarkable difference between the exchangeable calcium and magnesium results in a very low Ca/Mg ratio, which is thought one of the most probable causes of poor vegetation on serpentine soils (Proctor and Woodell 1975, Brooks 1987, Alexander et al. 1989). Alexander et al.(1989) demonstrated that the timber yield index for conifer stands in California was very closely related to the Ca/Mg ratio. The average Ca/Mg values for the deeper horizons in Groups 1 and 2 did not reach 0.01. This value is extremely low compared to serpentine soils in temperate regions, which form stunted serpentine flora (Proctor and Woodell 1975, Rabenhorst et al. 1982, Bonifacio et al. 1997). The soils in Group 4 seem to be favorable for tree growth, although their very low levels of exchangeable bases might constrain it. However, the Ca/Mg ratio for Group 4 (average 0.42) was not as low as in the other groups.

The exchangeable calcium and potassium levels were low in all the soils. The changes in exchangeable magnesium and cation exchange capacity (CEC) with depth were similar, reflecting the soils very high magnesium saturation, especially in Groups 1 and 2 (60–70%). The pH($H_2O$) values of 7.0 for the soils in Group 1 are higher than that of temperate serpentine soils (Rabenhorst et al. 1982, Alexander 1988, Bonifacio et al. 1997). The pH($H_2O$) tended to decrease from Group 1 to 4, in association with the leaching of bases.

## 7.3.2 Leaching of Mg and Si and Accumulation of Fe and Al

The residual accumulation of sesquioxides is another important pedogenic process in laterization (Kanno et al. 1965, Wildman et al. 1968). This process becomes pronounced as lateritic weathering proceeds. It is usually associated with the removal of silica and magnesium.

The distribution of magnesium, silicon, iron, and aluminum with depth in the soils studied is presented in Fig. 7.3. The magnesium and silicon contents decreased markedly from Group 1 to 4. In fresh serpentinite, the MgO and $SiO_2$ contents are about 40 and 45%, respectively, and fresh rocks collected in several profiles showed similar values. Since the MgO and $SiO_2$ contents of the soils in Group 1 were 8% and 37%, respectively, the MgO concentration in the fine earth fraction is one fifth of that in fresh rock, while the $SiO_2$ fraction is four fifths. These results indicate that a large portion of the magnesium is removed from the soils through leaching, while $SiO_2$, in contrast, remained. This also suggests that leaching of bases occurred first in the weathering process in the study area, and desilication followed. The $SiO_2$ contents of the soils of Groups 2 and 3 decreased to less than one-half that of the parent serpentinite. The MgO and $SiO_2$ contents of the Group 4 soils were no more than 1 and 9%, respectively. Based on the relatively low pH($H_2O$) for Group 4, it is

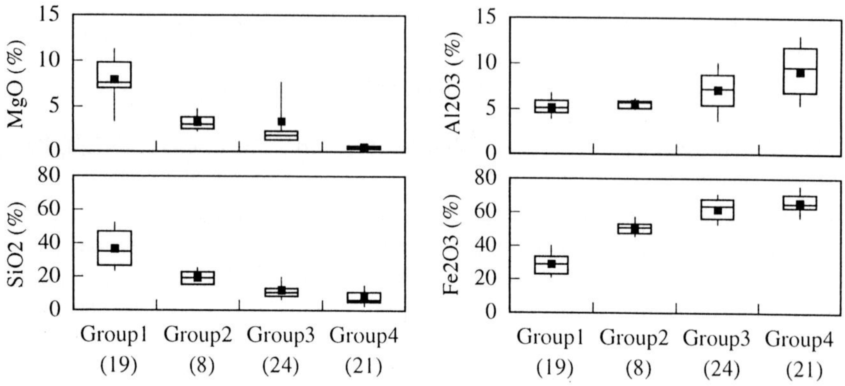


**Fig. 7.3.** Concentrations of total magnesium oxide, silica, iron oxide, and aluminum oxide of the soils in the four groups. Numbers in *parentheses* are the numbers of horizons. *Horizontal bars* dissecting the boxes represent the median between the whiskers (the range of values between 10th and 90th quartiles), *boxes* enclose the interquartile range, and *closed squares* denote means

possible that some silica might remain in clay minerals, and more might exist as amorphous $SiO_2$ in Group 4 as reported by Wildman et al. (1968).

The total iron increased remarkably from Group 1 (29%) to Group 4 (66%) (Fig. 7.3). Iron oxide accumulation occurs in other tropical soils, such as in the Philippines (Frasché 1941), South Kalimantan, Indonesia (Schellmann 1964), and New Caledonia (Lee et al. 1977). Serpentine soils can easily form a laterite crust in a warm humid tropical climate. However, attention should also be directed to the fact that non-lateritic Molisols (Group 1) containing high magnesium levels can also form in the same region.

## 7.3.3 Carbon, Nitrogen, and Phosphorous Status

Many workers suspect that low nutrient status hinders growth (e.g., Walker 1954, Griffin 1965, Proctor 1971), and many studies have reported that tropical soils are low in nutrients, especially phosphorous (Vitousek and Denslow 1986, Silver 1994).

The nutrient status of the epipedons and subhorizons in each group is presented in Fig. 7.4. The respective total carbon and nitrogen contents range from 4–9% and 0.5–2% for epipedons, and 0.3–0.5% and 0.05–0.2% for subhorizons. There were significant differences in the total carbon and total nitrogen between the groups. The available-N and total-P levels in Group 1 were markedly low in comparison with other groups. Brooks (1987) pointed out that serpentinite rock adsorbs soluble phosphorous. Although available-P was not determined for Group 1, judging from the presence of a considerable amount of serpentinite gravel in the Group 1 soils, the available-P level for Group 1 is assumed not to exceed those of Groups 2 to 4. It is

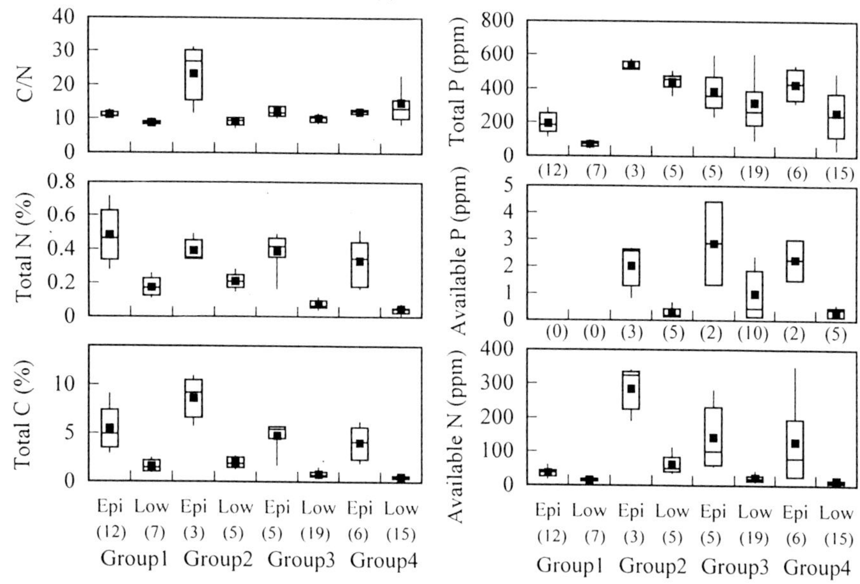


**Fig. 7.4.** Nutrient status at the epipedons and lower horizons of the soils in the four groups. Epi, epipedons; Low, B and C horizons. Numbers in *parentheses* are the numbers of horizons. The box plots are the same as in Fig. 7.3. Note that the numbers of horizons for available phosphorus are different from those for the other properties

presumed that it is the low levels of available nutrients in these serpentine soils, especially the Group 1 soils, that limits tree growth. The high solubility of phosphorous at pH 6.0 to 7.5 (Landon, 1984) may not be true for serpentine soils.

The serpentine soils were about twice as rich in total-C and total N content than the Ultisols under the lowland Dipterocarp forest around Bukit Soeharto Experiment Station, indicating that organic matter might be well stabilized in serpentine soils. However, the available nutrient status differs. The available-N content was lower in Group 1, and was similar or higher in Groups 2 to 4 than in Ultisols. The available-P levels in every serpentine soil were pronouncedly lower (0.3–2.9 ppm) than those of Ultisols, which are 23 ppm in the epipedons and 3 ppm in the lower horizons, on average (compiled from Ohta and Effendi 1992).

## 7.4 Soils on Serpentine Catena

The preceding section described prominent differences in the physico-chemical properties of serpentine soils in the four groups. These characteristics are likely to be closely related to drainage and slope stability in the pedogenesis of humid tropical serpentine soils. The schematic relationships between geomorphology and soils are illustrated in Fig. 7.5. The soils studied were divided into two principal groups by drainage. One group (Group 1, shown as 'I' in Fig. 7.5) might develop under poorer

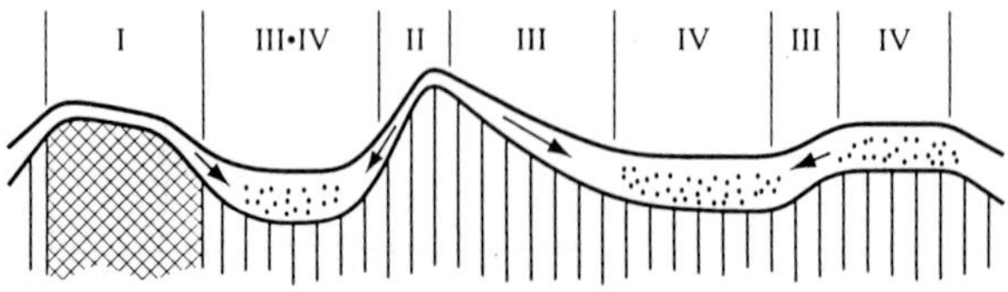


**Fig. 7.5.** Schematic relationship between soils, and geomorphology and drainage, on the serpentine catena developed in the humid tropics. I–IV correspond to Groups 1–4, respectively; *arrows* show soil mass movement; *dots*, iron oxide nodules; *vertical lines*, serpentinite bedrock; *cross-hatched lines*, poorly drained serpentinite bedrock

drainage on serpentinite bedrock with few cracks. Under these conditions, the leaching of magnesium and silicon might be depressed due to the slower drainage, and a high pH environment might develop from the abundant exchangeable magnesium derived from serpentine. These conditions might bring about the formation of 2:1 swelling clay minerals, such as smectite.

The soils in the other group (Groups 2 to 4, shown as 'II to IV' in Fig. 7.5) are primarily characterized by good drainage. The leaching of magnesium and silicon might proceed quickly due to rapid drainage, and weathering might be accelerated. In these groups, the steeper inclination causes mass movement on the soil surface, which is indicated by arrows in Fig. 7.5. Therefore, fresh soil material containing serpentine rock fragments becomes exposed in the shallower part of profiles located higher up on steeper slopes. The abundance of rock fragments in the Group 2 (II) surface soils and the Group 3 (III) subsurface horizons is explained by this difference in slope stability. The horizons with abundant rock fragments of any soil groups clearly showed high exchangeable magnesium levels and pH($H_2O$) values, and very low Ca/Mg ratio around 0.01. This implies that the serpentinite rocks in the profiles continue to supply a large amount of magnesium to the soil system, and that the excess exchangeable magnesium affects the principal characteristics of the soils. The pedogenic process may not proceed from the leaching stage to the oxidation stage if there are abundant rock fragments in the profiles. Under good drainage conditions (II – IV) on serpentine soil catena, it is assumed that slope stability influences the abundance of rock fragments, which in turn affects the chemical properties of the soil, especially the base status.

## 7.5 Conclusion

The soils on serpentine catena in the humid tropics contained a variety of soil orders, such as Molisols, Inceptisols, and Oxisols. The Oxisols can be described as soils that are infertile, but not toxic. Although the nutrient status of these soils is very low, weathering and leaching ameliorate their toxicity. No apparent hindrance to growth was seen at the site with thick Oxisols, where large trees more than 50 cm in diameter grew. The use of the land for agriculture may also be possible with fertilization. On the other hand, the Molisols and Inceptisols can be described as both infertile and toxic. This is mainly due to the continuous supply of magnesium from serpentine rocks. It would be difficult to change these areas into arable land, even with the use of fertilizers.

We should also pay attention to the biological diversity of this area. As mentioned in the introduction, the vegetation growing on serpentine soils is a pool of rare genetic resources, compared to that on Ultisols, which predominate in the humid tropics in Asia. These areas should be treated as conservation areas and given a high priority.

## References

Alexander EB (1988) Morphology, fertility and classification of productive soils on serpentinized peridotite in California (U.S.A.). Geoderma 41:337–351

Alexander EB, Adamson C, Zinke PJ, Graham RC (1989) Soils and conifer forest productivity on serpentinized peridotite of the Trinity ophiolite, California. Soil Sci 148:412–423

Bonifacio E, Zanini E, Boero V, Franchini-Angela M (1997) Pedogenesis in a soil catena on serpentinite in north-western Italy. Geoderma 75:33–51

Brooks RR (1987) Serpentine and its vegetation. Dioscorieds, Portland

Frasché DF (1941) Origin of the Surigao iron ores. Economic Geol 36:280–305

Griffin JR (1965) Digger pine seedling response to serpentinite and non-serpentinite soil. Ecology 46:801–807

Kanno I, Tokudome S, Arimura S, Onikura Y (1965) Genesis and characteristics of brown forest soils derived from serpentine in Kyushu, Japan. Part 2 Genesis and characteristics of Brown forest soils. Soil Sci Plant Nutr 11:141–150

Land Resource Development Centre (1987) Regional physical planning programme for transmigration Vol 1, Review of phase 1 results East and South Kalimantan. Foreign and Commonwealth Office, London (unpublished)

Landon JR (1984) Booker tropical soil manual. Longman Sci Tech, Essex England

Lee J, Brooks RR, Reeves RD, Boswell CR (1977) Plant-soil relationships in a New Caledonian serpentine flora. Plant and Soil 46:675–680

Ohta S, Effendi S (1992) Ultisols of 'Lowland Dipterocarp Forest' in East Kalimantan, Indonesia, II Status of carbon, nitrogen, phosphorus. Soil Sci Plant Nutr 38:207–216

Proctor J (1971) The plant ecology of serpentine. II Plant response to serpentine soils. J Ecol 59:397–410

Proctor J, Woodell SRJ (1971) The plant ecology of serpentine. I. Serpentine vegetation of England and Scotland. J Ecol 59:375–395

Proctor J, Woodell SRJ (1975) The ecology of serpentine soils. Adv Ecol Res 9:255–366

Rabenhorst MC, Foss JE, Fanning DS (1982) Genesis of Maryland soils formed from Serpentine. Soil Sci Soc Am J 46:607–616

Schellmann W (1964) Zur lateritischen verwitterung von sepentinit. Geol Jb 81:645–678

Silver WL (1994) Is nutrient availability related to plant nutrient use in humid tropical forests? Oecologia 98:336–343

Soil Survey Staff (1997) Keys to soil taxonomy. Soil Conservation Service USDA, Pocahontas, Blacksburg

Spence DHN (1970) Scottish serpentine vegetation. Oikos 21:22–31

Vitousek PM, Denslow JS (1986) Nitrogen and phosphorus availability in treefall gaps of a lowland tropical rainforest. J Ecol 74:1167–1178

Walker RB (1954) The ecology of serpentine soils. II. Factors affecting plant growth on serpentine soils. Ecology 35:259–266

Wildman WE, Jackson ML, Whittig LD (1968) Iron-rich montmorillonite formation in soils derived from serpentinite. Soil Sci Soc. Am Proc 32:787–794

# 8 Characteristics of Soils Developed on Volcanic Parent Materials in the Barong Tongkok Area

DARUL AKSA, NAGAHARU TANAKA, SEIICHI OHTA, and SYARIF EFFENDI

## 8.1 Introduction

Volcanic soils such as Andosols, which are the initial weathering product of volcanic ash, are well known for their distinctly superior fertility compared with other soils of the humid tropics (Burnhan, 1984). Volcanic activity has been very extensive on many of the islands of south-east Asia, and most particularly on the Sunda group (Sumatra, Java, Bali, Lombok, Sumbawa, and Flores) (Burnhan, 1984), and volcanic soils as Andosols are often found in this area. These soils have quite different properties from other soils such as Acrisols derived from the Tertiary deposits which are the major soils in the lowland of East Kalimantan. The Andosols are known to exhibit higher porosity and a lower bulk density, a "fluffy"consistency, higher capacities of water retention, cation exchange, and phosphate sorption, and an ability to form stable associations of mineral and organic matter which give their topsoils a characteristic dark color (Burnham, 1984).

At present there is no active volcano in East Kalimantan, and the major soils of the area are Acrisols derived from Neogene sedimentary rocks (Ohta and Effendi, 1992a), which are characterized by low porosity, high bulk densities, and a low capacity for water retention and cation exchange. However, in the central part of East Kalimantan there is an area where the geology is dominated by Neogene volcanic rocks and some extinct volcanoes are distributed there (Voss, 1982). Thus it is expected that soils derived from volcanic parent materials are distributed in the area. In fact, characteristically dark colored soils are distributed in the area, and the local people call them "Tana Linggu" distinguishing them from other soils for their higher productivity. Although the unique characteristics of these soils would have strong influences on the responses of forest ecosystem against various disturbances, the information on these soils is scarcely available. In this chapter, we introduce the characteristics of soils developed on volcanic parent materials in central East Kalimantan in comparison with the Acrisols derived from Neogene sedimentary rocks.

Ecological Studies, Vol. 140
Guhardja et al. (eds.) Rainforest Ecosystems of East Kalimantan: El Niño, Drought, Fire and Human Impacts

## 8.2 Study Area

The area studied, Barong Tongkok, is located in the middle course of the Mahakam River at south latitude between 0°08' and 0°20' and east longitude between 115°31' and 115°47' and about 170 km west of Samarinda. The area belongs to the physiographic region of the West Kutai Volcanic Region, and the geology is dominated by Neogene volcanic rocks (Voss 1982). In geomorphology the area is divided into three major units: the lava field, the volcanic shield, and the extinct volcano. The elevation ranged from approximately 80–230 m, 180–350 m, and 320–550 m in the lava field, volcanic shield, and extinct volcano, respectively. The climate of the region is classified into Aafw" or Aafw' in accordance with Koppen's system, tropical rainy (no months have mean rainfalls less than 60 mm), isothermal (the annual range of mean temperature is less than 5°C) climate with a hot summer (hottest month more than 22°C) and two rainfall maximums in the year. The mean annual rainfall of the study area is approximately 3000 mm.

In the area the Dayak people, an indigenous people in Borneo, dwell in the scattered villages of Mencimai, Asa, Barongtongkok, and others. They have lived on swidden agriculture for a long period (more than a few hundred years) and the soils and forest ecosystems of the area are assumed to have been subjected to human influences to various extents.

To discuss the soil characteristics and their distribution, soils from 18 different locations, 8 soils from the lava field, 4 from the volcanic shield and 6 from the extinct volcano area, were compared in soil morphology and physico-chemical properties with each other and also with Acrisols on the Tertiary deposits from Bukit Soeharto Education Forest (0°50' S, 117°10'E') and a logging concession area of the International Timber Corporation Indonesia (0°20'S–1°00'S, 116°30'E–117°00'E) (Ohta and Effendi, 1992a,b, Ohta et al. 1993).

## 8.3 Morphology and Physico-Chemical Properties

Three major types of soil were found in the area: Ferralsols, Nitisols, and Andosols. Ferralsols, an extremely weathered soil, had the largest extent in the area, and were found on the lava field except at a depression along the small streams, the volcanic shield and extinct volcano slopes except for the middle and the lower gentle parts of the slopes. The Ferralsols generally had a solum deeper than 2 m and no rock fragments. The morphological characteristics of the soils were representative of those for Ferralsols (FAO-UNESCO, 1988, Deckers et al. 1998) except for the soil color. In the soils distinct eluvi-illuvial features and cracks were not found except for very weakly developed clay skins in some sub-horizons. The surface horizons were characterized by darkish colors as blackish brown, dark brown or dark reddish brown, and the sub-horizons also by darkish colors of various hues reflecting the parent materials.

In physico-chemical properties, the Ferralsols were rich in clay, and were high in total porosity throughout the profiles (Fig. 8.1). The bulk densities of the soils were generally lower than those of the Acrisols (Fig. 8.2). The cation exchangeable capacity (CEC) per unit amount of clay, the concentrations of exchangeable base cations, and the base saturation percentages in CEC were also low. The values of CEC were also 4 to 5 times higher than that of effective CEC (ECEC) (Fig. 8.3). These chemical characteristics of the soils might be all attributed to low activity clay minerals such as kaolinite and oxides of Fe and Al, which are the predominant minerals in Ferralsols.

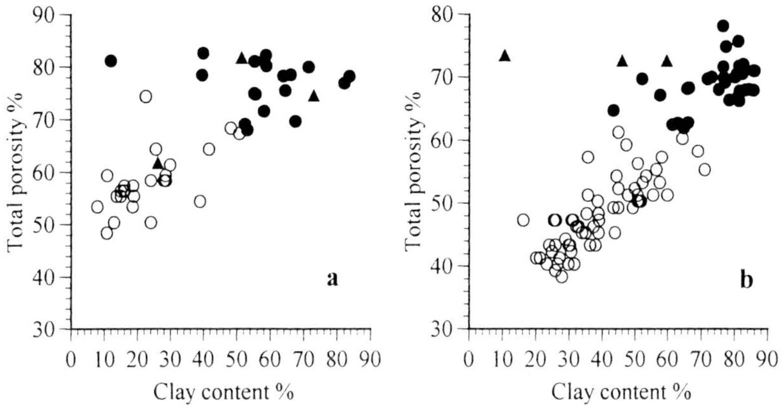


**Fig. 8.1.** The relationships between total porosity and clay content for the A horizon (**a**) and the lower B horizon (**b**) of Ferralsols (*solid circles*), Andosols (*solid triangles*) developed on volcanic parent materials, and Acrisols derived from Neogene sedimentary rocks (*open circles*)

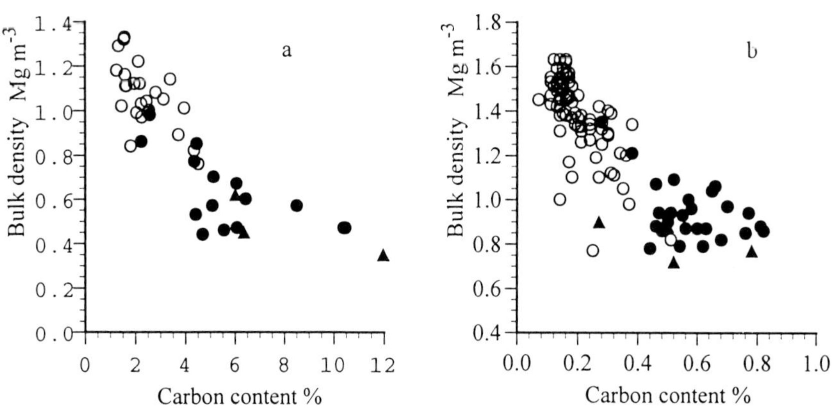


**Fig. 8.2.** The relationships between bulk density and carbon content for the A horizon (**a**) and the lower horizon (**b**), of Ferralsols and Andosols developed on volcanic parent materials, and Acrisols derived from Neogene sedimentary rocks. For explanation of symbols see Fig. 8.1

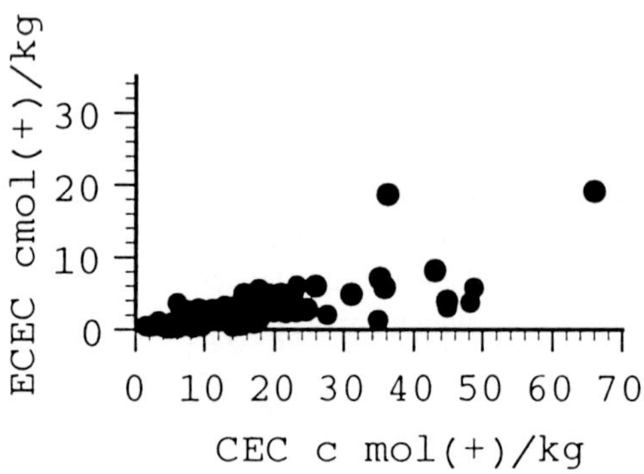


**Fig 8.3.** The relationship between cation exchangeable capacity (CEC) and effective CEC ECEC of soils developed on volcanic parent materials

On the middle slopes of the extinct volcanoes, Nitisols, which showed higher CEC per unit amount of clay than the Ferralsols, were distributed partially in association with the Andosols mentioned below. In the Nitisols, clay content was low in epipedons and subsoils displaying the maximum in the intermediate horizons. The soils were characterized by high bulk densities and low porosity, particularly of fine pores. Although the CEC per unit amount of clay was higher than the Ferralsols, the CEC of bulk soil and the concentration of exchangeable base cations were distinguishably lower than those of Ferralsols because of their lower clay contents.

These three different kinds of soil mentioned above were poor in nutrients as shown in their low levels of Av.P and exchangeable base cations. In contrast to these soils, quite morphologically different soils occurred on the lower gentle slopes of extinct volcanoes. Although these soils had brownish black epipedons with developed crumb structure and no elivi-illuviation as did the other types of soils, they had relatively shallow solum less than 2 m and much less weathered subrounded gravel. Plant roots also penetrated deeper into the horizons in these soils than in the other types of soils. In bulk density and porosity these soils were not significantly different from other types of soil, but they were clearly distinguished from others by their higher CEC per unit amount of clay, exchangeable base cations and base saturation percentages in ECEC. These soils satisfied the requirement to be classified into Andosols with andic properties.

In the case of mountains in Java island, at high elevation above 700 m, Andosols characterized by dark color and the presence of allophane dominate because of the regime of high constant rainfall and low temperature favor organic matter accumulation and impede the crystallization of allophane (Dudal and Soeprapiohardjo, 1960). In the area studied in East Kalimantan, however, the soil distribution pattern was different from that found in the island of Java. Andosols were not found at higher elevations but only in a limited area on the lower gentle slopes of extinct volcanoes at elevations around 450 m. These soil distribution patterns in the studied area might be associated with the difference in the weathering rate of parent material reflecting the geomorphological conditon. In the geomorphological unit where Andosols are distributed, weathering might have been retarded possibly due to slower water movement and supply of substances from upper slopes.

## 8.4 Comparison with Soils on the Tertiary Deposits

The volcanic soils have remarkably different physico-chemical characteristics in comparison with the Acrisols on the Tertiary deposits, which are the major soils in the surrounding region of East Kalimantan. In physical properties the volcanic soils were higher in clay content and porosity than the Acrisols, and the values were almost constant throughout the profiles, unlike the Acrisols (Fig. 8.1). On the other hand, the bulk density increased with depth but with lower magnitude than Acrisols. In carbon content the volcanic soils were significantly higher both in surface soils and subsoil than the Acrisols (Fig. 8.2). These findings suggest that the volcanic soils are better in water holding and drainage than the Acrisols.

In chemistry the Ferralsols were quite different from the Acrisols and also from the Andosols (Fig. 8.4). The Ferralsols were significantly lower in exchangeable base cations than the Acrisols except for the surface soils, and exchangeable Al was higher in the former than in the latter for all horizons. The contents of Total-C, Total-N and Av.P were, however, higher in the Ferralsols than in the Acrisols.

On the other hand, the Andosols derived from volcanic materials displayed higher exchangeable base cations contents except for K and lower exchangeable Al throughout the solum in comparison with the Acrisols.

Volcanic soils other than Andosols in the region were poorer in base cations than the Tertiary soils, though the level of exchangeable Al as a whole was lower in the former than the latter . These results indicate that all the soils derived from volcanic materials are not necessarily more fertile in chemistry than the soils on the Tertiary deposits, and they included quite different soils in characteristics and productivity. Comparison among the dark-colored surface soil horizons reveals that chemical fertility is similar among the Ferralsols and the Acrisols, while it is remarkably higher in the Andosols which are rich in mineral nutrients even in the subsoil.

Recently, the function of forests in carbon sequestration is a worldwide concern, and it is recognized that forest soil plays an important role in carbon storage. In the case of forest soils in Japan, it is well known that carbon storage of black soils derived from volcanic ash is higher than that of other soil types (Ohta, unpublished). The darker soil color of the soils studied suggests higher soil carbon storage in the Barong Tongkok area. The carbon storage in the 0–1.0 m layer of the soils studied was estimated to be 74–183 ton/ha, and the average was 134 ton/ha. The amount is much larger than the 81 ton/ha (0–1.5m) (Ohta and Effendi 1992b) found for Acrisols on the Tertiary deposit in the surrounding region. This amount is also larger than the results recorded in other tropical regions, e.g., 75 ton/ha in soils under Dipterocarp forest in Malaysia (Chan, 1982) and 117 ton/ha by Schlesinger (1986), and even larger than the 120 ton/ha in the soil of a mountain rainforest in New Guinea (Edward and Grubb, 1977). The vegetation of the study sites consists not only of intact natural forests but also includes secondary forest developed after repeated slash-and-burn cultivation. Nevertheless, the volcanic soils studies stored larger amounts of carbon than for soils of the humid tropics, and it is suggested that soils derived

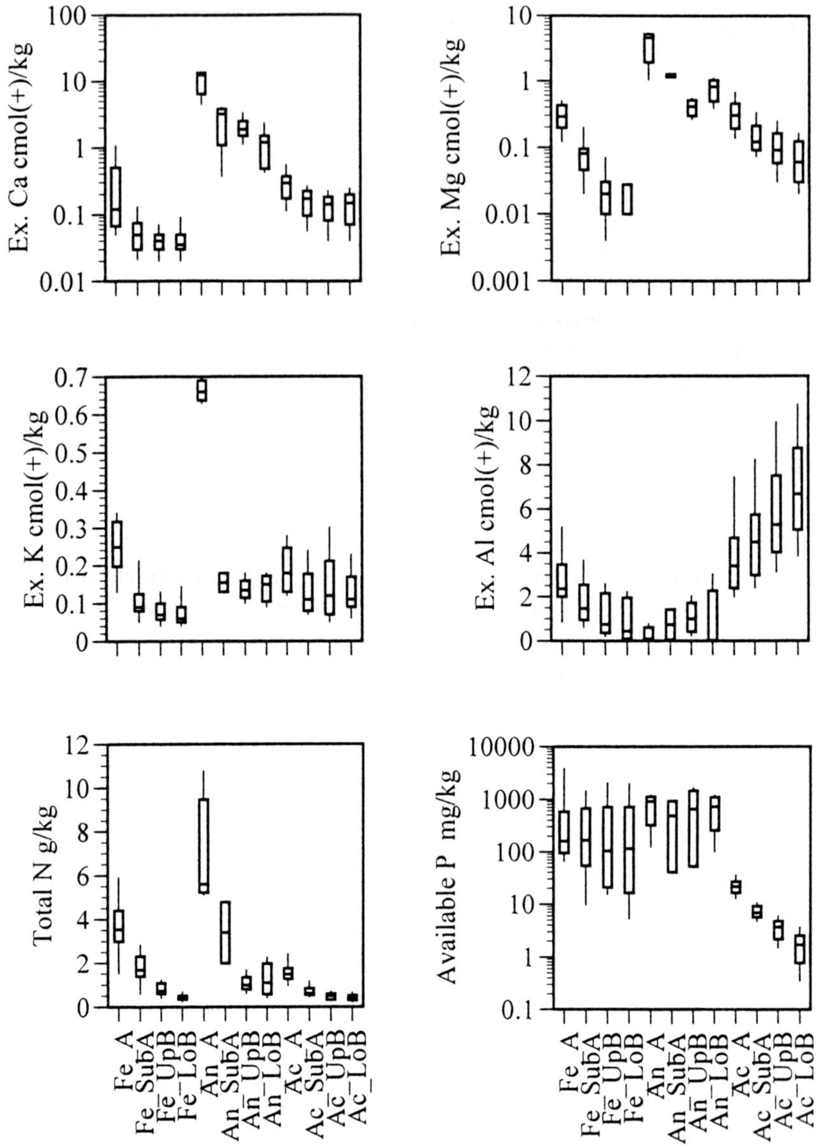


**Fig. 8.4.** The range of exchangeable (*Ex.*) cations, Ex. Al, Total N and available P in the surface horizon (*A*), sub-surface horizon (*SubA*), upper B horizon (*UpB*) and lower B (*LoB*) horizon of Ferralsols (*F*), Andosols (*An*) and Acrisols (*Ac*)

from volcanic parent materials have the capability of stabilizing soil organic matters as well as in temperate areas.

## 8.5 Conclusion

The soils developed on volcanic parent materials in Barong Tongkok area were classified into Ferralsols, Nitisols, and Andosols. These soils were quite different in characteristics and productivity, and were not necessarily more fertile in chemistry than the soils on the Tertiary deposits, except for the Andosols. In areas of agricultural fields, secondary forests bush and grassland, and soils assumed to have been heavily affected by slash-and-burn activities, dark-colored surface soils were often observed. In particular, on the lava field where slash-and-burn practices have been conducted repeatedly, the color of the topsoil was black to brownish black (10YR or 7YR in hue, 2 in value and 1 or 2 in chrome). Although the carbon contents of the surface horizons of these soils were highly similarly to the Andosols, the nutrient status as base cation level was poor in many cases (Fig. 8.5). It is possible the poorness in nutrients in the surface soils might be attributed to their inherent characteristics and also removal of nutrients from the soil system in the course of repeated slash-and-burn practices. Simultaneously, these nutrient poor black surface soils might have been developed by the accumulation of resistant organic matter through the supply of plant materials, including partially carbonized organic matter, under the presence of allophane and an amorphous aluminum-silicate mixture that forms stable organo-mineral complexes through repeated slash-and-burn practices. These findings also suggest that carbon accumulated in volcanic soils might not be depleted and functions as a stable carbon pool even under intensive disturbances.

From the results mentioned above it is highly possible that the soils called "Tana Linggu" by the Dayak local people are identical to the soils classified into Andosols in this study, and were in a relatively earlier stage of weathering. However, the reason why such lightly weathered soils are distributed in the same area together with strongly weathered soils has to be further investigated.

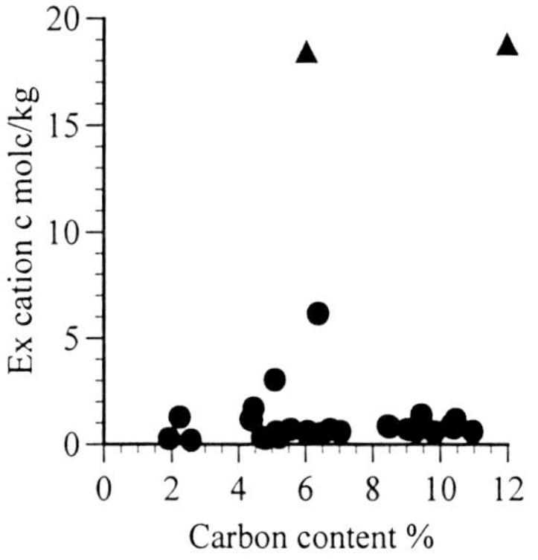


**Fig. 8.5.** The relationship between carbon content and total exchangeable (*Ex.*) cations of the black color surface horizon. *Solid circles*, Ferralsols; *solid triangles*, Andosols

## References

Burnham CP (1984) The forest environment: soils. In: Whitmore TC (ed) Tropical rainforests of the Far East. Oxford University Press, Oxford, pp 137–154

Chan Y (1982) Storage and release of organic carbon in peninsular Malaysia. J Environ Stud 18:211–222

Deckers JA, Nachtergaele FO, Spaargaren OC (eds) (1998) World reference base for soil resources, Introduction. Acco, Belgium

Dudal R, Soeprapiohardjo M (1960) Some considerations of the genetic relationship between Latosols and Andosols in Java, Indonesia. Trans. Seventh Int Congr Soil Sci 4:229–237

Edward PJ and Grubb PJ (1977) Studies of mineral cycling in a montane rainforest in New Guinea. I. The distribution of organic matter in the vegetation and soil. J Ecol 65:943–969

FAO–UNESCO (1988) Soil map of the world, revised legend. World Soil Resources Report 60, FAO, Rome

Ohta S, Effendi S (1992a) Ultisols of "lowland dipterocarp forest" in East Kalimantan, Indonesia, I. Morphology and physical properties. Soil Sci Plant Nutr. 38:197–206

Ohta S, Effendi S (1992b) Ultisols of "lowland dipterocarp forest" in East Kalimantan, Indonesia, II. Status of carbon, nitrogen and phosphorus. Soil Sci Plant Nutr 38:207–216

Ohta S, Effendi S, Tanaka N, Miura S (1993) Ultisols of "lowland dipterocarp forest" in East Kalimantan, Indonesia, III. Clay minerals, free oxides and exchangeable cation. Soil Sci Plant Nutr 39:1–12

Schlesinger WH (1986) Changes in soil carbon storage and associated properties with disturbance and recovery. In: The changing carbon cycle, a global analysis, Springer, Berlin Heidelberg New York, pp 194–220

Voss F (1982) Atlas of East Kalimantan, Indonesia. Transmigration Area Development Project (TAD). HWWA-Institute fur Wirtschaftsforschung, Hamburg

# Part 3

# Drought and Fire Impacts on Forest Ecosystems

# Part 3

# 9 Tree Species Composition of a Burned Lowland Dipterocarp Forest in Bukit Soeharto, East Kalimantan

PAULUS MATIUS, TAKESHI TOMA, and MAMAN SUTISNA

## 9.1 Introduction

Tropical lowland evergreen rainforests on the Malay Peninsula and islands of the Sunda Shelf are distinguishable from similar types of vegetation of this class found on other continents by their abundance of dipterocarp species (Whitmore 1984). The floristic composition in lowland dipterocarp forests is distinctive because of the very large numbers of tree species represented by a small number of mature individuals, and the general 'family dominance' of Dipterocarpaceae, especially in terms of basal area (Richards 1996). The species richness of lowland dipterocarp forest in Borneo Island, for instance, is illustrated by the species-area curve for a forest at Wanariset in East Kalimantan (Kartawinata 1981a).

The lowland dipterocarp forests in East Kalimantan have been rapidly degraded by human activities since the late 1960's (Kartawinata et al. 1981b; Chapter 1, this volume). Further degradation was caused by the large-scale forest fires in 1982–83 (e.g., Goldammer et al. 1996; Chapter 3, this volume). Large canopy gaps caused by the fires have been filled with a high density of pioneer tree species, especially *Macaranga gigantea* and *Macaranga triloba* (Matius and Okimori 1991, Goldammer et al. 1996). Thus, since the fire, the species composition of severely burned forest stands has become simpler. However, many species seemed to survive the forest fire in 1982–83. For example, Keßler et al. (1992) recorded 1200 different tree species from the Balikpapan-Samarinda area in East Kalimantan, most of which was affected by the fire in 1982–83.

This chapter deals with the tree species composition observed in a large sampling plot (ca. 9 ha) in a lowland dipterocarp forest which was burned by the 1982–83 fire. The recovery of the vegetation after the 1982–83 fire has been studied in relatively small plots, of up to 1 ha (Tagawa and Wirawan 1988, Matius and Okimori 1991, Nykvist 1996). However, as mentioned above, most species in lowland dipterocarp forests are represented by a small number of mature individuals. Thus, sampling small plots in pioneer-dominated stands is unlikely to detect all surviving species present in the burned forests. Therefore, further investigations have been undertaken in the large sample plot, which have confirmed that a large number of tree species survived the 1982–83 fire.

Ecological Studies, Vol. 140
Guhardja et al. (eds.) Rainforest Ecosystems of East Kalimantan: El Niño, Drought, Fire and Human Impacts

## 9.2 The Large-scale Study Plot in Bukit Soeharto

Bukit Soeharto Education Forest (BSEF) of Mulawarman University is situated between Samarinda and Balikpapan, in the lowland and coastal part of Kutai Regency, East Kalimantan, which was originally covered with lowland dipterocarp forests. Most forest stands in the BSEF had been subjected to selective cutting, partly by the local people and partly by commercial logging companies, until the area was designated a protected forest in 1978. Shortly after this, it was affected by the fire in 1982–83 (see Chapter 3, this volume). The severity of fire damage in the burned stands was related to the intensity of logging: the higher the previous logging intensity, the heavier the damage by the fire. The heavily damaged sites were dominated by pioneer *Macaranga* species, and the lightly damaged sites were dominated by surviving dipterocarp species in 1988, five years after the 1982–83 fire (Matius and Okimori 1991; Chapter 10, this volume). Thus, by 1997, as a result of selective logging and forest fire, the BSEF had become a mosaic of forest stands dominated by surviving dipterocarp trees and pioneer *Macaranga* trees established after the 1982–83 fire.

In May 1997, a 9-ha permanent plot was established in order to investigate the effects of the second felling in a logged-over dipterocarp forest, and subsequent recovery from the felling (see Chapter 19, this volume). The 9-ha plot was composed of nine 1-ha subplots (each 100 m × 100 m), arranged in a regular square of 3 × 3 plots. All trees greater than or equal to 10 cm in diameter at breast height (DBH) were tagged, identified and their dbh values were recorded. The present chapter deals with the tree species composition of the 9-ha plot at the initial census. After the initial tree census, an experimental logging operation was conducted in six subplots by Ruslim et al. (Chapter 19, this volume). These studies form part of a cooperative research program jointly organized by the Faculty of Forestry and PUSREHUT (Tropical Rain Forest Research Center) of Mulawarman University, CIFOR (Center for International Forestry Research) and JICA (Japan International Cooperative Agency).

## 9.3 Tree Density and Basal Area

Tree density and basal area (BA) in the 9-ha plot were lower than reported for the primary dipterocarp forests in East Kalimantan, but the values in some 1-ha subplots were comparable to those of the primary forests (Table. 9.1). There were large variations in the tree density and BA among the 1-ha subplots, and between individual subplots and the primary forests. The variations in tree density and BA indicate the necessity of sampling a large-scale plot, or a large number of small plots, for evaluating the tree density and BA in lowland dipterocarp forests, whether burned or unburned.

# 9.4 Species Composition

## 9.4.1 Number of Species

The number of species in the nine 1-ha subplots were less than the number reported for the primary forests in East Kalimantan (Table 9.1), ranging from 92 to 163 species $ha^{-1}$, at an average of 123 species $ha^{-1}$. The numbers of species in primary forests in East Kalimantan ranged from 160 to 276. Only two of the nine 1-ha subplots in the 9-ha plot described above had a number of species comparable to the lowest level (160 species $ha^{-1}$) found in the primary forests in East Kalimantan.

The species-area curve was clearly far from the asymptote for the 9-ha inventory, suggesting that that there are still many other, unrecorded species in the BSEF (Fig 9.1). There were 138 species represented by a single individual in the 9-ha plot, and the numbers in the 1-ha subplots ranged from 55 to 98. Furthermore, 157 of the total of 391 species were found at only one subplot. The quite low density of most species undoubtedly reduces the probability of finding them when sampling in a small area. In support of this assertion, 191 species were recorded in three other monitoring plots with a total area of 2.08 ha in the BSEF (Chapter 10, this volume), 91 of which were not found in the 9-ha plot.

**Table 9.1.** Tree density, basal area, and number of species in the 9-ha plot in Bukit Soeharto Education Forest (BSEF) and primary lowland dipterocarp forests in East Kalimantan.

| | BSEF[a] | | | | | | Berau[e] | | |
|---|---|---|---|---|---|---|---|---|---|
| | mean | max | min | Sebulu[b] | Wanariset[c] | Lempake[d] | mean | max | min |
| Tree density ($ha^{-1}$) | 429 | 524 | 363 | 592 | 541 | 445 | 531 | | |
| Basall area ($m^2$ $ha^{-1}$) | 27.2 | 32.7 | 23.8 | 33.9 | 29.7 | 33.7 | 31.5 | | |
| Number of Species ($ha^{-1}$) | 123 | 163 | 92 | 276 | 178 | 161 | 182 | 201 | 160 |

[a]Mean, maximum and minimum value of the nine 1-ha subplots.
[b]Sukardjo et al. (1990), [c]Kartawinata et al. (1981a), [d]Riswan (1987), [e]Sist and Saridan (1998).
For trees (DBH ≥ 10 cm)

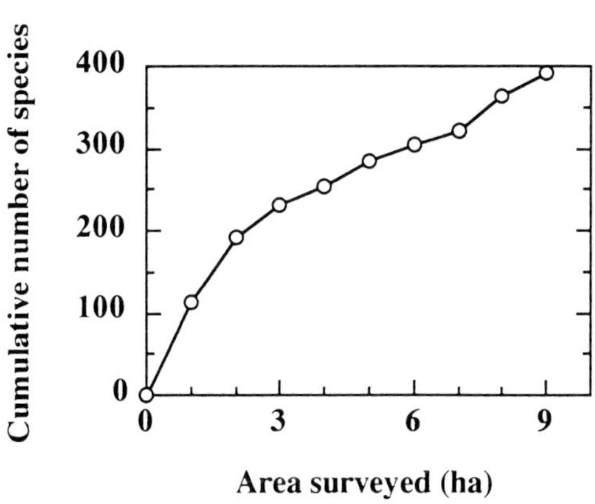


**Fig. 9.1.** Species-area curve for trees (DBH ≥ 10 cm) in the 9-ha plot

Recent studies in large sample plots have confirmed the generally high species richness in rainforests in Borneo Island. For example, in a mixed dipterocarp forest at Lambir hill, Sarawak, northern Borneo, 1198 and 1025 tree species were recorded for trees larger than 1 cm and 10 cm in DBH, respectively, in a 52-ha sampling plot (Yamakura et al. 1996). These values are much higher than those recorded in 50-ha plots at Pasoh in Peninsular Malaysia and Barro Colorado Island (BCI) in Panama (Condit et al. 1996). In Pasoh, 817 and 683 species were recorded for trees larger than 1 cm and 10 cm in dbh, respectively, and in BCI, the corresponding figures were 303 and 229 species, respectively (for 50 ha plots in both cases). Our record in the BSEF, of 482 tree species (DBH ≥ 10 cm) within an 11.08-ha sample area is already higher than the number of species recorded in the 50-ha sampling plot in BCI. The species-area curve of the 9-ha plot (Fig. 9.1) indicates that if we enlarge the sampling area we are likely to find many more species than so far recorded. The data from the 50- and 52- ha plots mentioned above also indicate that there are many more species in the 9-ha plot, and we could detect some of them by decreasing the lower DBH limit in the tree census. Clearly, however, even though the BSEF has suffered selective logging and forest fire, a large number of species have survived these severe disturbances.

### 9.4.2 The Dominant Families

The most dominant families in terms of tree density and BA were Euphorbiaceae and Dipterocarpaceae, respectively (Table 9.2). The tree density and BA of each family differed between the 1-ha subplots, but the dominance of Euphorbiaceae in terms of tree density, and Dipterocarpaceae (in BA) were commonly observed. In the entire 9-ha plot, 50 species of Euphorbiaceae were found, representing 15 genera, and 38 species of Dipterocarpaceae from 6 genera. The third most dominant

**Table 9.2.** Ten most dominant families in terms of tree density and basal area in the 9 ha plot[a]

| | Tree density ($ha^{-1}$) | | | | Basal area ($m^2$ $ha^{-1}$) | | |
|---|---|---|---|---|---|---|---|
| Family | mean | max | min | Family | mean | max | min |
| Euphorbiaceae | 194.3 | 236 | 135 | Dipterocarpaceae | 10.7 | 18.8 | 6.4 |
| Dipterocarpaceae | 52.2 | 85 | 27 | Euphorbiaceae | 6.1 | 9.4 | 2.7 |
| Lauraceae | 30.0 | 46 | 15 | Lauraceae | 3.5 | 4.7 | 1.5 |
| Sapotaceae | 28.0 | 52 | 9 | Sapotaceae | 0.9 | 1.4 | 0.5 |
| Myrtaceae | 12.6 | 29 | 4 | Moraceae | 0.6 | 1.0 | 0.3 |
| Moraceae | 9.8 | 15 | 6 | Bombacaceae | 0.5 | 1.7 | 0.1 |
| Myristicaceae | 9.7 | 15 | 3 | Leguminosae | 0.5 | 1.4 | 0.1 |
| Ebenaceae | 8.4 | 13 | 3 | Myrtaceae | 0.5 | 0.9 | 0.1 |
| Burseraceae | 7.3 | 15 | 3 | Burseraceae | 0.4 | 1.5 | 0.0 |
| Fagaceae | 7.2 | 13 | 4 | Myristicaceae | 0.3 | 0.5 | 0.1 |

[a]Mean, maximum and minimum values of the nine 1-ha subplots are shown.
For trees (DBH ≥ 10 cm)

family was Lauraceae, both in terms of tree density and BA, which consisted of 35 species from 12 genera. These three dominant families accounted for 64.5%, 74.9% and 39.1% of the tree density, BA, and number of species in the 9-ha plot, respectively. The dominance of these families in primary forests in East Kalimantan has been reported in several previous studies (Riswan 1987, Kartawinata et al. 1981a, Sukardjo et al. 1990, Sist and Saridan 1998).

## 9.4.3 Dominant Species

The most dominant species in terms of tree density and BA differed (Table 9.3). The pioneer *Macaranga* species were clearly dominant in tree density. In contrast, primary species such as *Eusideroxylon zwageri* (Lauraceae), and various dipterocarp species were prominent in BA. Furthermore, the tree density and BA of important pioneer and primary species varied between subplots (see the maximum and minimum values in Table 9.3).

The dominance of *Macaranga* spp. in terms of tree density was commonly observed. In the entire 9-ha plot, the three most dominant species in terms of tree density were all from the genus *Macaranga*, and together they accounted for 38% of the total tree density (Table 9.3). Such a high density of *Macaranga* trees has not been reported in the abovementioned studies of primary forests in East Kalimantan. The high density of a few pioneer species resulted in comparatively low species diversity in the burned forest. The Shannon-Wiener diversity index (H') and Pielow's eveness index (J') (Krebs 1985) in the 9-ha plot were 6.15 and 0.71, respectively. If we exclude the three dominant *Macaranga* species from caluculation, H' and J' increased to be 7.3 and 0.85, respectively. These values were lower than the values derived for the primary forests in Berau (Sist and Saridan 1998), where the number of species in each of three 4-ha plots ranged from 323 to 363. Even though the

**Table 9.3.** Ten most dominant species in terms of tree density and basal area in the 9 ha plot[a]

| Species | Tree density ($ha^{-1}$) | | | Species | Basal area ($m^2$ $ha^{-1}$) | | |
|---|---|---|---|---|---|---|---|
| | mean | max | min | | mean | max | min |
| *Macaranga gigantea* | 70.3 | 129 | 29 | *Eusideroxylon zwageri* | 2.63 | 3.57 | 1.09 |
| *Macaranga hypoleuca* | 55.1 | 93 | 15 | *Shorea laevis* | 2.62 | 7.54 | 0.01 |
| *Macaranga triloba* | 28.3 | 45 | 10 | *Macaranga gigantea* | 2.29 | 5.05 | 0.74 |
| *Eusideroxylon zwageri* | 15.7 | 26 | 7 | *Macaranga hypoleuca* | 1.63 | 2.70 | 0.60 |
| *Palaquium gutta* | 13.6 | 33 | 2 | *Shorea smithiana* | 1.22 | 4.12 | - |
| *Macaranga conifera* | 10.8 | 24 | 2 | *Macaranga triloba* | 1.19 | 2.14 | 0.19 |
| *Shorea laevis* | 10.8 | 32 | 1 | *Shorea johorensis* | 1.00 | 3.37 | 0.04 |
| *Shorea johorensis* | 5.6 | 14 | 1 | *Dipterocarpus humeratus* | 1.06 | 3.35 | - |
| *Palaquium rostratum* | 5.2 | 11 | 2 | *Dipterocarpus cornutus* | 0.91 | 2.53 | - |
| *Aporusa elmeri* | 4.6 | 10 | 1 | *Shorea lamellata* | 0.53 | 2.78 | - |

[a]Mean, maximum and minimum values of the nine 1-ha subplots are shown.
For trees (DBH ≥ 10 cm)

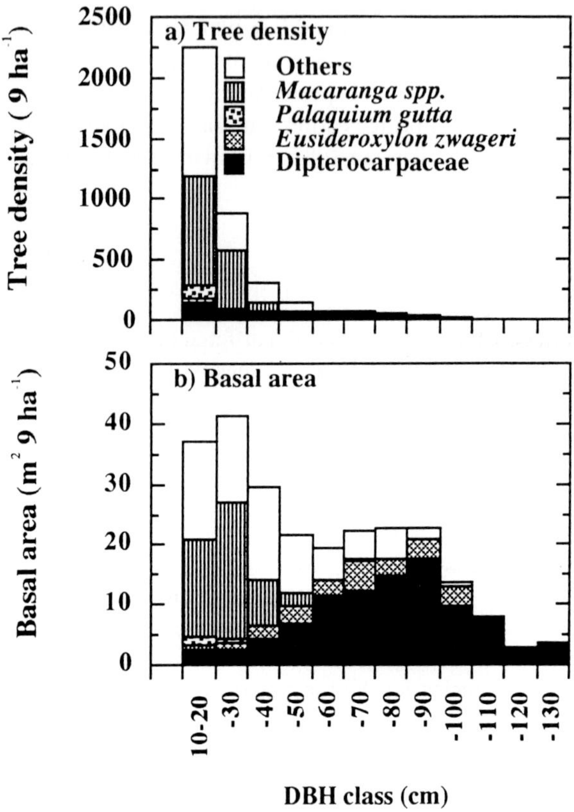


**Fig. 9.2a,b.** Distribution of trees in the 9-ha plot according to dbh classes. **a** Tree density; **b** Basal area (BA)

numbers of species in these 4-ha plots were lower than the corresponding numbers in the 9-ha plot, their lowest H' (7.36) and J' (0.88) figures (Sist and Saridan 1998) were higher than those in the 9-ha plot.

The most dominant species in terms of BA was a primary species, *Eusideroxylon zwageri*, followed by the dipterocarp species *Shorea laevis* (Table 9.3). These two species accounted for 19% of the total BA, but for only 6% of the total tree density in the 9-ha plot as a whole. The frequency distribution of tree density and BA according to dbh classes clearly indicates that trees of the pioneer *Macaranga* genus were restricted to the lower dbh classes, whereas dipterocarp trees and *Eusideroxylon zwageri* trees contributed strongly to the larger dbh classes (Fig. 9.2).

## 9.5 Conclusion

In 1997, the lowland dipterocarp forests in the BSEF, which were burned in the 1982–83 fires, had a high density of small pioneer *Macaranga* trees that established after the fire, and large dipterocarp and *E. zwagri* trees that survived the fire. Many additional species also survived the 1982–83 fires. Since tree densities of most species are quite low in these stands, it is unlikely that they will all be found by sampling only in a small plot. Recent discussions suggested that floristic composition is poorer in the eastern than other parts of Borneo, because occasional droughts have been more frequent in the eastern zone (Goldammer and Seibert 1990, Walsh 1996, Sist and Saridan 1998). However, the works of Kartawinata et al. (1981a) and Sukardjo et al. (1990) showed that the species richness of dipterocarp forests in lowland Kutai, East Kalimantan, were comparable to other parts of Borneo. A large number of species which are likely to be present in East Kalimantan have not even been listed in the basic inventory, and their roles in the forest ecosystem are entirely unknown. A large area will be needed in order to maintain the population of all the species present at low densities. Otherwise, almost all of them will become extinct, without leaving any record of their existence on the earth.

Fourteen years after the forest fires in 1982–83, further large-scale forest fires occurred in 1997–98, during the prolonged dry spells caused by a strong El Niño-Southern Oscillation (ENSO) event. Most land areas of the lowland part of Kutai Regency were burned by the fire, and the BSEF was no exception (see Chapter 3, this volume). Most of the forests in the BSEF were burned in 1998, just as they were in 1982–83 or worse (Chapter 10, this volume). The number of species in the 9-ha plot was 258 in November 1998, two-thirds of the numbers in May 1997. The severely burned sites have been dominated by a few pioneer species. The species composition of the forests in BSEF was simplified again and maintainance of tree species populations at low densities became difficult.

## References

Condit R, Hubbell SP, Lafrankie JV, Sukumar R, Manokaran N, Foster RB, Ashton PS (1996) Species — area and species — individual relationships for tropical trees: a comparison of three 50-ha plots. J Ecol 84:549–562

Goldammer JG, Seibert B (1990) The impact of droughts and forest fires on tropical lowland rain forest of east Kalimantan. In: Goldammer JG (ed) Fire in the tropical biota. Ecosystem processes and global challenges. Springer, Berlin, pp 11–31

Goldammer JG, Seibert B, Schindele W (1996) Fire in dipterocarp forests. In: Schlte A Schöne D (eds) Dipterocarp forest ecosystems: Towards sustainable management. World Scientific Publishing, Singapore, pp 155–185

Kartawinata K, Abdulhadi R, Partomihardjo T (1981a) Composition and structure of a lowland dipterocarp forest at Wanariset, East Kalimantan. Malay For 44:397–406

Kartawinata K, Adisoemarto S, Riswan S, Vayada AP (1981b) The impact of man on a tropical forest in Indonesia. Ambio 10:115–119

Keßler PJA, Sidiyasa K, Ambriansyah, Zainal A (1992) Checklist for a tree flora of the Balikpapan — Samarinda area, East Kalimantan, Indonesia. Tropembos Technical Series 8, The Tropenbos Foundation, Wageningen

Krebs JC (1985). Ecology: the experimental analysis of distribution and abundance. 3rd edn, Harper and Row, New York

Matius P, Okimori Y (1991) Floristic composition and dynamics of lowland dipterocarps forest largely burned in Bukit Soeharto protection forest East Kalimantan. BIOTROP Spec Publ 41:257–271

Nykvist N (1996) Regrowth of secondary vegetation after the 'Borneo fire' of 1982–1983. J Trop Ecol 12:307–312

Richards PW (1996) The tropical rain forest. 2nd edn. Cambridge University Press, Cambridge

Riswan S (1987) Structure and floristic composition of a mixed dipterocarp forest at Lempake, east Kalimantan. In: Kostermans AJH (ed) Proceedings of third round table conference on dipterocarps, Samarinda, 16–20 April 1985, UNESCO, Jakarta pp 437–457

Sist P, Saridan A (1998) Description of the primary lowland forest of Berau. In: Bertault J-G, Kadir K (eds). Silvicultural research in a lowland mixed dipterocarp forest of East Kalimantan. CIRAD-forest, FORDA, P.T. INHUTANI, Indonesia, pp 51–73

Sukardjo S, Hagihara A, Yamakura T, Ogawa H (1990) Floristic composition of a tropical rainforest in Indonesian Borneo. Bull Nagoya Univ For 10:1–44

Tagawa H, Wirawan N (eds) (1988) A research on the process of earlier recovery of tropical rainforest after a large scale fire in Kalimantan Timur, Indonesia. Kagoshima University Research Centre for the South Pacific. Occasional paper No. 14

Walsh RPD (1996) Drought frequency changes in Sabah and adjacent parts of northern Borneo since the late nineteenth century and possible implications for tropical rain forest dynamics. J Trop Ecol 12:385–407

Whitmore TC (1984) Tropical rain forests of the Far East, 2nd edn. Clarendon, Oxford

Yamakura T, Itoh A, Kanzaki M, Ohkubo T, Ogino K, Lee HS, Tang S, LaFrankie JV, Ashton PS (1996) Basical stastistics of species diversity in a large-scale plot at Lambir, Sarawak (in Japanese). Abstract for 7th annual meeting of the Japan Society of Tropical Ecology, p 48

# 10 Dynamics of Burned Lowland Dipterocarp Forest Stands in Bukit Soeharto, East Kalimantan

Takeshi Toma, Paulus Matius, Hastaniah, Yoshiyuki Kiyono, Ryuichi Watanabe, and Yasuyuki Okimori

## 10.1 Introduction

Two of the world's largest forest fires burned the lowland dipterocarp forests in East Kalimantan within a 15-year period. During the dry spell in 1982–83, wildfires in the Indonesian province of East Kalimantan affected about 3.6 million hectares of primary and secondary rainforest (Goldammer and Seibert 1990). From 1997 to 1998, severe droughts linked to another strong El Niño-Southern Oscillation (ENSO) event occurred, and resulted in a second immense forest fire in East Kalimantan (Chapter 3, this volume).

There were several studies on the effects of the 1982–83 fire and the subsequent recovery of the forest. The fire damage was greater in logged-over secondary forests than in primary forests (Riswan and Yusuf 1986, Tagawa and Wirawan 1988, Woods 1989). Logged and burned forests suffered severe canopy loss, and the fires killed most of the existing seedlings and saplings (Tagawa and Wirawan 1988, Woods, 1989). After the fire, the regenerated forests often had a different species composition from that of the original forest stands. Trees of pioneer species in the genus *Macaranga*, particularly *M. triloba* and *M. gigantea*, were definitely promoted by the 1982–83 fire, especially in a severely burned forest with large openings in the canopy (Goldammer and Seibert 1990, Matius and Okimori 1991, Goldammer et al. 1996). Despite the importance of this rare, catastrophic event on forests, few long-term monitoring studies of stand dynamics were made after the 1982–83 fire.

This chapter deals with the dynamics of burned forest stands at Bukit Soeharto, East Kalimantan, based on an annual tree census in permanent plots. Since 1988, an annual tree census has been conducted in the forest burned by the 1982–83 fires, revealing drastic changes in the forest, 1) due to *Macaranga* trees that became established after the 1982–83 fire and 2) induced by the droughts and fire in 1997–98. Tree community data, previous disturbance history, and the influence of disturbance on tree communities are important to evaluate and monitor the changes following disturbance. The information discussed in this chapter is critical for an understanding of the long-term stand-level dynamics in East Kalimantan, which has become wild fire prone.

Ecological Studies, Vol. 140
Guhardja et al. (eds.) Rainforest Ecosystems of East Kalimantan: El Niño, Drought, Fire and Human Impacts

## 10.2 Monitoring Plots with Different Intensities of Fire Disturbance

This study was conducted in the Bukit Soeharto Education Forest (BSEF) of Mulawarman University, where the original vegetation is lowland dipterocarp forest (Chapter 3, this volume). Before it was designated as a protected forest in 1978, the forest had been logged sporadically for *Dryobalanops* spp. and *Eusideroxylon zwageri* by the local people, and selectively logged for dipterocarp trees by timber companies (Matius and Okimori 1991). The large forest fires that occurred in East Kalimantan in 1982–83 burned the BSEF. Fire damage in previously logged areas was related to the intensity of logging: greater logging intensity resulted in more damage.

In 1988, five years after the 1982–83 fires, the BSEF was a mosaic of stands in which the top canopy was dominated by either large dipterocarp trees or pioneer *Macaranga* trees. The former stands had not been logged intensively, and the damage in the 1982–83 fires seemed to be light (Lightly disturbed stand, LDS). On the other hand, most of the large trees in the latter stands were logged before the 1982–83 fires and the fire damage seemed to be heavy (Heavily disturbed stand, HDS). There were also stands with intermediate levels of fire damage (Moderately disturbed stand, MDS).

Three monitoring plots were established in 1988 (Matius and Okimori 1991): Plot 1 (100 m × 100 m), Plot 2 (120 m × 60 m), and Plot 3 (60 m × 60 m) were established in a LDS, MDS, and HDS, respectively (Fig 10.1). All trees greater than or equal to 10 cm in diameter at breast height (DBH) were enumerated annually. The upper canopy of the LDS consisted of dipterocarp species, and the tallest tree exceeded 55 m. In the MDS, some of the top canopy consisted of *Macaranga* trees about 11 m in height, and the rest consisted of tall dipterocarp trees. In the HDS, the

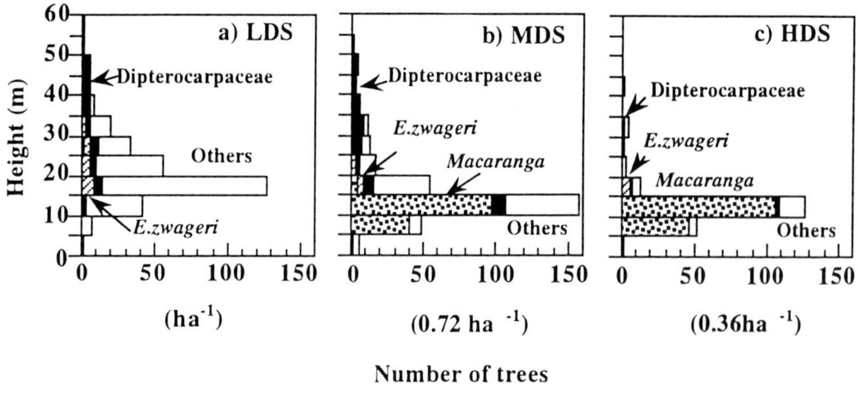


**Fig. 10.1a–c.** Height distribution of trees (diameter at breast height [DBH] ≥ 10 cm) in 1988: **a** lightly disturbed stand (*LDS*); **b** moderately disturbed stand (*MDS*); and **c** heavily disturbed stand (*HDS*)

continuous top canopy consisted of *Macaranga* trees about 11 m tall. A few large dipterocarp trees remained, indicating that this was once a tall dipterocarp stand.

## 10.3 Changes After the 1983 Fires

### 10.3.1 Tree Density and Basal Area

Tree density and basal area (BA) in the HDS and MDS changed significantly compared to LDS (Fig. 10.2). The greater changes observed in the HDS and MDS were due to *M. gigantea* and *M. triloba* trees that became established after the 1982–83 fire. The limited canopy openings in the LDS may inhibit the establishment of light-demanding pioneer species, such as *M. gigantea* and *M. triloba*. In the LDS,

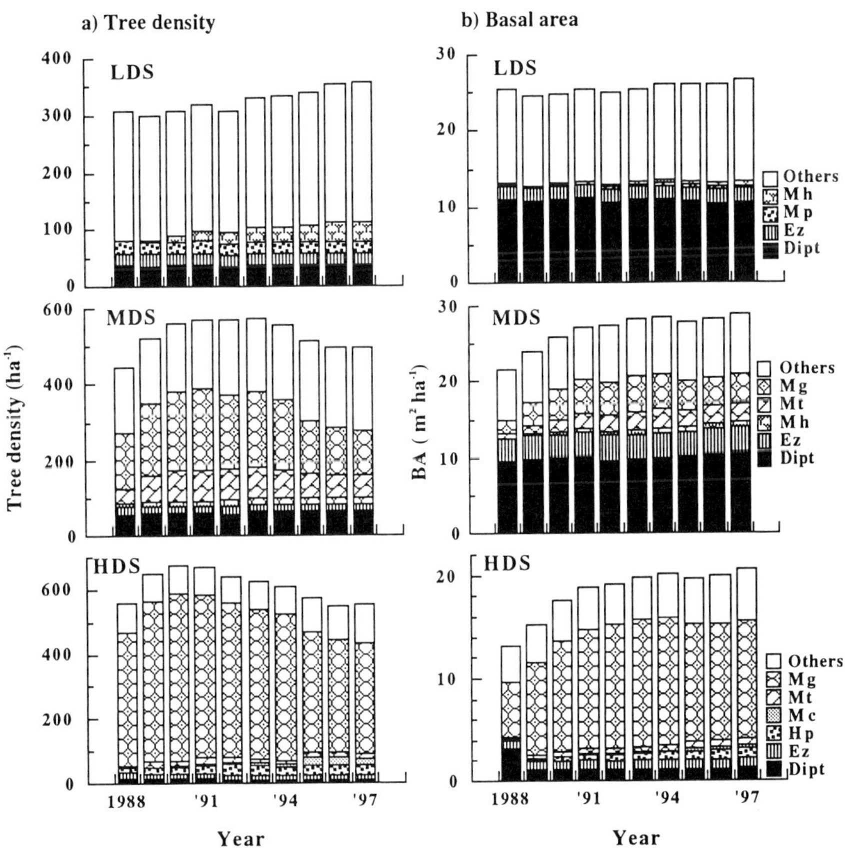


**Fig. 10.2a,b.** Trends in **a** tree density and **b** basal area in the monitored stands (modified from Toma et al. 1997). Species: Dipt: Dipterocarpaceae, Ez: *Eusideroxylon zwageri*, Mc: *Macaranga conifera*, Mg: *Macaranga gigantea*, Mh: *Macaranga hypoleuca*, Mt: *Macaranga triloba*, Mp: *Mallotus penangensis*, Hp: *Homalanthus populneus*

*Macaranga hypoleuca*, a pioneer species, started to appear in the tree-sized class (DBH ≥ 10 cm) in 1991, but far fewer trees of this species recruited than of *M. gigantea* and *M. triloba* in the MDS and HDS. This suggests that *M. hypoleuca* is relatively shade tolerant compared to *M. gigantea* and *M. triloba* (Davies et al. 1998). Changes in the density and BA of primary species, e.g. *Mallotus penangensis*, *E. zwageri*, and dipterocarp trees, were smaller than the changes in *Macaranga* trees.

The large BA of dipterocarp trees that survived the 1982–83 fires contributed to the large BA in the LDS and MDS compared to that in the HDS (Fig 10.2b). In 1988, the tree density in the HDS and MDS was already higher than that in the LDS because of the *Macaranga* trees; however, the BA was lower in the HDS and MDS than in the LDS. Beginning in 1991, the tree density of *M. gigantea* in the HDS and MDS started to decrease without a marked decrease in the BA of the stands. In 1997, the tree density in these two stands was still much higher than that in the LDS, while BA in the LDS was the same as in the MDS and much higher than that in the HDS, where there were few large dipterocarp trees.

## 10.3.2 Recruitment and Death

The peak of recruitment of *M. gigantea* and *M. triloba* to the tree-sized class (DBH ≥ 10 cm) occurred within five years of the 1983 fire. The rate of tree recruitment in the MDS and HDS was highest from 1988 to 1989, and decreased thereafter (Fig. 10.3). Early in the monitoring period, the tree recruitment rate was much higher in the MDS and HDS than in the LDS, but subsequently the rate in the MDS and HDS decreased to that in the LDS. The change was due to decreased recruitment of *M. gigantea* and *M. triloba* in the MDS and HDS.

Increased mortality in *Macaranga* trees occurred during occasional droughts in the BSEF. Higher mortality was observed from 1991 to 1992 and from 1994 to 1995 (Fig. 10.3), when severe droughts occurred in the BSEF (Chapter 2, this volume). The high mortality was due to the death of *Macaranga* spp. The proportion of trees that died between 1991 and 1992 that were *Macaranga* spp. was 87% and 80% in the MDS and HDS, respectively. All of the trees that died in the MDS and HDS between 1994 and 1995 were *Macaranga* trees.

The death of *Macaranga* trees seemed to be also due to natural thinning. Slow-

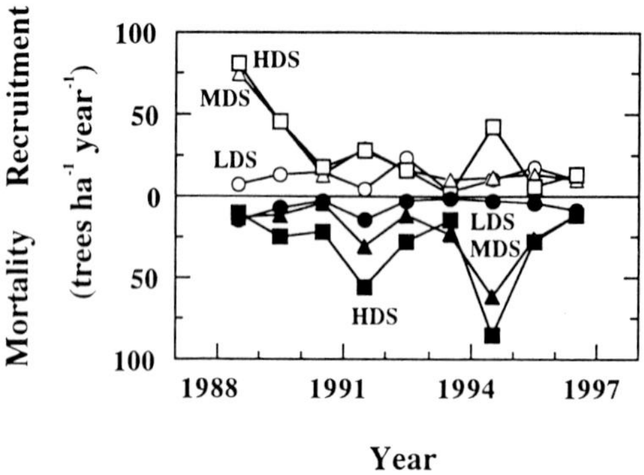


**Fig. 10.3.** Changes in recruitment and mortality rates of trees (DBH ≥ 10 cm) in each stand (after Toma et al. 1997)

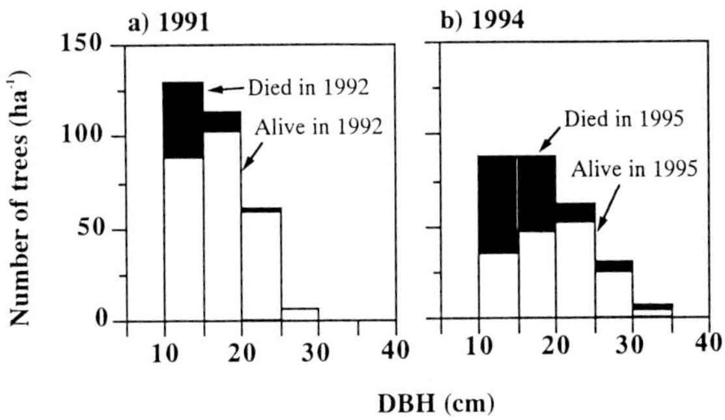


**Fig. 10.4a,b.** DBH frequency distribution for *Macaranga* trees in moderately disturbed stand (MDS) in **a** 1991 and **b** 1994. The *solid parts of bars* represent the trees that died before the next annual enumeration (modified from Toma et al. 1997)

growing, less vigorous individuals are not able to compete in either height or root growth, and eventually die without decreasing stand biomass (Peet and Christensen 1987). Large plants then have the advantage of being able to extract critical resources from the environment, so that mortality is concentrated in the smaller size classes (Peet and Christensen 1987). Tree density began to decrease without a marked decrease in BA in 1991 (Fig 10.2), and the dead *Macaranga* trees were biased to smaller size classes (Fig 10.4). Natural thinning of *Macaranga* trees started 8 years after their establishment in large sites opened up by the 1982–83 fires. The high mortality between 1994 and 1995 caused by severe drought resulted in marked loss of BA in HDS, which indicates that competition for limited water resources exceeded the natural thinning process in a normal wet year.

The DBH distribution pattern of *Macaranga* trees in the MDS and HDS changed to a bell-shaped distribution by 1997 (see the open columns in Fig. 10.4), which indicates that the regeneration of these pioneer species depended on large disturbances of the forest canopy. The number of trees in the lowest size class (10–15 cm DBH) in the MDS and HDS decreased from 1988 to 1997. The decrease in *Macaranga* trees in the smaller size class in the MDS and HDS was caused by growth into larger size classes, combined with a decrease in recruited trees and higher mortality in the smaller size class. Species with a bell-shaped DBH distribution are generally shade intolerant or disturbance dependent (Nakashizuka et al. 1992). Thus, if there were no large disturbances, such as forest fires, late successional species which can germinate and become established under a closed canopy would be expected to replace the *Macaranga* trees. The high mortality in *Macaranga* trees during occasional droughts would increase the chance for establishment of dipterocarp species, which flower and fruit after the onset of a drought (Chapter 11, this volume).

## 10.3.3 Aboveground Biomass and Productivity

The aboveground weight of a tree can be calculated from a series of allometric functions based on DBH (Ogawa and Kira 1977, Whitmore 1984). Tree height (H, m) can be estimated from DBH (D, cm), because a hyperbolic function approximates the relationship between H (m) and DBH (D cm) in LDS (Ogawa et al 1965). We calculated the weight of a tree using $D^2H$, which applies allometric coefficients obtained for non-*Macaranga* trees in a primary dipterocarp forest at Sebulu, East Kalimantan (Yamakura et al. 1986). For *Macaranga* trees, we used the allometric coefficients obtained for *M. gigantea* trees in the BSEF (Toma unpublished).

## DBH Growth Rate and Aboveground Biomass

The high growth rate of pioneer *Macaranga* trees did not persist for very long, less than 10 years (Fig. 10.5). The DBH growth rate of *Macaranga* trees was highest at the beginning of the monitoring period, and decreased thereafter. *Macaranga* trees had a higher DBH growth rate than dipterocarp trees at the beginning of the monitoring period, but later there was no clear difference.

Throughout the monitoring period, the LDS, which was dominated by dipterocarps, had the highest aboveground biomass of the three stands (Fig 10.6). The aboveground biomass in the LDS increased from 303 t $ha^{-1}$ in 1988 to 315 t $ha^{-1}$ in 1997. However, the aboveground biomass in the LDS was still much lower than that reported in primary dipterocarp forests in East Kalimantan, 509 t $ha^{-1}$ in Sebulu (Yamakura et al. 1986) and 460 t $ha^{-1}$ in Meratus (Ruhiyat 1996). Therefore, 14 years is inadequate for the recovery of aboveground biomass, even in the LDS.

The aboveground biomass of the *Macaranga* population increased rapidly until 1991 and the biomass then became nearly constant (Fig. 10.6a). The aboveground biomass of the *Macaranga* population was much less than those of LDS, MDS and the above mentioned primary forests. A large number of *Macaranga* trees with a high initial rate of increase in DBH (Fig 10.5) did not compensate for the loss in aboveground biomass by disturbances in the past. Further aboveground biomass accumulation is expected only after primary species replace the pioneer *Macaranga* species.

## Net Primary Productivity

Even in the stands dominated by pioneer species, net primary productivity (NPP) was lower than values reported in tropical primary forests (Fig. 10.6b; Table 10.1). The highest NPP in the MDS and HDS was observed when the DBH growth rate of

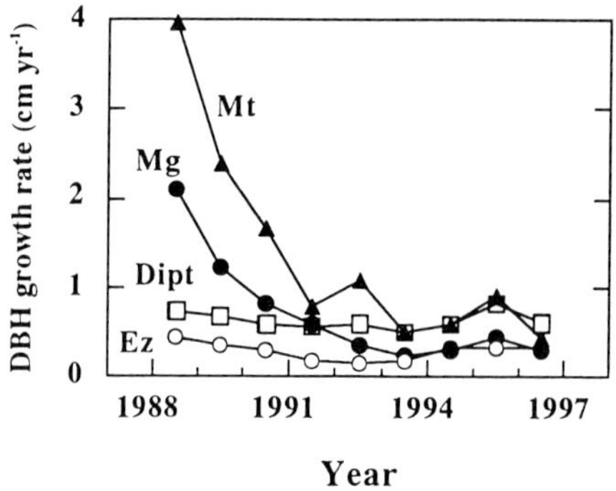


**Fig. 10.5.** DBH growth rate of the Dipterocarpaceae (Dipt), *Eusideroxylon zwageri*,(Ez), *Macaranga gigantea*,(Mg), and *Macaranga triloba*,(Mt) trees in the stands

*Macaranga* trees was highest (Fig 10.6) and they were comparable to tropical secondary and primary forests. The forest at Sipitang, in Sabah, Borneo, was also burned by the 1982–83 fires and dominated by *Macaranga* spp. in 1991 (Nykvist 1996). The estimated NPP of the forest in Sipitang from 1988 to 1991 was comparable to that in HDS during the same period. Over a longer time interval, the NPP in MDS and HDS became less (nearly half) that in primary forests because of the decrease in the DBH growth rate in *Macaranga* trees.

Occasional death of trees consumes the net production of the stands and the change in the aboveground biomass becomes negative. High rates of loss of aboveground biomass were observed in the LDS and HDS between 1988 and 1989 (Fig. 10.6c). This was due to the death of large dipterocarp trees, which was probably a residual effect of fire injury suffered in 1982–83 (Matius and Okimori 1991).

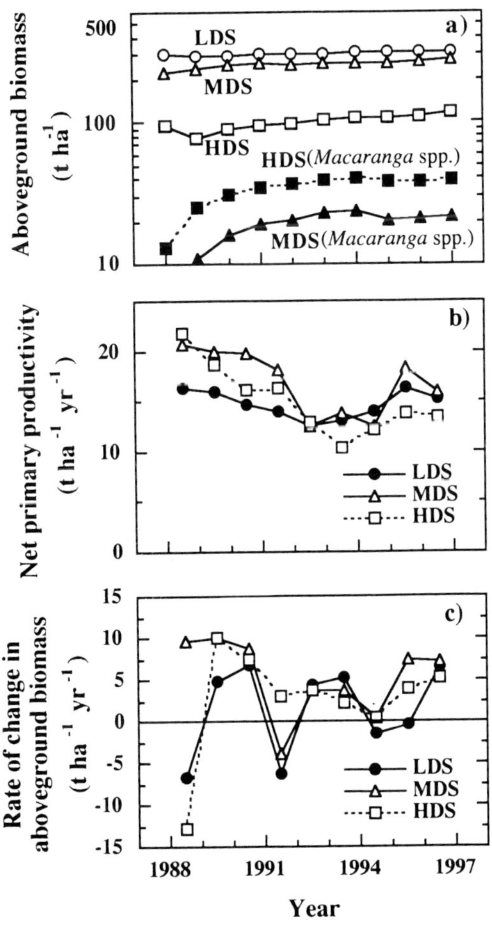


**Fig. 10.6 a,b.** Trend in **a** aboveground biomass, **b** net primary productivity (NPP), and **c** rate of change in aboveground biomass in the stands. The annual litterfall rate in LDS of 7.7 t ha$^{-1}$ yr$^{-1}$ in 1996 (Chapter 5, this volume) was applied to calculate the NPP of all the stands for the whole period

**Table 10.1.** Aboveground biomass (AB), aboveground biomass increment rate (ABI), litterfall rate (LR), and net primary productivity (NPP) in various tropical forests

| Forest type<br>Locality | AB<br>($t\ ha^{-1}$) | ABI<br>($t\ ha^{-1}\ yr^{-1}$) | LR<br>($t\ ha^{-1}\ yr^{-1}$) | NPP<br>($t\ ha^{-1}\ yr^{-1}$) |
|---|---|---|---|---|
| Burned by the 1982–83 fire | | | | |
| Bukit Soeharto, East Kalimantan | | | | |
| LDS (1988–97) | 309.3 | 7.0 | 7.7[f] | 14.7 |
| MDS (1988–89) | 234.1 | 12.9 | 7.7[f] | 20.6 |
| (1994–95) | 265.1 | 4.9 | 7.7[f] | 12.6 |
| HDS (1988–89) | 87.2 | 14.7 | 7.7[f] | 21.8 |
| (1993–94) | 105.5 | 2.1 | 7.7[f] | 10.4 |
| Sipitang, Sabah[a] | | | | |
| 1985–88 | 15.7 | 7.3[g] | 8.0[h] | 15.3 |
| 1988–91 | 44.3 | 11.8[g] | 8.0[h] | 19.8 |
| Primary forests | | | | |
| Pasoh, Malaysia[b] | 425 | 5.3 | 15.9 | 21.1 |
| Pinang Pinang, West Sumatra[c] | 414 | 10.3 | 9.4 | 19.8 |
| Gajabuih, West Sumatra[c] | 432 | 9.9 | 9.0 | 18.9 |
| Secondary forests | | | | |
| Memcimai, East Kalimantan[d] | | | | |
| (33 years after fallow) | 97.4 | 9.4 | 11.1 | 20.5 |
| Average of various forests in tropics[e] | | | | |
| (20 years old) | 140 | 7.0 | 8.0 | 15.0 |

LDS, MDS, and HDS, lightly, moderately, and heavily disturbed stand, respectively. Mean value in LDS, maximum and minimum values in MDS and HDS are shown.
[a]Nykvist(1996)
[b]Kira(1978)
[c]Yoneda et al.(1994)
[d]Kiyono and Hastaniah(1997)
[e]Brown and Lugo(1990)
[f]Measured in LDS in 1996(Chapter 5, this volume)
[g]Calculated from aboveground biomass in 1985, 1988 and 1991
[h]According to Brown and Lugo (1990)

## 10.4 Effect of the Drought and Fires in 1997 and 1998

### 10.4.1 Death of Trees Caused by Droughts and Fires

The droughts and fire in 1997–98 consumed the recovery process of the stands after the 1983 fire (Fig. 10.7). Between the annual tree census in August 1997 and 1998,

(Chapter 2, this volume). Surface fires occurred and burned most of the BSEF between the end of February and early March 1998 (Chapter 3, this volume). Tree density and BA in the stands decreased to be less than that in 1983 as a result of the droughts and fires in 1997–98 (Fig. 10.7).

The first drought resulted in high mortality only in *Macaranga* trees, while the second drought, together with the resulting surface fires, resulted in high mortality of both *Macaranga* trees and non-*Macaranga* species (Fig. 10.8). As the second drought progressed, many trees partly shed their leaves . However, even these weakened trees were still alive with leaves remaining before the surface fires burned the monitoring stands in late February and early March 1998. After the fire, many trees shed their leaves immediately, and most of these trees died. The direct cause of the death of non-*Macaranga* trees was the fire, not the drought. Drought without surface fire killed planted trees (Chapter 3, this volume) and trees in natural forests during the droughts in 1982–83 in East Kalimantan (Leighton and Wirawan 1986) and in Sabah (Woods 1989, Becker et al. 1998). Mortality caused by the 1983 drought alone was lower than that caused by drought together with fire (Leighton and Wirawan 1986, Woods 1989). However, we cannot quantitatively compare the drought and fire in 1998 as a cause of death, because all of our monitoring stands were burnt. A comparison of burned and unburned stands is necessary to compare the effects of drought and fire.

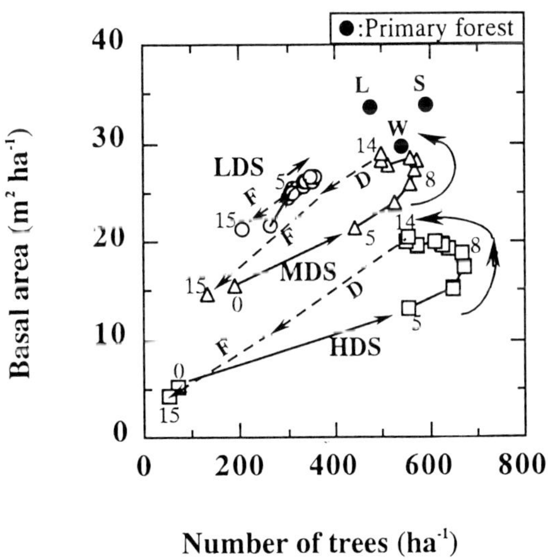


**Fig. 10.7.** Changes in the density and basal area (BA) of trees (DBH ≥ 10 cm) in the stands after the 1983 fire. *Arrows* represent the time course of change. The *numbers* indicate years since 1983. *Arrows with a dashed line* represent changes induced by drought in 1997 (*D*) and drought and fire in 1998 (*F*). The density and BA of the trees in 1983 was calculated from the 1988 DBH and the DBH growth rate from 1988 to 1989. Trees that died between 1983 and 1988 were ignored. The values of primary forests come from Wanariset (W, Kartawinata et al. 1981), Lempake (L, Riswan 1987), and Sebulu (S, Sukardjo et al. 1990) in Kutai Regency, East Kalimantan

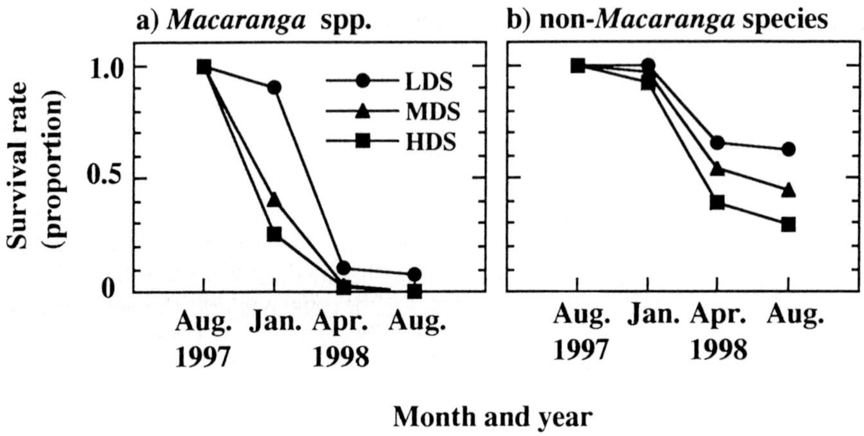


**Fig. 10.8a,b.** Survival rate of **a** *Macaranga* spp. and **b** non-*Macaranga* trees during the droughts and fire events in 1997–98. The first drought lasted from July 1997 to October 1997 and the second drought lasted from January to April 1997. Surface fires burned the stands at the end of February to early March in 1998

Tree mortality caused by the 1998 fire was related to the intensity of past disturbance. The stand that was heavily disturbed by the 1982–83 fire, dominated by *Macaranga* spp., had the highest mortality in the droughts and fire in 1997–98. Even if we exclude *Macaranga* trees, which had a higher mortality than other species, among the three stands the mortality of trees was highest in the HDS (Fig. 10.8b). From August 1997 to August 1998, the mortality of trees other than *Macaranga* spp. was 37, 56, and 70%, in the LDS, MDS, and HDS, respectively. In the MDS and HDS, these values exceeded the tree mortality caused by the 1982–83 fires in a primary (28.3%) and secondary (42.4%) forest in Lempake, East Kalimantan (Riswan and Yusuf 1986).

Dead trees were enumerated in August 1998, after normal rainfall recommenced in the middle of April 1998. Nevertheless, some newly-dead trees were found after the annual tree census in August 1998 had been completed. The initial estimates of tree mortality after a disturbance may not include the subsequent mortality of trees that are damaged by the disturbance but which die later when their often substantial reserves are exhausted (Walker 1995). Matius and Okimori (1991) observed a residual effect of fire injury to tree death five years after the 1983 fire. In the LDS, 13 trees died and fell in 1988–89 and 10 of these (73%) had burn scars from the 1982–83 fires (Matius and Okimori 1991). Such subsequent mortality will be examined by annual enumeration in the future.

### 10.4.2 Estimation of the Recovery Process After the 1998 Fire

When a canopy gap forms, the regeneration of primary and pioneer tree species usually starts from seedlings and seed banks, respectively (e.g. Whitmore 1984). As a result of the drought in 1997, many *Macaranga* trees died in the MDS and HDS, and the canopy in those forests became sparse. A sparse canopy might result in increased growth of the seedlings that grew under the closed *Macaranga* canopy. However, the 1998 fire burned and killed the seedlings of both pioneer and primary species. Mortality of seedlings due to the 1998 fires in LDS, MDS and HDS was 90, 81, and 90%, respectively. This would benefit pioneer species, which germinate from seed banks after a fire, since competing seedlings were eliminated.

*Macaranga* trees are expected to dominate both the MDS and HDS again. In September 1998, six months after the surface fire, there were about 100 000 and 60 000 newly germinated seedling $ha^{-1}$ in MDS and HDS, respectively. The ratio of *Macaranga* seedlings to total newly-germinated seedlings exceeded 80% in both of the stands. At the end of 1998, ten months after the fire, the MDS and HDS were covered with several pioneer species, herbs and ferns. Heights of *Macaranga* seedlings were the same or lower than the other species and most of the *Macaranga* seedlings were covered by other species. However, even if large open sites are covered with herbs and ferns for a few years after a fire disturbance or fallow, pioneer trees come to be the predominant species within 5 years (Mackie et al. 1986, Nykvist 1989; Chapter 16, this volume). Thus, we expect that even though the MDS and HDS are now covered with herb, ferns and shrub species, pioneer trees such as *M. gigantea* and *M. triloba* will overtop the other plants within 5 years.

Dipterocarp seedlings were observed in the above mentioned burned stands or fallows, and were thought to have come from seeds produced by the remaining mature trees near the stands (Mackie et al. 1986, Nykvist 1989, Chapter 16 this volume). In the LDS and MDS, trees belonging to primary species decreased but still remained after the 1998 fires, and they could be seed-bearing trees. Species replacement from *Macaranga* species to primary species is expected when the *Macaranga* population dominance starts to decrease in future. However in HDS, only a few matured trees of the primary species survived the 1998 fire and thus species replacement is seemingly hopeless compare to LDS and MDS.

## 10.5 Conclusion

Structure, species composition and stand dynamics in logged over forests are affected by logging intensity in the past. Since more heavily logged forest subsequently suffers greater damage due to fires, the effects of selective logging on those tree community characteristics are more emphasized when the fires occur. In this respect, long-term monitoring of forest with previous disturbance history is needed.

As of 1997 in lightly disturbed stands, the family Dipterocarpaceae was still predominant in terms of basal area and biomass because large dipterocarp trees that had survived the 1983 fires shared a large part of these values. The heavily disturbed stands were replaced by pioneer trees with a high density. Although these trees had a rapid growth rate, they did not compensate for the loss of basal area and biomass due to death of large dipterocarp trees. Up to 1997, our forest appeared to be moving toward recovery, i.e., successional species replacement from pioneer to primary species such as dipterocarps.

However, the fire in 1998 has again devoured the forests. The interval between the first and second fires, 15 years, was too short for recovery. Since the forest had not recovered enough, the damage due to the 1998 fire was more severe than that in 1983. Large trees that survived the 1998 fire were fewer than those that survived the 1983 fire. Seedling banks of primary species were destroyed. On the other hand, the 15-year period is likely to be long enough for pioneer species to prepare the next generation in seed bank. Therefore, the sites opened by the 1998 fire should be dominated by pioneer species quickly. However, because of the deficiency of juveniles and decreased density of seed-bearing trees, it will be extremely difficult for primary species to regenerate themselves, when the dominance of the pioneer population wanes in another 15 years.

# References

Becker P, Ong CL, Goh F (1998) Selective drought mortality of dipterocarp trees: no correlation with timber group distributions in Borneo. Biotropica 30:666–671

Brown S, Lugo AE (1990) Tropical secondary forests. J Trop Ecol 6:1–32

Davies SJ, Palmiotto PA, Ashton PS, Lee HS, Lafrankie JV (1998) Comparative ecology of 11 sympatric species of Macaranga in Borneo: tree distribution in relation to horizontal and vertical resource heterogeneity. J Ecol 86:662–673

Goldammer JG, Seibert B (1990) The impact of droughts and forest fires on tropical lowland rain forest of east Kalimantan. In: Goldammer JG (ed) Fire in the tropical biota. Ecological studies 84. Springer, Berlin, pp 11–31

Goldammer JG, Seibert B, Schindele W (1996) Fire in dipterocarp forests. In: Schlte A Schöne D (eds) Dipterocarp forest ecosystems: Towards sustainable management. World Scientific Publishing, Singapore, pp 155–185

Kartawinata K, Abdulhadi R, Partomihardjo T (1981) Composition and structure of a lowland dipterocarp forest at Wanariset, East Kalimantan. Malay For 44:397–406

Kira T (1978) Community architecture and organic matter dynamics in tropical lowland rain forests of Southeast Asia with special reference to Pasoh Forest, West Malaysia. In: Tomlinson PB, Zimmermann MH (eds) Tropical trees as living systems. Cambridge University Press, Cambridge, pp 561–590

Kiyono Y, Hastaniah (1997) Slash-and-burn agriculture and succeeding vegetation in East Kalimantan. PUSREHUT Spec Publ vol 6, Mulawarman University, Samarinda, Indonesia

Leighton M and Wirawan N (1986) Catastrophic drought and fire in Borneo rain forests associated with the 1982–3 El Niño southern oscillation event. In: Prance GT (ed) Tropical rain forests and the world atmosphere, Westview, Boulder, Colorado, pp 75–102

Mackie C, Jessup TC, Vayda AP, Kartawinata K (1986) Shifting cultivation and patch dynamics in an upland forest in East Kalimantan. In: UNESCO-MAB Regional Workshop on Impact of Man's Activities on upland forest Ecosystem. Serdang, Malaysia, February, 1986, pp 486–518

Matius P, Okimori Y (1991) Floristic composition and dynamics of lowland dipterocarps forest largely burned in Bukit Soeharto protection forest East Kalimantan. BIOTROP Spec Publ No 41:257–271

Nakashizuka T, Iida S, Tanaka H, Shibata M, Abe S, Masaki T, Niiyama K (1992) Community dynamics of Ogawa Forest Reserve, a species rich deciduous forest, central Japan. Vegetatio 103:105–112

Nykvist N (1996) Regrowth of secondary vegetation after the 'Borneo fire' of 1982–1983. J Trop Ecol 12:307–312

Ogawa H, Kira T (1977) Methods of estimating forest biomass. In Shidei T and Kira T (eds) Primary productivity of Japanese forests, JIBP Synthesis 16. University of Tokyo Press, Tokyo, pp 15–25

Ogawa H, Yoda K, Ogino K, Kira T (1965) Comparative ecological studies on three main types of forest vegetation in Thailand. II. Plant biomass. Nature and life in southeast Asia 4:49–80

Peet RK, Christensen NL (1987) Competition and tree death. Bioscience 37:586–595

Riswan S (1987) Structure and floristic composition of a mixed dipterocarp forest at Lempake, East Kalimantan. In Kostermans AJH (ed) Proc of 3rd round table conference on dipterocarps, Samarinda, 16–20 April 1985, UNESCO, Jakarta, pp 437–457

Riswan S, Yusuf R (1986) Effect of forest fire on trees in the lowland dipterocarp forest of East Kalimantan, Indonesia. BIOTROP Special Publication 25:155–163

Ruhiyat D (1996) Biomass estimation of tropical rain forest in East Kalimantan. Rimba Kalimantan (Bull Fac For Mulawarman Univ) 1:42–57 (In Indonesian with English abstract)

Sukardjo S, Hagihara A, Yamakura T, Ogawa H (1990) Floristic composition of a tropical rain forest in Indonesian Borneo. Bull Nagoya Univ For 10:1–44

Tagawa H, Wirawan N (eds) (1988) A research on the process of earlier recovery of tropical rainforest after a large scale fire in Kalimantan Timur, Indonesia. Kagoshima University Research Centre for the South Pacific. Occasional paper No. 14

Toma T, Matius P, Hastaniah, Kiyono Y, Watanabe R, Okimori Y (1997) Floristic composition and stand dynamics in logged-over and burnt lowland dipterocarp forests in Bukit Soeharto, East Kalimantan. In: Suhartoyo H, Fatawi M (eds) Proc 2nd Intern Symp Asean Trop For Manage, PUSREHUT Univ Mulawarman, Samarinda pp 198–206

Walker LR (1995) Timing of post-hurricane tree mortality in Puerto Rico. J Trop Ecol 11:315–320

Whitmore TC (1984) Tropical rain forests of the Far East, 2nd edn. Clarendon, Oxford

Woods P (1989) Effect of logging, drought, and fire on structure and composition of tropical forests in Sabah, Malaysia. Biotropica 21:290–298

Yamakura T, Hagihara A, Sukardjo S, Ogawa H (1986) Above ground biomass of tropical forest stands in Indonesian Borneo. Vegetatio 68:71–82

Yoneda T, Ogino K, Kohyama T, Tamin R, Syhbuddin, Rahman M (1994) Horizontal variance of stand structure and productivity in a tropical foothill rain forest, West Sumatra, Indonesia. Tropics 4:17–33

# 11 Flowering and Fruiting Phenologies of Dipterocarps in a Rainforest in Bukit Soeharto, East Kalimantan

YOSHIYUKI KIYONO and HASTANIAH

## 11.1 Phenological Observations at Bukit Soeharto Education Forest

Flowering and fruiting periodicity varies from tree to tree, even within a single dipterocarp species. However, much of the knowledge about flowering phenology in tropical trees is based on broad observations at the community level (Bawa and Krugman 1991). To examine the variation of individual trees, observations were recorded of the flowering and fruiting phenologies of dipterocarps in Bukit Soeharto Education Forest (BSEF), in East Kalimantan, for the 6 years from 1992 to 1998. This chapter deals with interspecific and inter-individual variation in relation to variation in diameter at breast height (DBH) and with interspecific variation in the reproductive success of mature trees. In addition, the influence of the severe drought and fires in 1997–98 on the survival of dipterocarp trees is discussed.

In December 1992, we selected 782 trees from 24 dipterocarp species (Table.11.1, Fig. 11.1), whose entire crown could easily be observed from one of five forest roads in BSEF: Ibu-1, Ibu-2, Enggang-A, Enggang-B, and Landak. Individual trees were observed for flowering and fruiting every two to four weeks. In 1987–88, Okimori and Permono (1991) investigated the flowering and fruiting phenologies of 18 trees in 11 dipterocarp species in BSEF: *Dryobalanops lanceolata, Dipterocarpus cornutus, Dipterocarpus humeratus, Shorea ovalis, Shorea smithiana, Shorea gibbosa, Shorea parvifolia, Shorea pauciflora, Shorea laevis, Shorea lamellata*, and *Anisoptera costata*. They found that the flowering periods of individual trees were between one and three months long, and their fruiting periods were between 42 and approximately 200 days. Therefore, our regular observations were unlikely to miss episodes of flowering or fruiting.

In our study, the 12 species that included 20 or more trees when observed in 1992 were deemed major species, and the other 12 species that included less than 20 trees were deemed minor species. This classification reflects the actual distribution of dipterocarp tree species in BSEF. The number of individual trees observed, the range of DBH, and the annual death rate are shown for each dipterocarp species (Table 11.1).

Ecological Studies, Vol. 140
Guhardja et al. (eds.) Rainforest Ecosystems of East Kalimantan: El Niño, Drought, Fire and Human Impacts

**Table 11.1.** Dipterocarp trees observed

| Species | Number of trees in 1992 | DBH Max. (cm) | DBH Min. (cm) | Death rates 1992/97 (%/y[a]) | Death rates 1997/98 (%/y[a]) | Death rates 1992/98 (%/y[a]) |
|---|---|---|---|---|---|---|
| *Shorea laevis* | 270 | 180 | 25 | 0.1 | 72.0 | 19.2 |
| *Shorea smithiana* | 72 | 195 | 30 | 0.3 | 57.7 | 13.6 |
| *Dipterocarpus cornutus* | 67 | 160 | 30 | 0.0 | 40.3 | 8.2 |
| *Shorea pauciflora* | 47 | 160 | 40 | 0.4 | 50.0 | 11.2 |
| *Shorea johorensis* | 44 | 210 | 34 | 0.5 | 55.8 | 13.1 |
| *Shorea parvifolia* | 38 | 125 | 29 | 0.0 | 52.6 | 11.7 |
| *Shorea seminis* | 38 | 130 | 22 | 0.5 | 54.1 | 12.5 |
| *Dipterocarpus humeratus* | 30 | 165 | 35 | 1.4 | 32.1 | 7.3 |
| *Shorea leprosula* | 28 | 110 | 40 | 0.7 | 48.1 | 10.9 |
| *Cotylelobium lanceolatum* | 24 | 105 | 25 | 0.8 | 4.3 | 1.4 |
| *Shorea gibbosa* | 20 | 180 | 41 | 0.0 | 65.0 | 16.1 |
| *Shorea ovalis* | 20 | 125 | 35 | 0.0 | 55.0 | 12.5 |
| *Dryobalanops beccarii* | 17 | 140 | 25 | 2.5 | 93.3 | 37.6 |
| *Dipterocarpus tempehes* | 10 | 85 | 35 | 2.1 | 11.1 | 3.7 |
| *Dipterocarpus confertus* | 9 | 115 | 32 | 0.0 | 22.2 | 4.1 |
| *Dipterocarpus gracilis* | 9 | 130 | 40 | 0.0 | 44.4 | 9.3 |
| *Hopea mengarawan* | 9 | 45 | 15 | 4.9 | 85.7 | 30.7 |
| *Shorea bracteolata* | 8 | 115 | 45 | 0.0 | 62.5 | 15.1 |
| *Shorea lamellata* | 7 | 130 | 50 | 0.0 | 42.9 | 8.9 |
| *Vatica javanica* var. *scaphifolia* | 6 | 120 | 35 | 0.0 | 33.3 | 6.5 |
| *Anisoptera costata* | 4 | 160 | 80 | 5.6 | 0.0 | 4.7 |
| *Hopea rudiformis* | 3 | 65 | 18 | 0.0 | 100.0 | 100.0 |
| *Parashorea* sp. | 1 | 30 | 30 | 0.0 | 0.0 | 0.0 |
| *Shorea parvistipulata* | 1 | 50 | 50 | 0.0 | 100.0 | 100.0 |
| Total | 782 | | | 0.4 | 56.9 | 13.4 |

DBH, runk diameter at breast height in 1992
[a]Percent per year

## 11.2 Flowering and Fruiting Fluctuations

It was obvious that dry spells induced flowering. Between 1992 and 1998, many dipterocarp trees flowered simultaneously and fruited in 1993–94, 1994–95, and 1997–98 (Fig. 11.2). Every flowering event started at least one to two months after the onset of a relatively prolonged dry spell. The flowering patterns of the 12 major species can be classified into three groups: the *Dipterocarpus* type (two species), the *Shorea* type (nine species in three subgroups), and the *Cotylelobium lanceolatum* type (one species).

In the *Dipterocarpus* group, *D. cornutus* and *D. humeratus* flowered three times between 1992 and 1998: in 1993–94, 1994–95, and 1997–98 (Fig. 11.3). The ratio of flowering trees to the total number of observed trees in each species was smallest in

1994–95 and largest in 1997–98. Similar fluctuations were also observed in minor *Dipterocarpus* species: *D. confertus, D. gracilis,* and *D. tempehes*.

In the *Shorea* group, *Shorea gibbosa* also flowered three times between 1992 and 1998: in 1993–94, 1994–95, and 1997–98 (Fig. 11.3). The ratio of flowering trees to the total number of trees was smallest in 1993–94 and largest in 1997–98. Similar fluctuations occurred in *S. leprosula, S. parvifolia,* and *S. seminis*. On the other hand, most *S. johorensis, S. laevis,* and *S. pauciflora* did not flower in 1993–94, but flowered in 1994–95 and 1997–98. A few *S. ovalis* and *S. smithiana* flowered in 1994–95, but most flowered in 1997–98. These nine *Shorea* species can be further

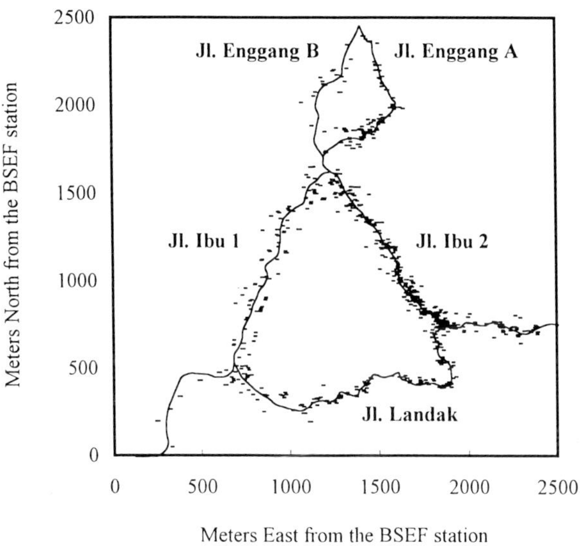


**Fig. 11.1.** Location of the dipterocarp trees observed. *BSEF*, Bukit Soeharto Education Forest

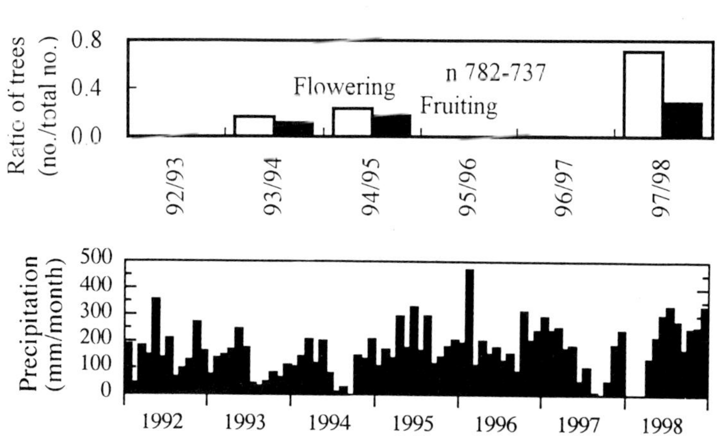


**Fig. 11.2.** Ratios of flowering and fruiting trees to the total number ($n = 782–737$) of observed dipterocarp trees and the monthly precipitation at Bukit Soeharto. A total of 782 trees were observed in December 1992, and 737 remained in April 1998. Rainfall ata cited from Chapter 2 in this volume

classified into three groups. The first group, *S. gibbosa, S. leprosula, S. parvifolia,* and *S. seminis* started flowering one to two months after a dry spell commenced. The second group, *S. johorensis, S. laevis,* and *S. pauciflora* started flowering two to six months after the onset of a dry spell. The third group, *Shorea ovalis* and *Shorea smithiana*, started flowering three to five months after dry spells started. The first group contains species considered sensitive to a dry spell, while the second group includes less sensitive species, and the species in the third group appear relatively insensitive. In these *Shorea* groups, the proportion of flowering trees increased in the order 1993–1994, 1994–95, and 1997–98.

Finally, of the 12 major species, *Cotylelobium lanceolatum* showed peculiar flowering responses (*Cotylelobium lanceolatum* type). More than half of the *Cotylelobium lanceolatum* trees flowered in 1994–95, and about one-third flowered in 1997–98 (Fig. 11.3). More precisely, flowers of *Cotylelobium lanceolatum* were found in May–June 1994 and in February 1995. In the former period, flowering started before the beginning of the 1994–95 dry spell. Therefore, the first flowering appeared to be induced by the 1993–94 dry spell that started ten months earlier. The 1994–95 dry

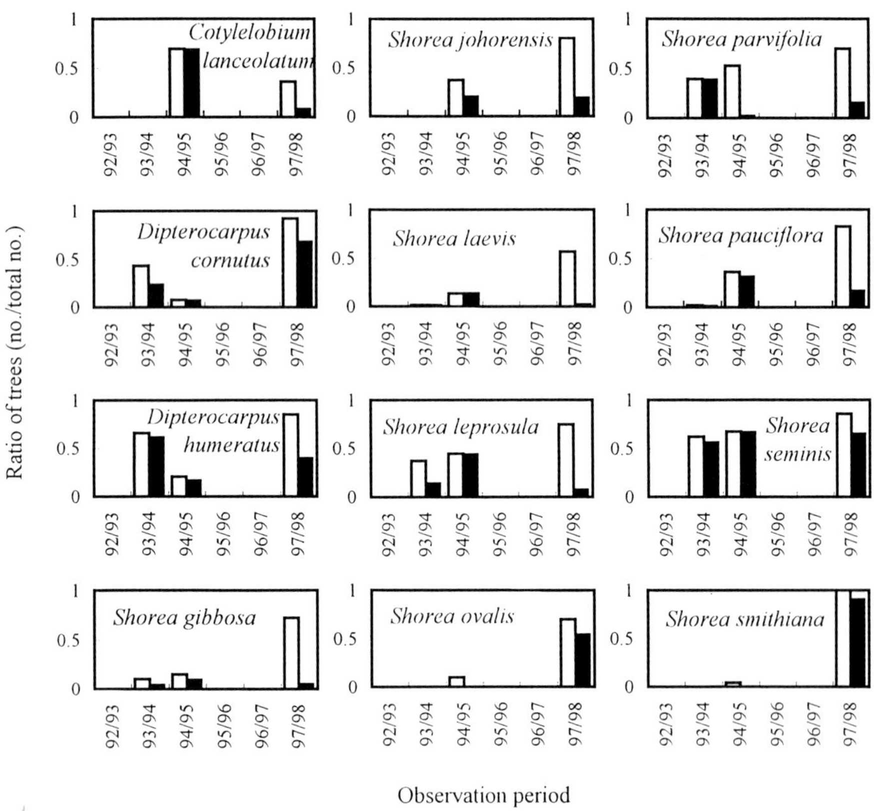


**Fig. 11.3.** Ratios of flowering and fruiting trees to the total numbers of the major dipterocarp species observed. *Open bars,* ratios of flowering trees; *closed bars,* ratios of fruiting trees. Some *Cotylelobium lanceolatum* trees were still in flower in November 1998

spell might have induced the second flowering in February 1995, seven months after the dry spell started. Does *Cotylelobium lanceolatum* need as long as seven or ten months to start flowering? If so, then according to this hypothesis *Cotylelobium lanceolatum* should have flowered only rarely early in the interspecific mass-flowering period in 1997–98. It finally joined the interspecific mass flowering in July 1998, one year after the dry spell started.

Of the 12 minor species, four *Dipterocarpus* species, *Anisoptera costata, Dryobalanops beccarii, Hopea mengerawan, Shorea bracteolata*, and *Vatica javanica* var. *scaphifolia* seemed to flower in response to dry spells, because they started flowering one to six months after the dry spells started. On the other hand, *Shorea lamellata*, another of the minor species, had a peculiar flowering response. This species flowered in 1994–95 and 1996–97. In the first period, flowers were found in June 1994, one month before the beginning of the 1994–95 dry spell, while in the latter period, no dry spell had occurred in the previous two years. *Shorea lamellata* appears not to need the cue of a dry spell to flower. Many researchers have reported that droughts cause mass-flowerings of dipterocarps (e.g., Kostermans 1992). Furthermore, Ashton et al. (1988) speculated that such dipterocarps might have arisen initially in the seasonal tropics. However, the flowering of *Shorea lamellata* requires further explanation.

## 11.3 Reproductive Success

Even non-fruiting trees can become parents by donating pollen to other trees that fruit. In this paper, reproductive success is defined as the frequency of becoming a parent, and was calculated for each of the 12 major tree species from the flowering and fruiting record from December 1992 to May 1998. The frequency of fruiting events ($F_{min}$) represents the minimum frequency of becoming a parent, while the frequency of fruiting plus non-fruiting events ($F_{max}$) is equivalent to the maximum frequency of becoming a parent. Therefore, the actual frequency of becoming a parent (reproductive success) is between $F_{min}$ and $F_{max}$. The coefficients of variation (CV; standard deviation divided by the mean) for $F_{min}$ and $F_{max}$ were calculated for each species, to determine the range of deviation of reproductive success within species (Table 11.2).

The following results were obtained: (1) The smaller the ratio of trees becoming a parent, the larger the CV. (2) Species that flowered and fruited frequently (e.g., *Shorea leprosula*, *Shorea seminis*, *Dipterocarpus cornutus*, and *Dipterocarpus humeratus*) had larger CVs than species that flowered and fruited less frequently (e.g., *Shorea smithiana* and *Shorea ovalis*) when their ratio of trees becoming a parent were the same. (3) The CVs of *Shorea laevis*, *S. gibbosa*, *S. johorensis*, and *S. pauciflora* were greatly increased by low pollination success, since more than half of the flowering trees failed to fruit (Fig. 11.3). (4) The smallest CV was found in *Shorea smithiana*, while *Cotylelobium lanceolatum* had the largest CV. *S. laevis*, *S. gibbosa*, *S. johorensis*, *S. pauciflora*, *S. parvifolia*, and *S. leprosula* all had larger CVs than *Shorea smithiana*. Assuming that the CVs remain at the observed levels, it

**Table 11.2.** Within species deviation of reproductive success

| Species | CV[a] F min[b] | CV[a] F max[c] | Species | CV[a] F min[b] | CV[a] F max[c] |
|---|---|---|---|---|---|
| *C. lanceolatum* | 0.99 (0.57) | 0.99 (0.57) | *S. leprosula* | 1.15 (0.52) | 0.62 (0.85) |
| *D. cornutus* | 0.78 (0.70) | 0.53 (0.90) | *S. ovalis* | 0.90 (0.55) | 0.65 (0.70) |
| *D. humeratus* | 0.76 (0.72) | 0.59 (0.83) | *S. parvifolia* | 1.16 (0.47) | 0.69 (0.82) |
| *S. gibbosa* | 2.00 (0.20) | 0.85 (0.70) | *S. pauciflora* | 1.20 (0.45) | 0.56 (0.87) |
| *S. johorensis* | 1.57 (0.33) | 0.55 (0.86) | *S. seminis* | 0.60 (0.82) | 0.49 (0.89) |
| *S. laevis* | 2.32 (0.16) | 0.81 (0.65) | *S. smithiana* | 0.38 (0.88) | 0.17 (0.97) |

[a]Standard deviation divided by the mean
[b]Frequency of fruiting in December 1992–May 1998
[c]Frequency of fruiting and non-fruiting. Values parenthesized are the ratios of trees becoming parents

is likely that species with a higher deviation have differentiated genetically from an original species. This should occur more rapidly in species with a lower deviation.

## 11.4 DBH Ranges of Flowering and Fruiting Trees

The flowering and fruiting behaviors of dipterocarp trees in three DBH classes were analyzed: DBH ≤ 49 cm, 50 cm ≤ DBH ≤ 100 cm, and DBH ≥ 101 cm (Fig. 11.4). In *Cotylelobium lanceolatum, Dipterocarpus cornutus, Dipterocarpus humeratus*, and *Shorea parvifolia*, the larger the DBH class the higher was the ratios of flowering and fruiting trees to the total number of observed trees. On the other hand, the influence of DBH on flowering and fruiting was ambiguous in *Shorea laevis*, *S. pauciflora, S. seminis*, and *S. smithiana*. In *Shorea gibbosa, S. johorensis*, *S. leprosula*, and *S. ovalis*, trees with a DBH less than 50 cm flowered rarely, if ever.

## 11.5 Death in the Drought and Fires of 1997 and 1998

Of 782 trees, 452 died between 1992 and 1998. Of these deaths, 96% occurred in the dry spell and fires of 1997–98, which were linked with an El Niño-Southern Oscillation event (see Chapter 2, this volume) (Table 11.1). In 1997–98, *Dryobalanops beccarii* had a remarkably high death rate (93%/year). Of the 12 major dipterocarp species, *Shorea* species had higher death rates (48.1–72.0%/year) than *Dipterocarpus* species (32.1–40.3%/year) in 1997–98. *Cotylelobium lanceolatum* had the lowest death rate (4.3%/year). This species flowered peculiarly late, as mentioned above. It has the highest sprouting and root-suckering ratios after slash-and-burn (see Chapter 17, this volume), suggesting that it may survive repeated forest fires and contribute to forming the secondary vegetation at Bukit Soeharto.

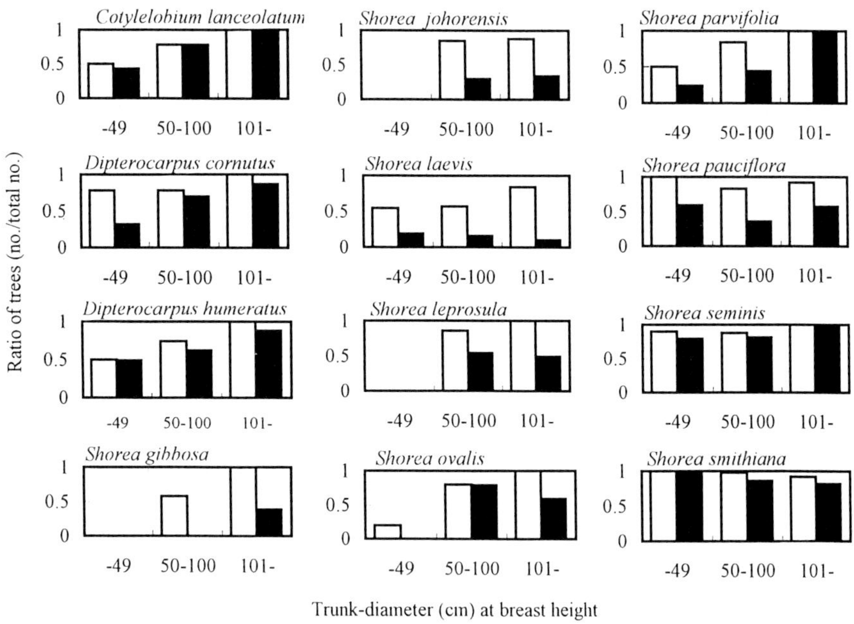


**Fig. 11.4.** Diameter at breast height (DBH) classes and the ratios of flowering and fruiting trees to the total number of trees observed from December 1992 to May 1998. *Open bars,* ratios of flowering trees; *closed bars,* ratios of fruiting trees

In April 1998, one month after the particularly large fires in BSEF from February to March 1998 (see Chapter 3, this volume), the death rates in 11 of the major species were higher in the Ibu-1, and Enggang-A and -B areas than in the Ibu-2 and Landak areas. *Shorea smithiana* was the only major species that was an exception. A population reduction in any species would increase its homozygosity and lower its reproductive output (Ledig 1986).

## 11.6 Conclusion

The flowering and fruiting of 782 trees representing 24 species of dipterocarps were observed for 6 years at Bukit Soeharto Educational Forest in East Kalimantan, Indonesia. Dipterocarp trees flowered simultaneously three times between 1992 and 1998. All 12 major species and 9 minor species apparently flower in response to relatively prolonged dry spells. Flowering in one minor species, *Shorea lamellata*, was not related to dry spells. The deviation in reproductive success varied between and within species. The ratio of flowering and fruiting trees to the total trees, the frequency of flowering and fruiting, and the pollination success all influence the deviation. Species with a high deviation may have rapidly differentiated genetically from an original species. The effects of DBH size on flowering and fruiting were clear in 8 of the 12 major species. In 4 major species, only trees over 100 cm in DBH acted as seed parents. *Cotylelobium lanceolatum* is considered fire and drought resistant, while the other major species of *Shorea* and *Dipterocarpus* appeared sensitive to drought.

## References

Ashton PS, Givnish TJ, Appanah S (1988) Staggered flowering in the Dipterocarpaceae: new insights into floral induction and the evolution of mast fruiting in the aseasonal tropics. Am Nat 132:44–66

Bawa KS, Krugman SL (1991) Reproductive biology and genetics of tropical trees in relation to conservation and management. In: Gomez-Pompa A, Whitmore TC, Hadley M (eds) Rainforest regeneration and management, pp 119–136

Kostermans AJGH (1992) A handbook of the Dipterocarpaceae of Sri Lanka. Yayasan tumbuh-tumbuhan yang berguna di Asia (Foundation for useful plants of tropical Asia). Vol. III. Wildlife Heritage Trust of Sri Lanka, Jakarta

Ledig TH (1986) Heterozygosity, heterosis and fitness in outbreeding plants. In: Soule ME (ed) Conservation biology: The science of scarcity and diversity. Sinauer Associates, pp 77–104

Okimori Y, Permono R (1991) Tree phenology of lowland dipterocarp forest in East Kalimantan. Fourth round-table conference on dipterocarp, BIOTROP Special Publication 41: 515–533

# 12 Leaf Gas Exchange and Canopy Structure in Wet and Drought Years in *Macaranga conifera*, a Tropical Pioneer Tree

ATSUSHI ISHIDA, TAKESHI TOMA, and MARJENAH

## 12.1 How Does Drought Affect the Leaf Gas Exchange Process?

Despite a high annual rainfall in the rainforest region, unusually prolonged and severe droughts occurred in eastern Borneo from June 1982 to May 1983 (Goldammer and Seibert 1990) and from June 1997 to April 1998 (Toma et al., Chapter 2, this volume). The prolonged droughts were linked to particularly strong El Niño-Southern Oscillation (ENSO) events. In the 1997–98 ENSO, there was no precipitation period from January 2 to April 15 (for 104 days), 1998 (Fig. 12.1). Although soil water potential was 0–45 kPa at a soil depth of 10 to 60 cm from May 1989 to April 1990 (normally wet years), it was 60–66 kPa from January to February 1998. When the soil water potential decreased (within 2 months of a period of no precipitation), both net photosynthetic rate ($P_n$) and water vapor stomatal conductance ($g_s$) decreased in the uppermost canopy leaves of a climax tree, *Dipterocarpus cornutus* Dyer (Fig. 12.1). Similar data were also found in the top canopy leaves of pioneer tree species, *Macaranga conifera* (Zoll.) Muell. Arg. and *Macaranga gigantea* (Reichb. f. & Zoll.) Muell. Arg.

Death of tissue or even individual whole plants was sometimes found during such severe droughts, especially in pioneer canopy trees. Leaf fall and reduced $g_s$ appear to reduce water loss at the whole canopy level. The drought appeared to bring not only changes in the canopy structure but also changes in the carbon gain process within the whole canopy. However, there are few published data about the effects of prolonged droughts associated with ENSO events on leaf gas exchange and canopy structure in tropical rainforests. This chapter deals with a specific instance of a general issue — how does drought affect the leaf gas exchange process in tropical rainforest trees?

We compared the characteristics of leaf gas exchange and canopy structure between a month in a normally wetter year (December, 1996) and a month in a drier year (January, 1998) in *Macaranga conifera* (Zoll.) Muell. Arg, a pioneer tree species. The physiological and morphological characteristics of leaves generally vary with canopy position, because different canopy positions are associated with different light environments. The variations of leaf characteristics within the canopy con-

Ecological Studies, Vol. 140
Guhardja et al. (eds.) Rainforest Ecosystems of East Kalimantan: El Niño, Drought, Fire and Human Impacts

tribute to the effective carbon gain of the whole canopy (e.g., Field 1983; Hirose and Werger 1987; Hollinger 1996). How physiological processes vary within plant canopies and how single-leaf measurements can be scaled up to the whole-plant level are important issues in plant ecophysiology (Caldwell et al. 1986; Ehleringer et al. 1986; Brooks et al. 1996). Thus, we attempted to scale up the processes at the different levels, to examine the effects of prolonged drought on carbon gain process within a canopy of *Macaranga conifera*.

At the single-leaf level, we will discuss what microclimatological and physiological factors limit carbon gain for the uppermost canopy leaves in the plant, especially in a drought year. Because a part of this research effort was directed in particular to examine the significance of leaf inclination angles on leaf gas exchange, the measurements of diurnal time changes in gas exchange and chlorophyll fluorescence were done in naturally east-facing leaves (EL), west-facing leaves (WL), and leaves artificially fixed in a horizontal position (HL) in the uppermost canopy leaves of an adult tree of *Macaranga conifera*, whose top canopy height was about 20 m.

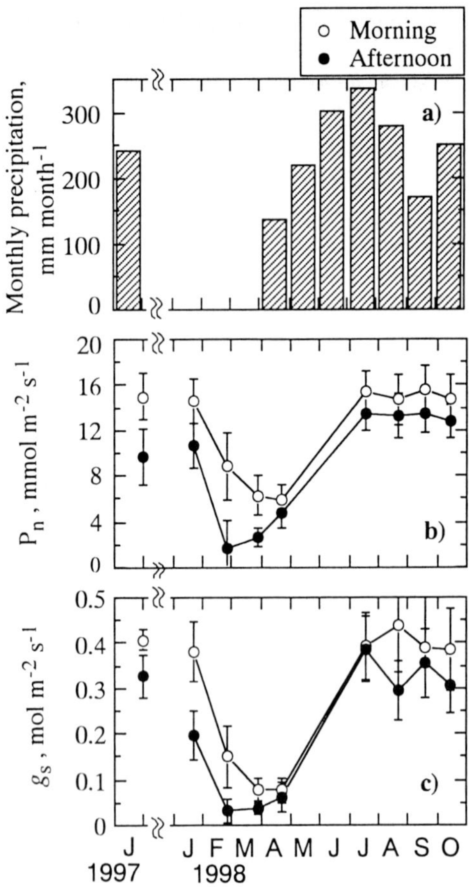


**Fig. 12.1a–c.** Seasonal changes in **a** monthly precipitation, **b** net photosynthetic rate ($P_n$), and **c** water vapor stomatal conductance ($g_s$) in the morning (*open circles*) and at midday (*closed circles*) in the uppermost canopy leaves of a climax tree, *Dipterocarpus cornutus*. *Vertical lines*, ± 1 SD

## 12.2 Canopy Structure and Leaf Characteristics

The cumulative leaf area index (LAI) was 5.1 in the wet year. Because the value of LAI has ranged from 5.2 to 11.4 in rainforests (Yamakura et al. 1986), the low LAI (thin canopy) in *Macaranga conifera* is probably characteristic of canopies of pioneer tree species with a high demand for light. The cumulative LAI was 2.3 in the drought year (Fig. 12.2). Thus, the decrease in LAI was about 50%. Because the variation of LAI linked with seasonal changes in soil moisture was only 15% in a rainforest in the Amazon (Roberts et al. 1996), the decrease in LAI in *Macaranga conifera* was relatively large. The canopy width decreased from 2.4 m in the wet year to 1.6 m in the drought year. The decrease in LAI was mainly due to the death of whole twigs that originated from lower positions in the trunk, and would function by reducing water loss from the whole canopy.

The leaves in the upper canopy layer showed near-vertical inclination angles in both the wet and drought years (Fig. 12.3). The median inclination angles in the canopy leaves of the upper, medium, and lower canopy-layers in the wet year were 58°, 42°, and 38°, respectively; those in the drought year were 77°, 80°, and 60°, respectively. Thus, the leaf inclination angle tended to increase in the drought year.

Leaf physiological properties were related to the ratio of photosynthetic photon flux density (PPFD) within the canopy to PPFD of full sunlight (relative PPFD; RPPFD) (Fig. 12.4). The drought influenced the amount of Nitrogen (N) and the arrangement of N within leaves between light harvesting components such as chlorophyll and $CO_2$ fixation components such as Rubisco. Photosynthetic N use efficiency (the ratio of maximum $P_n$ under high $CO_2$ and high PPFD to N content) was more or less maintained at constant values within the canopy, and was higher in the wet year (mean ± 1 SD, 0.215 ± 0.032 mmol $s^{-1}$ $molN^{-1}$) than in the drought year (0.188 ± 0.026 mmol $s^{-1}$ $molN^{-1}$) (t-test, $P<0.001$). The total amount of N within the whole canopy leaves decreased (60%) in the drought year, probably as a result of

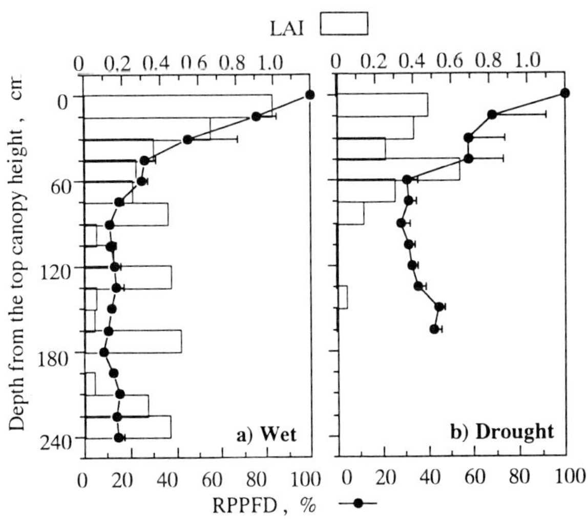


**Fig. 12.2a,b.** Vertical changes of leaf area index (*LAI*) in each 15-cm depth slice of the canopy and photosynthetic photon flux density relative to full sunlight (*RPPFD*) in **a** a wet year (December 1996) and **b** a drought year (January 1998) for *Macaranga conifera*. *Horizontal lines*, 1 SD of RPPFD (modified from Ishida et al 1999a)

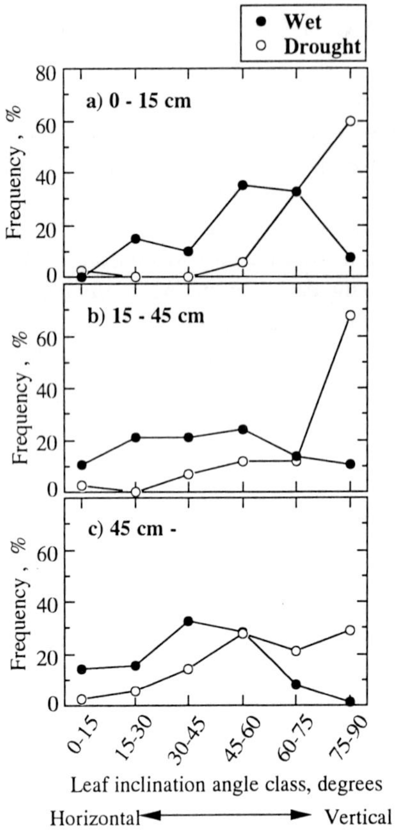


**Fig. 12.3a–c.** Distributions of leaf inclination angles with respect to the horizontal plane at **a** the upper canopy-layer (0–15 cm from the top of the canopy), **b** the middle canopy-layer (15–45 cm from the top of the canopy), and **c** the lower canopy-layer (> 45 cm), in wet (December 1996) and drought years (January 1998) for *Macaranga conifera*

reduced water uptake from the ground. This suggests that this tree suffered not only from water stress but also from N stress during the drought year. The lamina chlorophyll/N ratio within the canopy was higher, when PPFD relative to full sunlight (relative PPFD: RPPFD) was below 20%. The high chlorophyll/N ratio in the shaded leaves can be expected to increase the efficiency of light capture, suggesting a dynamic rearrangement of N from $CO_2$ fixation components to light harvesting components within the leaf (e.g., Evans 1989). The lamina area-based and mass-based chlorophylls at a given RPPFD were lower in the drought year than in the wet year. The average absorbance of PPFD in the top canopy leaves was 0.919 in the normal wet year and 0.878 in the drought year, because of the decreases in area-based chlorophyll. The decreases in the amount of chlorophyll and in chlorophyll/N ratio may be beneficial in a drought year, as it could result in a decrease in the absorption of excess-radiation against photoinhibition (Adams and Demmig-Adams 1995).

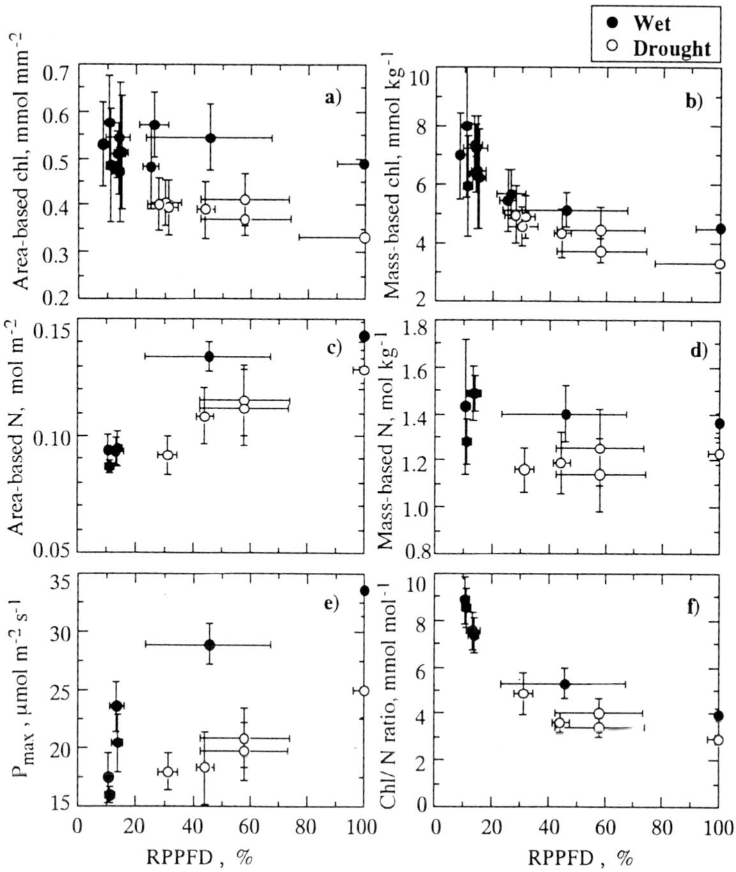


**Fig. 12.4a–f.** Relationships between photosynthetic photon flux density relative to full sunlight (*RPPFD*) and: **a** area-based and **b** dry mass-based chlorophyll (*chl*), **c** area-based and **d** dry mass-based nitrogen (*N*), **e** maximum photosynthetic rate ($P_{max}$) under high PPFD and high $CO_2$ concentrations, and **f** chl/N ratio, in wet (December 1996) and drought years (January 1998) for *Macaranga conifera* (modified from Ishida et al 1999a). *Vertical and horizontal lines*, ⊥ 1 SD

# 12.3 Leaf Gas Exchange of the Uppermost Canopy Leaves

## 12.3.1 Photosynthesis and Conductance Responses

When PPFD at the leaf surface exceeded 900 µmol $m^{-2}$ $s^{-1}$ (under near light-saturating conditions), a curvilinear relationship between $P_n$ and $g_s$ was found; there was no conspicuous difference in the relationship between the wet and the drought years (Fig. 12.5a). A reduced $g_s$ at midday has been reported in the uppermost canopy leaves in other lowland rainforest trees. For example, daytime $g_s$ began to decrease when the leaf-to-air vapor pressure deficit ($\Delta W$) exceeded 2.5–3.5 kPa in *Dryobalanops aromatica* Gaertn. f. (Dipterocarpaceae) (Ishida et al. 1996) and *Dialium pachyphyllum* Harms (Caesalpiniaceae) (Koch et al. 1994), whereas daytime $g_s$ began to decrease in the uppermost leaves of *Macaranga conifera* at a relatively small $\Delta W$ level. When $\Delta W$ exceeded about 1 kPa in the drought year, $g_s$ began to decrease (Fig. 12.5b). On the other hand, when $\Delta W$ exceeded about 1.5 kPa in the wet year, $g_s$ began to decrease. Therefore, the stomatal sensitivity to $\Delta W$ differed between the wet and drought years.

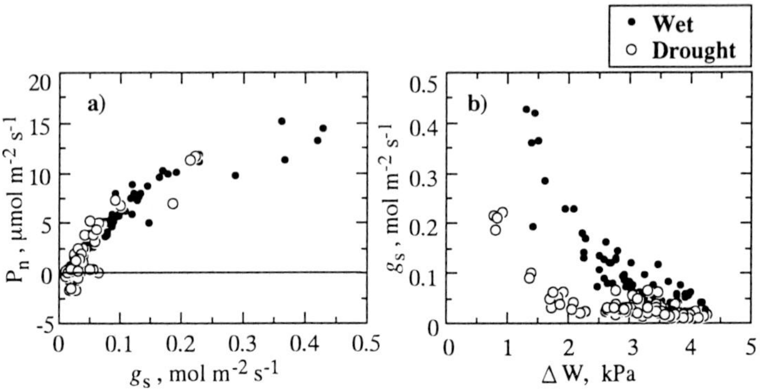


**Fig. 12.5a,b.** Relationships between **a** water vapor stomatal conductance ($g_s$) and net photosynthetic rate ($P_n$) and **b** $g_s$ and leaf-to-air vapor pressure deficit ($\Delta W$), in wet (December 1996) and drought years (January 1998) for *Macaranga conifera* (modified from Ishida et al 1999a,b). These data were obtained at relatively high light conditions (PPFD > 900 µmol $m^{-2}$ $s^{-1}$)

**Table 12.1.** Gas exchange and chlorophyll fluorescence characteristics in the uppermost canopy leaves of *Macaranga conifera* in wet (December 1996) and drought (January 1988) years [a]

| | Wet conditions | | | Drought conditions | | |
|---|---|---|---|---|---|---|
| Characteristics[c] | EL[b] | HL | WL | EL | HL | WL |
| Daily maximum $P_n$ | 13.2 a | 11.4 a | 6.8 b | 10.1 c | 6.4 b | 5.8 b |
| Daily maximum $g_s$ | 239 a | 357 b | 339 b | 198 c | 81 d | 146 e |
| Daily maximum *E* | 3.8 a | 4.4 a | 3.8 a | 1.5 b | 1.7 b | 1.9 b |
| Daily minimum ΔF/Fm' | 0.20 a | 0.08 b | 0.12 c | 0.10 c | 0.05 b | 0.08 c |
| Daily maximum ETR | 117 a | 114 a | 109 a | 103 a | 78 b | 81 b |
| Daily maximum NPQ | 5.0 a | 5.0 a | 3.7 b | 4.0 ab | 4.7 ab | 5.0 a |

[a]Values are mean of six leaves. Means followed by the same letter (a-e) after data indicate no significant difference between values in each row according to a t-test. Data from Ishida et al (1999a,b).
[b]EL, east-facing leaves; WL, west-facing leaves; HL, leaves artificially fixed in a horizontal position.
[c]$P_n$, net photosynthetic rate (µmol $m^{-2}$ $s^{-1}$); $g_s$, water vapor stomatal conductance (mmol $m^{-2}$ $s^{-1}$); *E*, transpiration rate (mmol $m^{-2}$ $s^{-1}$) ; ΔF/Fm', PSII quantum yield; ETR, electron transport rate through PSII (µmol $m^{-2}$ $s^{-1}$).

If leaf water potential or turgor potential remains stable when the soil water potential decreases, stomata must begin to close at smaller values of $\Delta W$ in drier conditions than in wetter conditions, to limit transpirational water loss and to supply the water to leaves (e.g., Ishida et al. 1992; Ryan and Yoder 1997). Because the turgor of leaves was maintained throughout the day even in the drought year, the difference in sensitivity of stomata to $\Delta W$ between the two years would act to effectively keep water within leaves. This supports the hydraulic limitation hypothesis (Ryan and Yoder 1997) that old trees stop growing in height as they become taller, because of an increase in soil-to-leaf hydraulic resistance. Seasonal changes in stomatal sensitivity to $\Delta W$ between wet and drought seasons have also been reported for other species (e.g., Meinzer 1982; Meinzer et al. 1993; Prior et al. 1997).

## 12.3.2 Leaf Gas Exchange Related to Leaf Orientation and Inclination Angle

In the uppermost canopy leaves, the leaf orientation and inclination angle affected the leaf gas exchange processes. The daily maximum value (13.2 ± 2.7 μmol $m^{-2}$ $s^{-1}$)

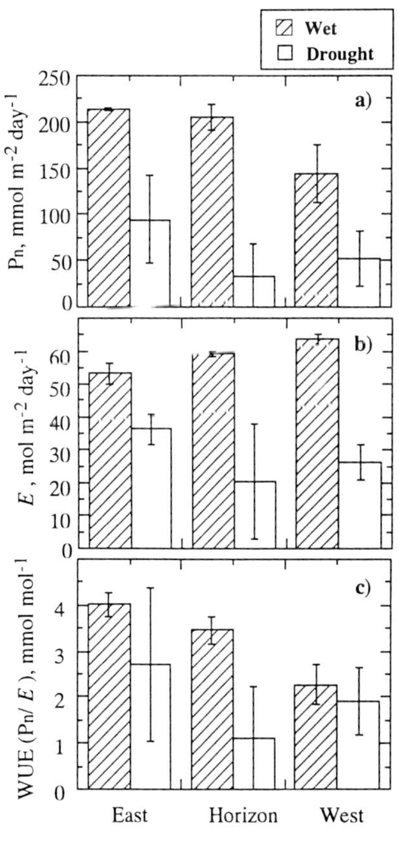


**Fig. 12.6a–c.** Leaf area based cumulative **a** net photosynthetic rate ($P_n$), **b** transpiration rate ($E$), and C water use efficiency ($P_n/E$) for 0800–1800 h in east-facing leaves (*EL*), horizontally-fixed leaves (*HL*), and west-facing leaves (*WL*) in uppermost canopy leaves in wet (*hatched column*; December 1996) and drought years (*open column*; January 1998) for *Macaranga conifera*. *Vertical lines*, ± 1 SD

in $P_n$ was found in EL and HL in the wet year, whereas the daily maximum value (10.1± 1.5 μmol $m^{-2}$ $s^{-1}$) was found in EL in the drought year (mean ± 1 SD.; Table 12.1). The daily minimum values of PSII (photosystem II) quantum yield in the light (ΔF/Fm') were lower in HL than in EL or WL in both the wet and the drought years This indicates that HL suffered from reduced photochemical capacity from heat and light loads at midday. The daily minimum values in electron transport rate through PSII (ETR) were also found in HL in the drought year (data not shown).

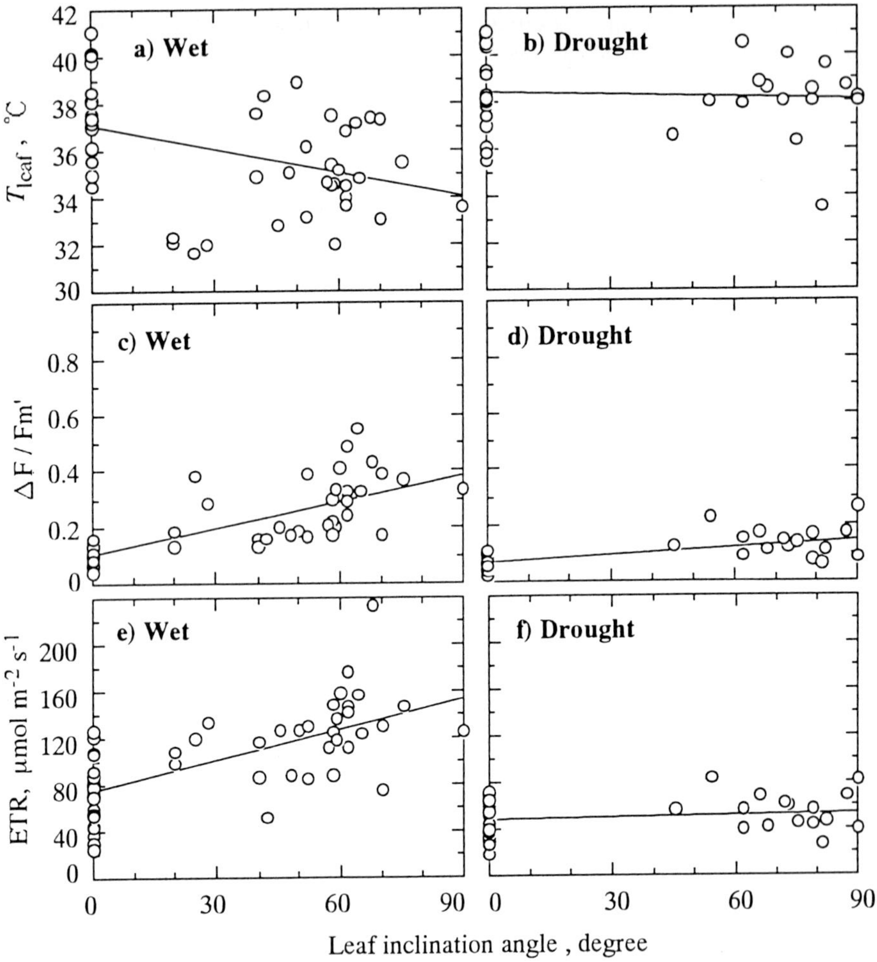


**Fig. 12.7a–f.** Relationships between leaf inclination angle and: **a,b** leaf temperature ($T_{leaf}$), **c,d** quantum yield of PSII (ΔF/Fm'), and **e,f** electron transport rate through PSII (ETR) of the uppermost canopy leaves for *Macaranga conifera*. These data were obtained at relatively high light conditions (PPFD > 600 μmol $m^{-2}$ $s^{-1}$) at midday in the upper canopy-layer (0–15 cm from the top of the canopy), **a,c,e** in a wet year (December 1996) and **b,d,f** a drought year (January 1998) (modified from Ishida et al 1999a,b). Leaves that had an angle of 0 were artificially fixed in a horizontal position

The daily cumulative carbon gain, water loss, and water use efficiency (WUE; the ratio of carbon gain to water loss) for 0800–1800 h in the top canopy leaves were lower in the drought year than in the wet year (Fig. 12.6). WUE in this plant was lower in the drought year than in the wet year, indicating that the drought had a severe effect on carbon gain and water balance processes in the top canopy leaves.

The leaf inclination angle had significant effects on leaf temperature ($T_{leaf}$) as well as the photochemical and the leaf gas exchange processes. In the wet year, when the leaves were exposed to relatively high PPFD (> 600 µmol m$^{-2}$ s$^{-1}$), both ΔF/Fm' and ETR decreased and $T_{leaf}$ increased, as the leaf inclination angle decreased (Fig. 12.7). Similar phenomena have also been reported for another tropical herb (He et al. 1996). In the wet year, since the maximum ETR was found in top canopy leaves with about 60° in leaf inclination angle, the leaf inclination angle was consistent with the average value within the leaves as shown in Fig. 12.3, suggesting an optimal leaf inclination angle for carbon gain. Whereas in the drought year, $T_{leaf}$ was high and ΔF/Fm' and ETR were low regardless of the leaf inclination angle, suggesting high heat and radiation stresses.

## 12.3.3 Assessment of the Limiting Factors in Photosynthesis

Because ETR would be mainly partitioned between $CO_2$ assimilation and $O_2$-dependent metabolism such as photorespiration (Edwards and Baker 1993; Di Marco et al. 1994; Ishida et al. 1999b), simultaneous measurements of chlorophyll fluorescence and $P_n$ can provide information on the use of light energy during the carbon assimi-

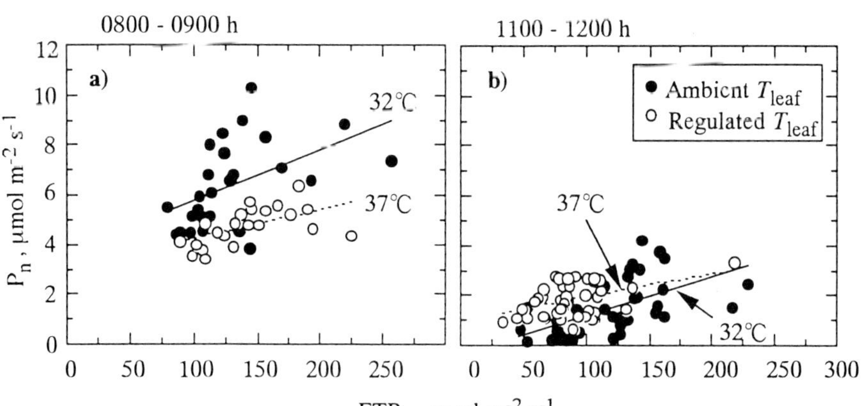


**Fig. 12.8a,b.** Relationships between electron transport rate through PSII (*ETR*) and net photosynthetic rate ($P_n$) in the uppermost canopy leaves of *Macaranga conifera*, when PPFD at the leaf surface exceeded 300 µmol m$^{-2}$ s$^{-1}$. The measurements were made at ambient $T_{leaf}$ conditions and regulated $T_{leaf}$ conditions at **a** 0800–0900 h and **b** 1100–1200 h in a drought year (December 1998) (modified from Ishida et al 1999b)

lation processes. On a clear day in the drought year, $T_{leaf}$ was lower in the morning (about 32°C) than at midday (about 37°C). Daytime $P_n$ at a given ETR was significantly larger in the early morning than at midday (Fig. 12.8). In order to clarify the effects of $T_{leaf}$ on the relationships between $P_n$ and ETR, we increased $T_{leaf}$ in the morning to the same level at midday (37°C) by heating, and decreased $T_{leaf}$ at midday to the same level in the morning (32°C) by cooling. The $P_n$ at a given ETR in the morning was decreased significantly (by about 29%) by increasing $T_{leaf}$ (ANCOVA, $P<0.01$), indicating that the low $T_{leaf}$ in the early morning was important for maintaining high $P_n$. On the other hand, $P_n$ at a given ETR at midday was increased significantly (by about 22%) by decreasing $T_{leaf}$. Under the same $T_{leaf}$ conditions (37°C or 32°C), $P_n$ at a given ETR was significantly lower at midday than in the morning (ANCOVA, $P<0.001$), indicating a high photorespiration rate at midday. Thus, $P_n$ at midday was mainly limited by a high photorespiration rate, because of low $g_s$ and the resulting reduced Ci (leaf intercellular $CO_2$), rather than by a high $T_{leaf}$. The high photorespiration rate at midday would consume excess-light energy and would protect photosystems from chronic photoinhibition (Jones 1983; Park et al. 1996; Osmond et al. 1997).

The combination of high $T_{leaf}$ and high PPFD is known to reduce PSII photochemical activity (Ludlow and Björkman 1984; Ludlow 1987; Gamon and Pearcy 1990; Epron 1997; Königer et al. 1998). Gamon and Pearcy (1989) concluded that excess $T_{leaf}$ at midday, not high PPFD, was the principle cause of reduced $P_n$ in *Vitis california* Benth. In the top canopy leaves in *M. conifera*, the low $T_{leaf}$ in the early

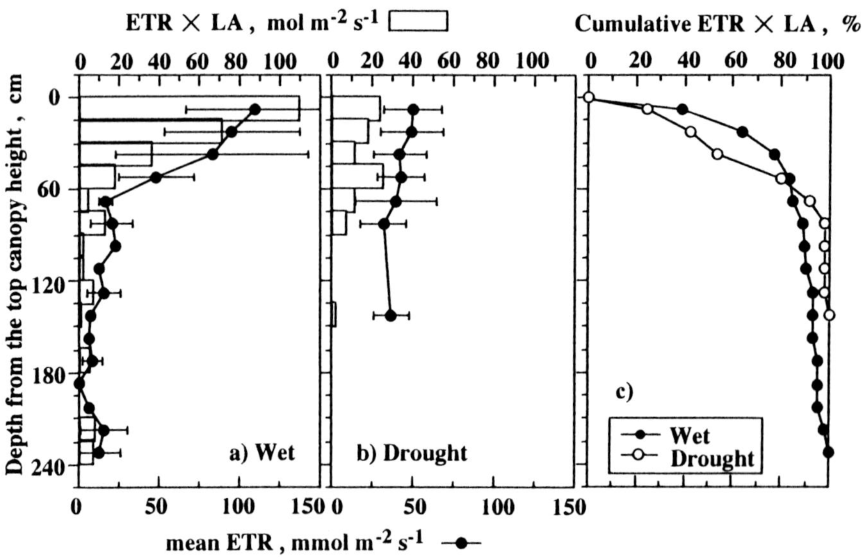


**Fig. 12.9a–c.** Vertical changes of the multiple individual electron transport rate through PSII and individual leaf area (*ETR × LA*) and mean ETR in each 15-cm thick slice of the canopy, in **a** wet (December 1996) and **b** drought years (January 1998) in *Macaranga conifera*. *Horizontal lines*, show 1 SD of ETR. **c** Cumulative ETR × LA from the top of the canopy, in wet (December 1996) and drought years (January 1998)

morning (when $g_s$ was high) was important for maintaining high $P_n$, probably because it reduced mitochondrial-respiration and photorespiration. Overall, our results suggest that the midday depression of $P_n$ in the drought year was mainly due to not only reductions in photochemical processes but also increased stomatal limitation, and the effect of $T_{leaf}$ on $P_n$ changed with changing $\Delta W$ and $g_s$ during the daytime.

## 12.4 Chlorophyll Fluorescence Within Canopy Leaves

Because good relationships were found between ETR and $P_n$ as shown in Fig. 12.8, the product of ETR and leaf area in each leaf (ETR × LA) would be a good indicator of the carbon gain of individual leaves. The cumulative ETR × LA at midday within

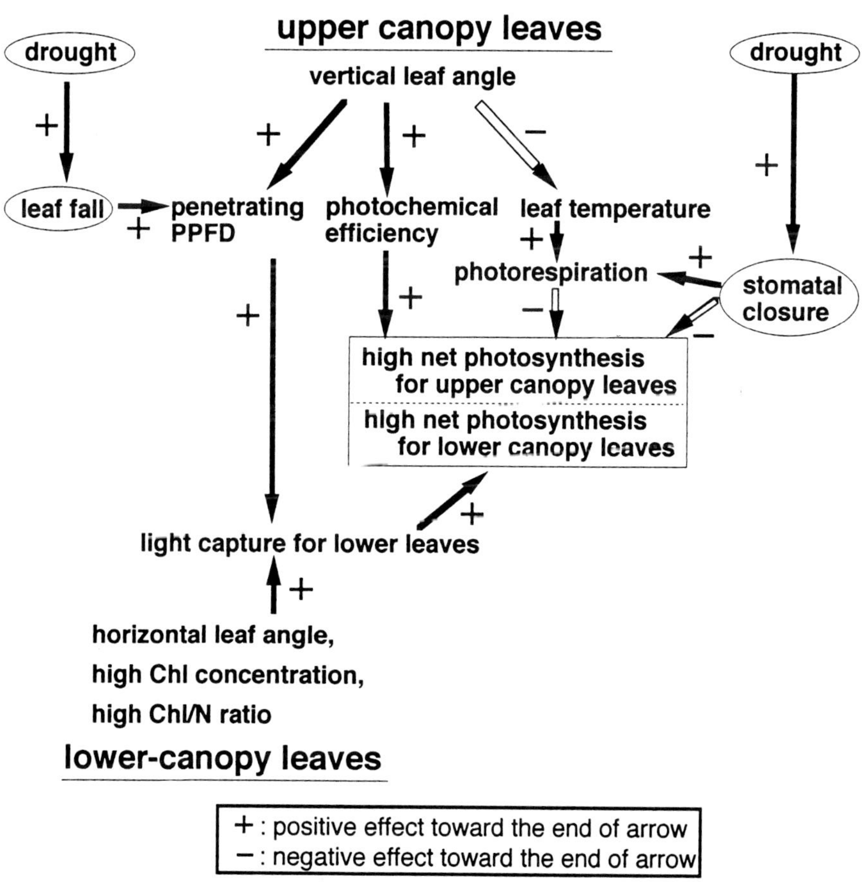


**Fig. 12.10.** Schematic summary of the present study. *Solid arrows* indicate positive (+) effects and *open arrows* indicate negative (−) effects, in the direction of the arrow. See text for details

the canopy based on 1 $m^2$ ground area were 281 mol $s^{-1}$ in the wet year and 96 mol $s^{-1}$ in the drought year. To examine the carbon gain process within the canopy, the vertical profiles in cumulative ETR × LA in each 15-cm depth slice of the canopy at midday, are shown in Figure 12.9. Only 50% of the total ETR × LA in the drought year existed in the 0–30 cm layer from the top of the canopy, while 70% of the total ETR × LA in the wet year existed in this layer. Thus, the carbon budget would shift toward lower layers within the canopy from the wet year to the drought year, probably because of pronounced light penetration within the canopy in the drought year.

## 12.5 Conclusion

The drought associated with an ENSO event brought reduced $P_n$, $g_s$, and LAI. Although Sá et al. (1996) showed that $g_s$ was less affected by a seasonal decrease in soil water contents in rainforest trees in the Amazon, both $P_n$ and $g_s$ decreased with decreasing soil water potential in early and late successional canopy trees in East Kalimantan. Remarkable decreases were found 1–2 months after a period of no precipitation linked to a prolonged ENSO event. Leaf fall during the drought period was found. The decrease of LAI was about 50% in *Macaranga conifera*, and the decrease was larger than the variations (15%) of LAI linked with seasonal changes of soil moisture at a rainforest in the Amazon (Roberts et al. 1996).

Although many individual canopy trees of *Macaranga* taxa (species of this genus are typical pioneer trees in Southeast Asia) died during the drought period of 4 months in 1999, the death of a whole plant was rare in climax dipterocarp trees (Toma et al., Chapter 10, this volume). This indicates that the damage due to the drought was greater in pioneer trees than in late successional trees. Although pioneer trees can better tolerate hot and bright conditions than can climax dipterocarp trees (Ishida et al., Chapter 24, this volume), probably pioneer trees are suited to wet conditions. Thus, these observations indicate that the drought linked to ENSO events affects succession in rainforests.

The integration of carbon gain processes from the single-leaf level to the whole-plant level is shown in a schematic diagram (Fig. 12.10). In a wet year, the near-vertical leaf angle in the top canopy leaves increased the photochemical efficiency and suppressed an increase in $T_{leaf}$. In the early morning, the low $T_{leaf}$ helped to maintain a high $P_n$. On the other hand, at midday, $P_n$ decreased because of the low $g_s$ rather than the high $T_{leaf}$, probably associated with a reduced Ci. Thus, the midday depression of $P_n$ would be due not only to a decrease in photochemical activity but also to an increase in stomatal limitation. A near-vertical leaf angle in the uppermost canopy leaves would increase the penetration of light within the canopy. Especially in a drought year, because of an increase in leaf inclination angles, the carbon budget shifts toward the lower layers within the canopy. The leaf characteristics in the lower-canopy leaves such as a horizontal leaf angle, high chlorophyll concentration, and high chlorophyll/N ratio would increase the capacity of light capture, contributing to high carbon gain within the whole-plant. The prolonged drought linked to an ENSO event affected the ecophysiological processes at the single leaf and whole canopy levels in canopy trees, and could potentially affect global climate change, such as $CO_2$ levels and the water flux between forests and air.

## References


Adams III WW, Demmig-Adams B (1995) The Xanthophyll cycle and sustained thermal energy dissipation activity in *Vinca minor* and *Euonymus kiautschovicus* in winter. Plant Cell Environ 18:117–127

Brooks JR, Sprugel DG, Hinckley TM (1996)The effects of light acclimation during and after foliage expansion on photosynthesis of *Abies amabilis* foliage within the canopy. Oecologia 107:21–32

Caldwell MM, Meister H-P, Tenhunen JD, Lange OL (1986) Canopy structure, light microclimate and leaf gas exchange of *Quercus coccifera* L. in a Portuguese macchia: measurements in different canopy layers and simulations with a canopy model. Trees 1:25–41

Di Marco G, Iannelli MA, Loreto F (1994) Relationship between photosynthesis and photorespiration in field-grown wheat leaves. Photosynthetica 30:45–51

Edwards GE, Baker NR (1993) Can $CO_2$ assimilation in maize leaves be predicted accurately from chlorophyll fluorescence analysis? Photosynth Res 37:89–102

Ehleringer JR, Pearcy RW, Mooney HA (1986) Recommendations of the workshop on the future development of plant physiological ecology. Bull Ecol Soc Am 67:48–58

Epron D (1997) Effects of drought on photosynthesis and on the thermotolerance of photosystem II in seedlings of cedar (*Cedrus atlantica* and *C. libani*). J Exp Bot 48:1835–1841

Evans JR (1989) Photosynthesis and nitrogen relationships in leaves of $C_3$ plants. Oecologia 78:9–19

Field CB (1983) Allocating leaf nitrogen for the maximization of carbon gain: leaf age as a control on the allocation program. Oecologia 56:341–347

Gamon JA, Pearcy RW (1989) Leaf movement, stress avoidance and photosynthesis in *Vitis california*. Oecologia 79:475–481

Gamon JA, Pearcy RW (1990) Photoinhibition in *Vitis california*: interactive effects of sunlight, temperature and water status. Plant Cell Environ 13:267–275

Goldammer JG, Seibert B (1990) The impact of droughts and forest fires on tropical lowland rainforest of East Kalimantan. In: Goldhammer JG (ed) Fire in the tropical biota. Ecosystem processes and global challenges. (Ecological Studies, vol. 84). Springer, Berlin Heidelberg New York, pp 11–31

He J, Chee CW, Goh CJ (1996) 'Photoinhibition' of *Heliconia* under natural tropical conditions: the importance of leaf orientation for light interception and leaf temperature. Plant Cell Environ 19:1238–1248

Hirose T, Werger MJA (1987) Maximizing daily canopy photosynthesis with respect to the leaf nitrogen allocation pattern in the canopy. Oecologia 72:520–526

Hollinger DY (1996) Optimality and nitrogen allocation in a tree canopy. Tree Physiol 16:627–634

Ishida A, Yamamura Y, Hori Y (1992) Roles of leaf water potential and soil-to-leaf hydraulic conductance in water use by understorey woody plants. Ecol Res 7:213–223

Ishida A, Toma T, Marjenah (1999a) Leaf gas exchange and chlorophyll fluorescence in relation to leaf angle, azimuth, and canopy position in a tropical pioneer tree, *Macaranga conifera*. Tree Physiol 19:117–124

Ishida A, Toma T, Marjenah (1999b) Limitation of leaf carbon gain by stomatal and photochemical processes in the top of the canopy of *Macaranga conifera*, a tropical pioneer tree. Tree Physiol 19:467–473

Ishida A, Toma T, Matsumoto Y, Yap SK, Maruyama Y (1996) Diurnal changes in leaf gas exchange characteristics in the uppermost canopy of a rainforest tree, *Dryobalanops aromatica* Gaertn. f. Tree Physiol 16:779–785

Jones HG (1983) Plants and Microclimate. Cambridge University Press, London

Koch GW, Jeffrey SA, Goulden ML (1994) Diurnal patterns of leaf photosynthesis, conductance and water potential at the top of a lowland rainforest canopy in Cameroon: measurements from the Radeau des Cimes. Tree Physiol 14:347–360

Königer M, Harris GC, Pearcy RW (1998) Interaction between photon flux density and elevated temperatures on photoinhibition in *Alocasia macrorrhiza*. Planta 205:214–222

Ludlow MM (1987) Light stress at high temperature. In: Photoinhibition. Kyle DJ, Osmond CB, Arntzen CJ (eds) Elsevier, Amsterdam, pp 89–109

Ludlow MM, Björkman O (1984) Paraheliotropic leaf movement in Siratro as a protective mechanism against drought-induced damage to primary photosynthetic reactions: damage by excessive light and heat. Planta 161:505–518

Meinzer FC (1982) The effect of vapor pressure on stomatal control of gas exchange in Douglas fir (*Pseudotsuga menziesii*) saplings. Oecologia 54:236–242

Meinzer FC, Goldstein G, Holbrook NM, Jackson P, Cavelier J (1993) Stomatal and environmental control of transpiration in a lowland tropical forest tree. Plant Cell Environ 16:429–436

Osmond CB, Badger M, Maxwell K, Björkman O, Leegood RC (1997) Too many photons: photorespiration, photoinhibition and photooxidation. Trend Plant Sci 2:119–121

Park Y-L, Chow WS, Osmond CB, Anderson JM (1996) Electron transport to oxgen mitigates against photoinactivation of photosystem II *in vivo*. Phototynth Res 50:23–32

Prior LD, Eamus D, Duff GA (1997) Seasonal and diurnal patterns of carbon assimilation, stomatal conductance and leaf water potential in *Eucalyptus tetrodonta* saplings in a wet-dry savanna in Northern Australia. Aust J Bot 45:241–258

Roberts JM, Cabral OMR, da Costa JP, McWillian A-LC, Sá TDdeA (1996) An over view of the leaf area index and physiological measurements during ABRACOS. In: Gash JHC, Nobre CA, Roberts JM, Victoria RL (eds) Amazonian deforestation and climate. John Wiley and Sons, Singapore, pp 287–306

Ryan MG, Yoder BJ (1997) Hydraulic limits to tree height and tree growth. BioScience 47:235–242

Sá TDdeA, da Costa JdePR, Roberts JM (1996) Forest and pasture conductances in south Pará, Amazon. In: Gash JHC, Nobre CA, Roberts JM, Victoria RL (eds) Amazonian deforestation and climate. John Wiley and Sons, Singapore, pp 241–263

Yamakura T, Hagihara A, Sukardjo S, Ogawa H (1986) Above-ground biomass of tropical rainforest stands in Indonesian Borneo. Vegetatio 68:71–82

# 13 Responses of Dipterocarp Seedlings to Drought Stress

Satoshi Ito, Yoshihiko Nishiyama, and Wawan Kustiawan

## 13.1 Need for Evaluation of Drought Tolerance in the Wet Tropics

Tropical rainforests are among the wettest terrestrial habitats on the surface of the Earth (Walter 1971). However, even in these wet environments, dry periods of brief duration commonly occur on a diurnal, seasonal and annual basis. Especially in East Kalimantan, prolonged droughts linked with strong El-Niño southern oscillation events occurred in 1982–83 and 1997–98 for the last two decades (see Chapter 2, this volume). Aside from these severe droughts, during the 1990s there were two short drought periods in 1991 and 1994 (see Chapter 2, this volume). Thus, frequent drought events occurred even in the rainforest region in East Kalimantan. As a consequence, plants growing in these habitats may have been often exposed to moderate drought stress. When there was no precipitation for one to two months from January 1998, reduced stomatal conductance and net photosynthetic rate were found in the top canopy leaves of pioneer *Macaranga* and climax dipterocarp trees in East Kalimantan (see Chapter 12, this volume). Further, during the course of practical forest management, reforested young stands planted after logging may suffer tissue water-deficit because of rapid evaporation from the soil surface and the undeveloped juvenile root system. Thus, it is important to investigate the drought tolerance of tropical rainforest trees for successful reforestation.

Kartawinata et al. (1980) reported that *Shorea leprosula* and *S. parvifolia* showed high tolerance to water stress (in areas of relatively low precipitation and a long dry season). Their report suggested that there are ecophysiological differences in terms of mechanism of drought tolerance between species even in the wet environments of tropical rainforests. Recently, several studies have reported on the drought tolerance of tropical rainforest trees (e.g., Robichaux and Peacy 1980, Oberbauer 1982, Medina 1983, Myers et al. 1987). However, there is less information on the drought tolerance of dipterocarp species, which are the most important in the Asian tropics.

In this chapter, we deal with the responses of five dipterocarp seedlings to soil water-deficit, and determine their growth in a series of irrigation experiments. Generally, environmental stresses often affect not only growth and physiological functions but also dry matter allocation, which has strong physiological significance for

Ecological Studies, Vol. 140
Guhardja et al. (eds.) Rainforest Ecosystems of East Kalimantan: El Niño, Drought, Fire and Human Impacts

water use and dry matter economy. Further, changes in dry matter allocation may have some significance for other functions of seedling architecture, because different functions often act under trade-off relationships in terms of carbohydrate allocation (Kramer 1983). Functional changes in seedling architecture are important in predicting the long-term effects of stresses, because they may affect the potential of further growth. Thus, we compare and discuss the dry matter allocation and functional changes of architecture induced by water-deficit in order to interpret the overall effects of droughts.

## 13.2 Plant Materials and Irrigation Treatments

Species tested by irrigation treatments were *Shorea parvifolia*, *S. leprosula*, *S. ovalis*, *Dryobalanops aromatica*, and *Hopea mengarawan*, which are common dipterocarps in the lowland tropical rainforests of East Kalimantan. Current-year-seedlings of each species collected from forests in East Kalimantan were transferred to plastic pots in September 1990 to grow at field capacity and without particular artificial shading in a greenhouse. After a growth period of two months, five levels of irrigation (50, 100, 150, 200 and 250 ml per pot at intervals of two days) were applied to 30 seedlings for each species from November 5 in 1990 to January 16 in 1991. The initial average sizes of seedlings and soil water regime promoted by irrigation are shown in Table 13.1 and Table 13.2, respectively.

**Table 13.1.** Initial average size of seedlings subjected to irrigation treatments

| Species | Height (cm) | Basal Diameter (mm) | Number of branches |
|---|---|---|---|
| *S.leprosula* | 55.7 (7.0)[a] | 3.62 (0.73) | 1.2 (1.1) |
| *S.parvifolia* | 46.7 (7.7) | 3.42 (0.72) | 2.2 (1.5) |
| *D.aromatica* | 49.0 (8.6) | 4.37 (0.70) | 2.1 (1.5) |
| *H.mengarawan* | 36.9 (7.1) | 3.29 (0.56) | 5.0 (1.6) |
| *S.ovalis* | 37.0 (8.9) | 4.04 (0.76) | 0.2 (0.5) |

[a] Values are the averages of 150 seedlings for each species measured at the beginning of irrigation treatments on November 5 in 1990. Figures in parentheses indicate standard deviations.

**Table 13.2.** Average soil water status during irrigation treatments

| | Irrigation regime (ml/2days) | | | | |
|---|---|---|---|---|---|
| | 50 | 100 | 150 | 200 | 250 |
| pF value of soil[a] | 2.39 (0.23)[b] | 2.21 (0.20) | 1.99 (0.16) | 1.93 (0.15) | 1.92 (0.11) |

[a] Soil water status was measured at intervals of two hours in the day time (06:00-20:00) from Feb.12 to Feb.28 under the same irrigation regimes as the experiments. Tensio-meters were applied for pots of *D.aromatica* seedlings. For each treatment, three replications were made of five- or six-day series of measurements for each seedling.

[b] Values are means of all 104 measurements for each treatment. Figures in parentheses indicate standard deviations.

## 13.3 Growth Response to Drought Stress

Evidence of strategic differentiation of competitors and stress-tolerators in relation to water resource was suggested even in the wet tropics as interspecies differences in growth responses in our short-term water–deficit experiments. Temporal changes of relative growth rates of seedling height (RHGR), calculated on the basis of measurements at intervals of two weeks (Fig. 13.1), showed the clearest interspecies differences among growth measures, such as height, basal diameter, and number of branches (data not shown). Although water–deficit affected the height growth of all species, we observed high susceptibility and temporal sensitivity of *S. leprosula* to severe

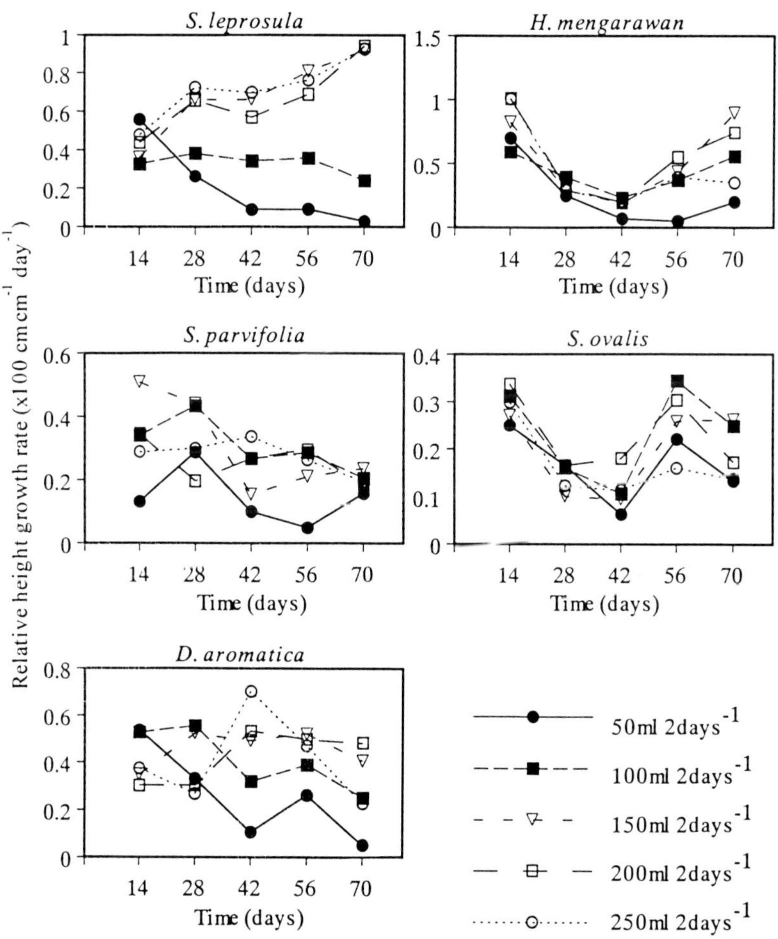


**Fig. 13.1.** Changes in relative height growth rate (RHGR) of five dipterocarp seedlings during irrigation treatments

stress, showing an early decline (from the 28th day) in RHGR in 50 ml 2days$^{-1}$. Interestingly, *S. leprosula* maintained high RHGR in moderately stressed conditions (150–200ml 2day$^{-1}$). In *D. aromatica*, an early decline of RHGR was also observed from the 42nd day in 50 and 100ml treatments, while lesser declines were found in 150 and 200ml 2day$^{-1}$ treatments through the experimental period. These two species are more or less susceptible to severe water-deficit, although they have higher RHGR compared with the other three species. On the other hand, *S. parvifolia* decreased its RHGR from the 42nd day in the 50ml treatment. *S. ovalis* had the lowest initial RHGR among the five species, and showed no clear trend of RGHR along the gradient of water-deficit. These three species, which grow relatively slowly, appear to have a low responsiveness to water-deficit stress both in degree of decline and temporal sensitivity, as do stress tolerators.

Contrary to results reported by Kartawinata et al. (1980), where high tolerance of *S. leprosula* and *S. parvifolia* to water stress was observed, our experiment indicated a similar or lower RHGR for *S. leprosula* and *S. parvifolia* under dry conditions compared with other species. These different results may be due to the difference in degree of soil water-deficit. In particular, the large decline of *S. leprosula* under severe water-deficit stress indicates the importance of susceptibility to severe stress. Further, gradual recovery of growth that was observed at the end of the experiment for *S. parvifolia* and *H. mengarawan* implies the possibility of physiological adaptation (Fitter and Hay 1981, Kramer 1983) to drought. Thus, long-term monitoring of growth and physiology using a wide range of standardised stress treatment conditions might be needed to interpret species tolerance, when we consider the influence of severe droughts such as those linked with strong El-Niño southern oscillation events (see Chapter 2, this volume).

## 13.4 Changes in Dry Matter Allocation

The interspecies differences in shoot growth responses partly resulted from variations in dry matter allocation induced by water-deficit stresses. The shoot/root ratio (Fig. 13.2a), calculated on the basis of ten seedlings of average size sampled from each species in each treatment at the end of experiment, shows typical adaptive changes of dry matter allocation in the susceptible species. The shoot/root ratio is a measurement having strong physiological significance for water-use by plants (Richards 1976), i.e., a low shoot/root ratio indicates an advantage in maintaining water balance. The shoot/root ratios of the two susceptible species, *S. leprosula* and *D. aromatica*, were ca. 1.5 times of the other three species under unstressed conditions, declining to the same level as the other species under stress. This change may be an evidence of adaptation in avoiding tissue desiccation, as described for temperate species (Jones and Zur 1984, Steinberg et al. 1990).
A possible reason for the susceptibility of these two species to water-deficit may be due to the high shoot/root ratio which developed under unstressed conditions, while low or constant ratios shown by the other three species may indicate relatively higher tolerance capacity to water-deficit stress. However, Osonubi and Davis (1978) re-

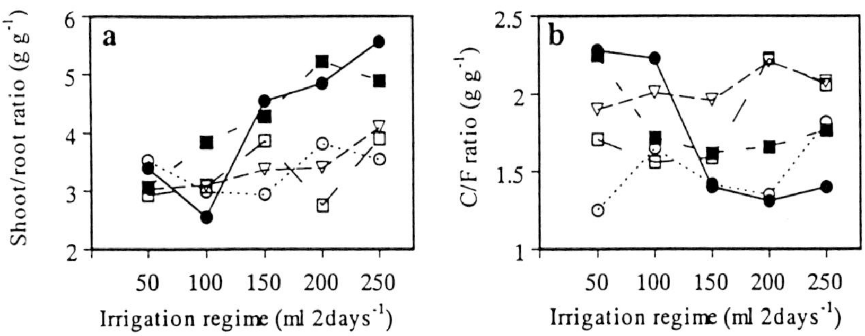


**Fig. 13.2a,b.** Dry matter allocation of five dipterocarp seedlings under water-deficit stress. **a** Top/root ratio (the ratio of the mass of the above-ground part to the underground part). **b** *C/F* ratio (the ratio of the mass of nonphotosynthetic organs to photosynthetic organs). *Closed circle*, S. Leprosula; *open square*, H. mengarawan; *trianglate*, S. parvifolia; *open circle*, S. oralis; *closed square*, D. aromatica

ported a high shoot/root ratio (3.0) in drought tolerant oak species. Hence, we suggest that the tolerance of tropical rainforest trees may be low compared with plants of drier regions.

The decrease of shoot mass of two susceptible species under water-deficit was a consequence of both decreased leaf mass and increased root mass (data not shown). These changes influence dry matter economy through a change of C/F ratio (Culm/Foliage ratio: a mass ratio of non-photosynthetic organs to photosynthetic organs, Iwaki 1958). High C/F ratios of *S. leprosula* and *D. aromatica* under water-deficit (Fig. 13.2b) indicate a disadvantage in dry matter economy because of the large amount of quota of non-photosynthetic organs (which only consume carbohydrates) to unit foliage mass. Finally, the dry matter allocation of the susceptible species under water-deficit stress changes to provide an improved water balance by sacrificing the advantage in dry matter economy.

## 13.5 Functional Changes of Seedling Architecture

The changes of dry matter allocation in susceptible species have significance for other functions of seedling architecture, which may affect their potential for further growth. These functions should be evaluated at least from hydraulic, supportive (mechanical stability), and assimilative aspects. We evaluated the functional changes of architecture induced by water-deficit using seedling dimensions measured for the same samples used in dry matter allocation analyses.

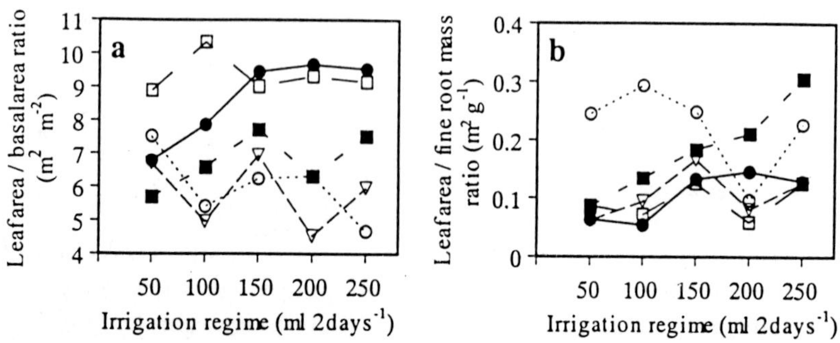


**Fig. 13.3a,b.** Hydraulic indices of seedling architecture of five dipterocarp species under water-deficit stress. **a** Leaf area/stem basal area ratio. **b** Leaf area/fine root mass ratio. Symbols and lines are the same as in Fig. 13.2

## 13.5.1 Hydraulic Architecture

The changes of dry matter allocation and seedling dimensions of susceptible species modify their hydraulic architecture advantage for maintaining water balance (Fig. 13.3) similar to that detected in the shoot/root ratio (Fig. 13.2). Hydraulic architecture is basically evaluated as the water conductivity of conductive organs for unit leaf area or leaf mass (Zimmermann 1983). We adopted two indices for rough estimation; 1) leaf area/basal area ratio and 2) leaf area/fine root mass ratio, both of which indicate hydraulic deficiency as the ratio of plant parts for transpiration and water absorption or conduction. Among the two susceptible species, *S. leprosula* had a low leaf area/basal area ratio, which was improved by water-deficit conditions (Fig. 13.3a). On the other hand, *D. aromatica* had a low leaf area/fine root mass ratio under water-deficit stress (Fig. 13.3b).

## 13.5.2 Supportive Architecture

Supportive architecture is important in toleration of mechanical stresses and disturbances such as wind blow, and has been known to normally differ due to status of competition (King 1981, 1990) and suppression (cf. Kohyama 1987). In addition to these factors related to light resource availability, water-deficit can also affect supportive architecture.

Relative height (stem height (*H*) / stem diameter at 10 % height of seedlings (*D*) ratio), which indicates supportive deficiency (Ito 1992, c.f., King 1981, 1990), was high in the most susceptible species, *S. leprosula* (Fig. 13.4a). High relative height of this species suggests a competitive architecture (Kohyama 1987) with low efficiency in mechanical support, making a contrast with low values for *S. ovalis*, the slow growing species. A similar difference was observed for stem density index (stem weight/$D^2H$). Stem density index may indicate rough supportive efficiency (Ito 1992) because stem strength is assumed to be proportional to bulk density (Panshin and de Zeeuw 1980). In the comparison of this index, *S. leprosula*, *S. parvifolia*, and

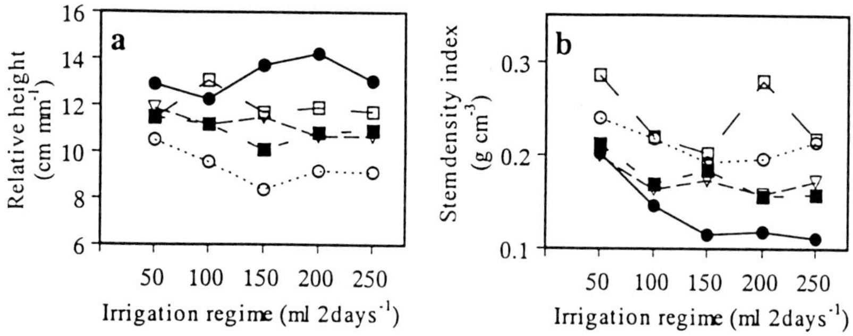


**Fig. 13.4a,b.** upportive indices of seedling architecture of five dipterocarp species under water-deficit stress. **a** Relative height (stem height [*H*] / stem diameter at 10 % of stem height [*D*]). **b** Stem density index (stem weight / $D^2H$). Symbols and lines are the same as in Fig. 13.2

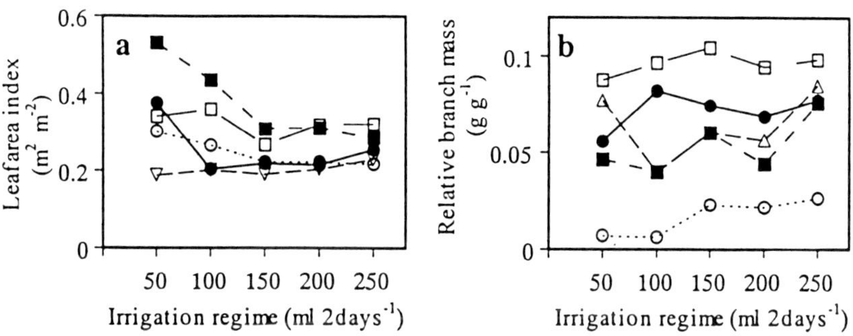


**Fig. 13.5a,b.** Assimilative indices of seedling architecture of five dipterocarp species under water-deficit stress. **a** Leaf area index (*LAI*). **b** Relative branch mass (the ratio of mass of branches to the whole shoot). Symbols and lines are the same as in Fig. 13.2

*D. aromatica* had lower stem density indices than the other two species over the whole range of irrigation regimes (Fig. 13.4b), suggesting a deficiency of mechanical support as for competitive architecture (Kohyama 1987). The stem density is influenced by vessel ratio in addition to stem taper, cell size and thickness of cell wall. Hence, there is a possibility that these three species have architectures that may be advantageous for water conduction rather than mechanical support. The stem density index showed a common tendency to increase under drought stress (Fig. 13.4b). This was especially clear for *S. leprosula*, the most susceptible species to water-deficit stress, which indicated shifting to non-competitive architecture.

## 13.5.3 Assimilative Architecture for Energy Capture

Assimilative architecture, evaluated as the efficiency of light capture, may be important especially for the degraded dry matter economy (i.e., high C/F ratios) of stressed seedlings. However, changes of assimilative architecture due to water-deficit occur in a negative way to reduce transpiration demand owing to reduced energy capture capacity. Leaf area index (*LAI*, leaf area/crown projection area), which indicates deficiency in light capture by self-shading of leaves, increased in *S. leprosula* under severe water-deficit stress (50 and 100ml treatments, Fig. 13.5a). *D. aromatica* also increased its leaf area index in the 50ml treatment, partly as a result of decreased crown projection area due to suspension of leaves. In *S. leprosula*, the relative branch mass (branch mass/shoot mass), which indicates efficiency of light capture (cf. Kohyama 1987), also declined in the 50ml treatment (Fig. 13.5b).

## 13.6 Conclusion

These experiments emphasize the importance of taking into account susceptibility to drought in the selection of sites and species for reforestation. Both *S. leprosula* and *D. aromatica* originally have the growth characteristics of a competitor. However, they lost their competitive characteristics due to water-deficit stress changing allocation patterns. Among these two species, *S. leprosula* is a candidate for planting on sites of moderate drought stress because of its better growth performance, although it is not suitable for severe drought sites. *S. parvifolia* is recommended for drier sites owing to its tolerance to severe drought. On the other hand, *S. ovalis* seemed to be a stress-tolerator showing the lowest susceptibility to water-deficit stress in our experiment, and is recommended for drier sites owing to its tolerance to severe drought. Further, the leaves of *S. ovalis* had a higher tolerance capacity for photoinhibition under hot and bright conditions than those of *Hopea* and *Dryobaranops* (Ishida et al. Chapter 26, this volume). However, we still cannot determine if this species is a drought tolerator, since its optimum water regime may exceed that of our best watering treatment. Ashton (1983) described this species as usually growing in moist places in valleys and low-lying ground. Our data show that the low growth rate in *S. ovalis* compared to that of the other species also supports this possibility. Tests of *S. ovalis* and other species at a wider range of water regimes are needed for better selection of planting species and sites.

Most changes in dry matter allocation and functions of architecture of the susceptible species seemed to reduce the potential for further growth as a competitor. This was a result of the primary change to improve the water balance under drought and may have a significant influence on long-term growth performance. On the other hand, their plasticity, i.e., the changeable dry matter allocation and functions of architecture suggest the possibility that they can recover their growth potential when they are released from drought. Thus, the duration and frequency of dry periods may be important in discussing effects of drought on inter-species relationships between stress-tolerators and competitors as a consequence of long-term growth performance in natural processes.

Among the measures examined in this chapter, the shoot/root ratio may be a good indicator of the susceptibility of a species to drought stress. The proportion of leaf area to fine root mass is also an important factor in predicting the growth and survival of seedlings, as fine root mass was the most variable factor affected by water-deficit stress in these experiments. As growth of fine roots is often influenced by mycorrhizal infections, it will be important to examine the role of mycorrhiza under varying levels of drought stress. Because a linkage between prolonged drought and large wild fires is found in East Kalimantan (see Chapter 3, this volume), understanding interspecies variations for drought tolerance processes in their physiology and morphology is important for the rehabilitation of deforested sites, especially in forest areas degraded due to forest fires and severe slash-and-burn agriculture.

## References


Ashton PS (1983) Dipterocarpaceae. In: Van Steenis CGGJ (ed) Flora Malesiana. Martinus Nijihoff / Dr W Junk, The Hague, pp 237–552

Fitter AH, Hay RKM (1981) Environmental physiology of plants. Academic, New York

Jones WJ, Zur B (1984) Simulation of possible adaptive mechanism in crop subjected to water stress. Irrigation Sci 5:251–264

Ito S (1992) Dry matter partitioning in seedlings and sprouts of several deciduous broad-leaved tree species: with special reference to the functions of tree architecture. Bull Kyushu Univ Forests 66:19–30

Iwaki H (1958) The influence of density on the dry matter production of *Fagopyyrum esculemtum*. Jap J Bot 16:210–226

Kartawinata K, Riswan S, Soedjita H (1980) The floristic change after disturbance in lowland dipterocarp forests in East Kalimantan. In: Furtado JI (ed) Tropical Ecology and Development, Int Soc Tropical Ecology, Kuala Lumpur, pp 47–54

King DA (1981) Tree dimensions: maximizing the rate of height growth in dense stands. Oecologia 51:351–356

King DA (1990) The adaptive significance of tree height. Amer Natural 1135:809–823

Kohyama T (1987) Significance of architecture and allometry in saplings. Funct Ecol 1:399–404

Kramer PJ (1983) Water relations of plants. Academic, NewYork

Medina E (1983) Adaptations of tropical trees to moisture stress. In: Golly FB (ed) Tropical rainforest ecosystem: structure and function. Elsevier, Amsterdam, pp 225–237

Myers BJ, Robichaux RH, Unwin GL, Craig IE (1987) Leaf water relations and anatomy of a tropical rainforest tree species vary with crown position. Oecologia 74:81–85

Oberbauer SF (1982) Water relations of *Pentaclethra macroloba*, a wet tropical forest tree. Bull Ecol Soc Amr 63:178

Osonubi O, Davis WL (1978) Solute accumulation in leaves and roots of woody plants subjected to water stress. Oecologia 32:323–332

Panshin AJ, de Zeeuw C (1980) Textbook of wood technology, Vol.1, 4th edn, McGraw-Hill, pp 200–236

Richards D (1976) Root-shoot interactions: a functional equilibrium for water uptake in peach [*Purnus persica* (L.) Batsch]. Ann Bot 41:279–281

Robichaux RH, Peacy RW (1980) Environmental characteristics, field water relations, and photosynthetic responses of $C_4$ Hawaiian Euphobia species from contrasting habitats. Oecologia 47:99–105

Steinberg SL, Miller Jr JC, Mcfarland MJ (1990) Dry matter partitioning and vegetative growth of young peach trees under water stress. Aust J Plant Physiol 17:23–36

Walter H (1971) Ecology of tropical and subtropical vegetation. Van Nostrand Reinhold, New York

Zimmermann M H (1983) Xylem structure and the ascent of sap. Springer, Berlin Heidelberg New York, pp 66–82

# 14 The Effect of Droughts and Fires on Coleopteran Insects in Lowland Dipterocarp Forests in Bukit Soeharto, East Kalimantan

HIROSHI MAKIHARA, HARUO KINUURA, KATSURO YAHIRO, and CHRISTOPHORUS SOEYAMTO

## 14.1 Introduction

The lowland dipterocarp forest in Borneo is one of the most species-rich ecosystems in the world (e.g., Whitmore 1984). The dipterocarp ecosystem is home to hundreds of thousands of animal and plant species, which form an intrinsic network of interactions (Schulte 1996). There are thought to be a large number of species of insects living in the lowland dipterocarp forest. We have been investigating the insect profiles in the lowland dipterocarp forest in Bukit Soeharto Education Forest (BSEF) of Mulawarman University, East Kalimantan, and have confirmed the species richness of insects in the lowland dipterocarp forest. For example, more than 550 species of the family Cerambycidae (Longicorn beetles) were found in a restricted area (ca. 1000 ha) of our study site in 1998 (Makihara 1999). This number is more than half of the total number of cerambycid species in all of North America (Linsley and Chemsak 1997).

Drought and fire affected the forest in BSEF in 1998. When the pronounced El Niño-Southern Oscillation (ENSO) event occurred in 1997–98, two droughts occurred in East Kalimantan, from July to October 1997 and from January to April 1998 (see Chapter 3, this volume). There was normal rainfall in BSEF between the two droughts, in November and December 1997. Wildfires swept BSEF repeatedly over a one-month period beginning in the middle of February, and affected the entire forest until normal rainfall recommenced in the middle of April 1998. These fires were mostly ground fires in the forest. The effects of the droughts and fires were more severe in the secondary forests than in the near-primary forest and the fire killed most of the smaller trees (see Chapter 10, this volume).

We investigated the insect population profiles for 10 months, beginning in a normal humid period, through the period of drought and fire, and on into the following normal rainy period. There have been many reports on the effects of forest drying or fire on insect populations (e.g., Werner 1983, Holliday 1984, 1992, McCullough et al. 1998). Almost all of these studies, however, have been conducted as late as 1, 2, 5, or even 10 years after a fire, whereas we describe the sequential changes in the number of species and individuals, as the drought became pronounced, during repeated wild fires, and after normal rainfall recommenced. This chapter deals with

Ecological Studies, Vol. 140
Guhardja et al. (eds.) Rainforest Ecosystems of East Kalimantan: El Niño, Drought, Fire and Human Impacts

**Fig. 14.1a–i.** Species frequently caught by Malaise traps. **a** *Celanephes parallelus* (Carabidae), body length 6.5 mm; **b** *Xylothrips flavipes* (Bostrychidae), body length 7.5 mm; **c** *Xylotrechus australis* (Cerambycidae), body length 11 mm; **d** *Pterolophia melanura* (Cerambycidae), body length 12 mm; **e** *Pterolophia banksi* (Cerambycidae), body length 12 mm; **f** *Genyocerus* sp.2 (Platypodidae), body length 3 mm; **g** *Genyocerus abdominalis* (Platypodidae), body length 3 mm; **h** *Platypus parallelus* (Platypodidae), body length 4.5 mm; **i** *Hypothenemus* sp.1 (Scolytidae), body length 2 mm

the effects of drought and fire on changes in the number of species and individuals in five families of coleopteran insects that are found in different habitats: predacious ground beetles (Carabidae), dry wood eaters (Bostrychidae), timber beetles (Cerambycidae), and wood borers (Platypodidae and Scolytidae). Figure 14.1 shows representative insects from each of these families.

## 14.2 Sampling Procedure

A Malaise trap in Town's configuration is a house-like structure, about 2 m long, with fine netting into which flying insects wander and then move upwards into a collecting bottle. Four Malaise traps were set in windy, open sites on ground paths (forest floor) in a near-primary dipterocarp forest, and a 60 m tower was installed. Three traps were set on the 60 m tower: at ground level, and at heights of 20 m and 40 m. Two traps were set on a 30 m tower that was installed in a secondary stand of *Macaranga* growing in a location that burned in 1982–83. One trap was at ground level and the other 20 m above the ground. In the stands where the towers were installed, the top canopy level was at heights of 40 and 20 m near the 60 m and 30 m towers, respectively. All the traps were first set on 29 or 30 December 1997 and samples were collected every week for 10 months, until 2 November 1998. The first fire occurred in BSEF and the surrounding area for about 10 days in late February, and the fire passed through the trapping sites on 26–27 February. The second fire was quite small, and occurred at scattered sites in BSEF about one month after the first fire. This fire occurred near the trap sites on April 1–2. The traps were moved just before the fires reached them, and then reinstalled once the fire danger was past. Specimens of Carabidae, Bostrychidae, Cerambycidae (including Disteniidae), Platypodidae, and Scolytidae were identified as precisely as possible, and the numbers of individuals in each species were noted for each trap.

## 14.3 Effects of Drought and Fire on Various Coleopteran Insects

### 14.3.1 Carabidae (Predacious Ground Beetles)

The nine Malaise traps caught 627 individuals representing 40 species of carabids between January and October 1998. Carabids occur in a wide variety of habitats: on the ground near the edge of streams or ponds, or in swamps, and above the ground on tree trunks, logs, or leaves (Lawrence and Britton 1991). Although Malaise traps are not normally used to catch ground dwellers such as carabids, they are still useful to estimate relative population trends. Moreover, no carabids were caught in pitfall traps, which are normally used for sampling this family. The most frequently caught species was *Celanephes parallelus* Schmidt-Gobel (Fig. 14.1a), and 414 individuals were caught, representing 60.0% of the total number of carabids. This species occurs widely, and is found in southern Japan, Southeast Asia, Micronesia, Polynesia, New Guinea, and Australia (Ohkura 1985).

The carabid population was affected by the fires and had still not recovered nine months afterwards (Figs. 14.2, 14.3). This is probably because most of the carabids, which are not wood-boring insects or deep soil-inhabiting insects, were burned to death in the fires. More carabid species were caught in the four Malaise traps on the forest floor before the fires than after the fires (Fig. 14.2a). More than 10 species were caught in a single week before the fires, but after the fires the numbers of species and individuals were low. The number of individuals caught at ground level increased just after the first fire (Figs. 14.2a, 14.3b,c) due to an increase in *Celanephes parallelus* (Figs. 14.2b, 14.3c). A marked increase in the numbers of both species and individuals was observed for July to August in the four Malaise traps on the forest floor (Fig. 14.2 a), but was not seen at the towers (Fig. 14.3).

*Celanephes parallelus* was the most frequently caught species just after the first fire. *Celanephes parallelus* is an arboreal species, which is found in the shrub crown. According to our observations, this beetle prefers to live in layers of living and dead leaves from shrubs, and was never found in the crowns of tall trees, although they seem to be strong flyers. The adults spend most of the time climbing around the base of trees in search of prey such as caterpillars. Therefore, most of the individuals of this species were caught at ground level, and few were found in the traps installed 20 m and 40 m above the ground (Figs. 14.2b, 14.3c). *Celanephes parallelus* was not observed in the 20 and 40 m traps just after the first fire, when their numbers in the ground traps increased dramatically. We assume that their habitat and prey were lost

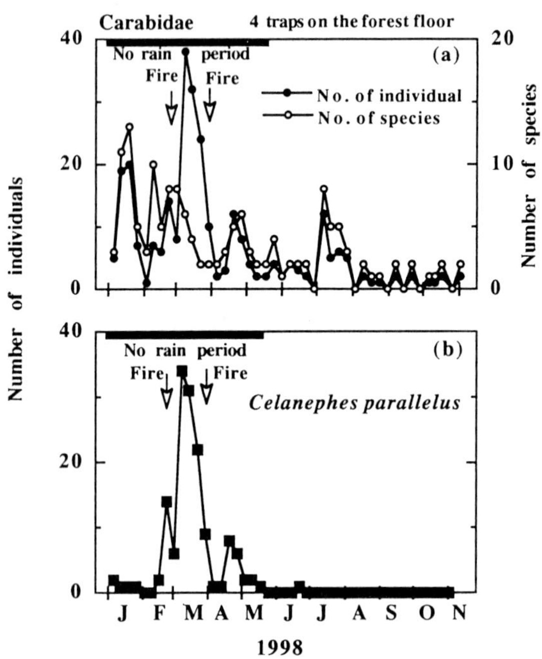


**Fig. 14.2a,b.** Changes in **(a)** the numbers of species and individuals of Carabidae, and **(b)** the number of individuals of *Celanephes parallelus*. The insects were caught in the four Malaise traps set on the forest floor

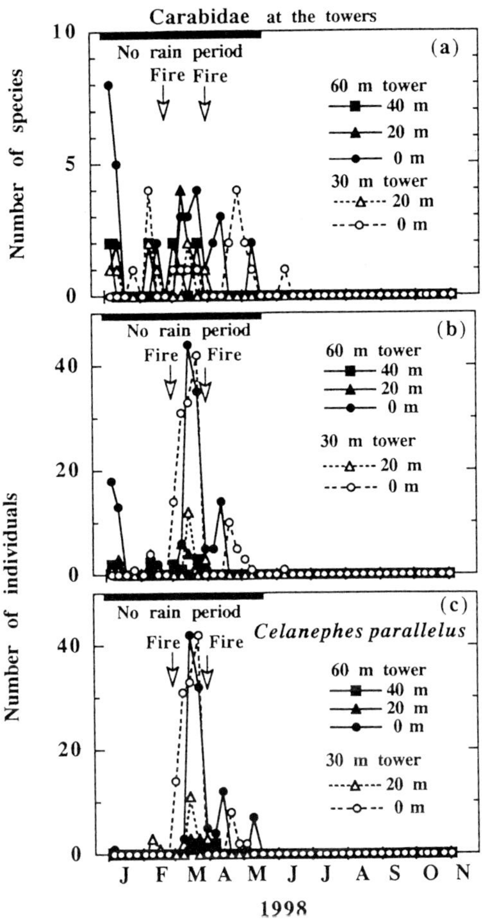


**Fig. 14.3a–c.** Changes in the numbers of **(a)** species and **(b)** individuals of Carabidae, and **(c)** the number of individuals of *Celanephes parallelus*. The insects were caught in the Malaise traps set on the towers

in the fire, and that the beetles were caught in the Malaise traps on the ground while escaping from the fire. Although some adults of this species escaped the first fires and survived in the burnt area for a short term, almost all likely died eventually because of insufficient food and loss of habitat as a result of the fires.

## 14.3.2 Bostrychidae (Dry Wood Eaters)

Four species of Bostrychidae, represented by 107 individuals, were caught in the nine Malaise traps between January and October 1998. More than 90% (98 beetles) were *Xylothrips flavipes* (Illiger) (Fig. 14.1b). There are few ecological differences between the species in this family, so we refer mainly to this species. Most bostrychids bore into moribund or freshly felled trees, dry, dead sapwood, or other dry plant material containing starches and sugars (Lawrence and Britton 1991).

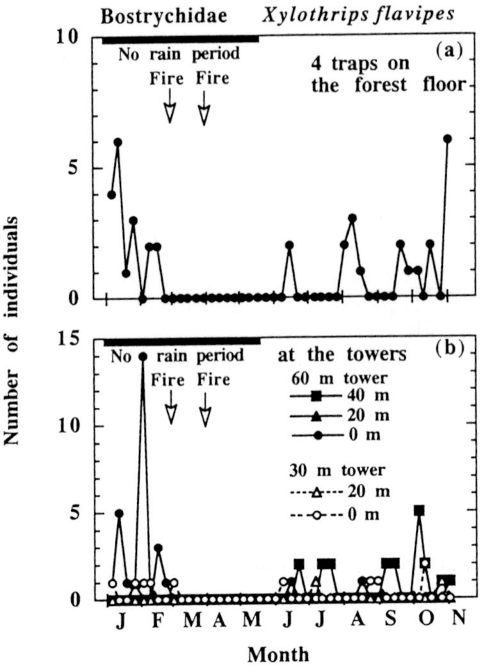


**Fig. 14.4a,b.** Changes in the number of *Xylothrips flavipes* (Bostrychidae) caught in the Malaise traps set on **(a)** the forest floor and **(b)** towers

No beetles from this family were captured in the four traps on the forest floor for 3 months after the first fire, but the numbers returned to normal within 6 months (Fig. 14.4a). The trap at ground level on the 60 m tower caught as many as 14 beetles in early February, while none were caught in the 20 or 40 m traps during the same period. Three months after the fires, however, many were trapped at 40 m (Fig. 14.4b). This may be due to an increase in newly-dead branches in tall trees weakened by the fires, while the small dead trees near the ground were too dry for this family to eat and live in.

Compared with other groups such as the Cerambycidae, only the Bostrychidae recovered their abundance within six months after the fires. Since members of this family can lay their eggs in a wide range of woods, from fresh, moist wood to hard, dry wood, the family is very resistant to dry conditions. Some larvae in wood probably survive fires, and the next generation of adults has plenty of oviposition sites. These conditions might contribute to the species' immunity to fire damage.

## 14.3.3 Cerambycidae (Timber Beetles)

A total of 2241 cerambycid beetles representing 190 species were caught in the nine Malaise traps between January and October 1998. Adult cerambycids are active fliers, and feed on flowers, foliage, or bark. Many are diurnal or nocturnal. The larvae are phytophagous, and usually feed internally on the bark, phloem, sapwood, or heartwood of a variety of trees and shrubs (Lawrence and Britton 1991).

The four traps on the forest floor captured large numbers of species and individuals in early January 1998, when the forest was still humid, but the numbers decreased progressively with the drought (Fig. 14.5a). The numbers of species and individuals caught dropped drastically soon after the first fire, and the numbers remained low until rainfall recommenced. The numbers of species and individuals increased slightly in June, two months after the fires, decreased again in July, and then began to increase gradually, but remained lower than the January levels (Fig. 14.5a). During the wet periods, two *Pterolophia* species, *Pterolophia melanura* (Pascoe) (Fig. 14.1d) and *P. banksi* Breuning (Fig. 14.1e) were much more common than the other cerambycid species. These two species were trapped in lesser numbers in February when it turned dry, none were captured in March and April, and only one individual was caught in May. On the contrary, many *Xylotrechus australis* (Catelnau et Gory) (Fig. 14.1c) were trapped near the first fire and for one month after the fires, when almost no other cerambycids were captured (Fig. 14.5b). *Xylotrechus australis* is widespread in Papua New Guinea and Indonesia beyond the Wallace line, although its distribution is not as broad as that of *Celanephes parallelus*.

The number of species captured at the 30 m tower decreased after the fires, and species were trapped at ground level before they were trapped at a height of 20 m (Fig. 14.6a). Within 3 months of the fires, the numbers at ground level recovered to the level found before the fires, and slightly more individuals were caught after September than before the fires (Fig. 14.6b).

The numbers of species and individuals caught in the trap at the base of the 60-m tower were the same as those caught in the four other traps on the forest floor. The

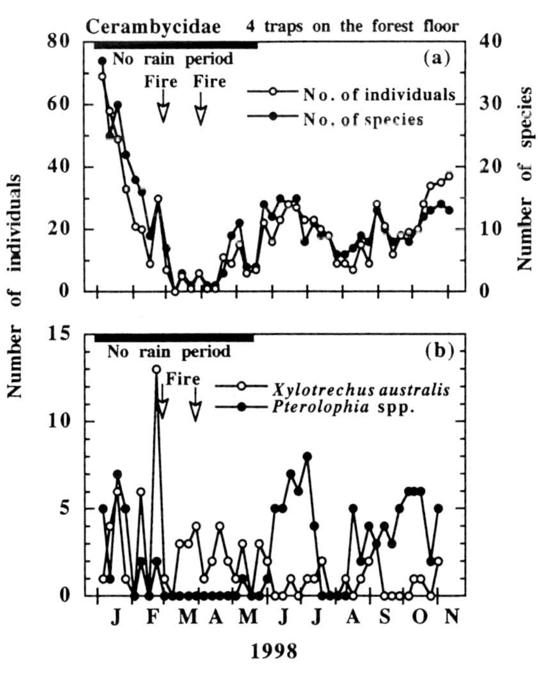


**Fig. 14.5a,b.** Changes in **(a)** the numbers of species and individuals of Cerambycidae and **(b)** the numbers of individuals of *Xylotrechus australis* and *Pterolophia* spp caught in the four Malaise traps set on the forest floor

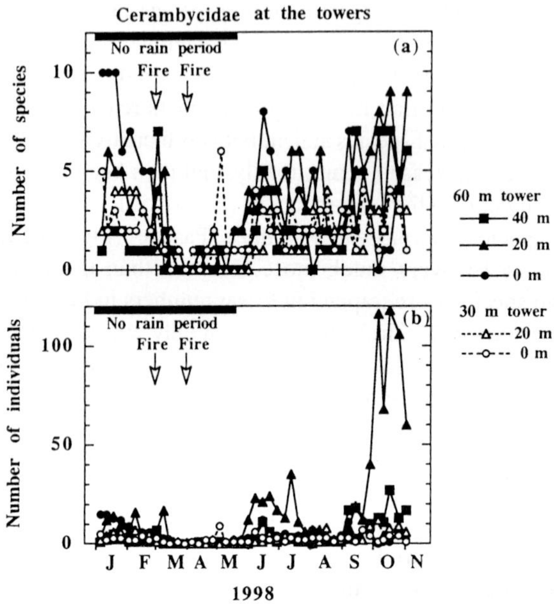


**Fig. 14.6a,b.** Changes in the numbers of **(a)** species and **(b)** individuals of Cerambycidae caught in Malaise traps set on the towers

number of species trapped at a height of 40 m on the 60 m tower increased when the first fire reached the base of the tower (Fig. 14.6a). These species were those that escaped the ground fire. During the second fire, no increase in the number of cerambycids at 40 m was observed, probably because the first fire killed many of the insects.

In October 1998, the number of individuals collected 20 and 40 m above the ground jumped to more than 10 times the level seen in early January 1998, just after the drought started (Fig. 14.6b). Although the sampling occurred in a limited area, this phenomenon was observed for a number of species (Fig. 14.6a). In contrast, the numbers of both species and individual cerambycids at ground level had not fully recovered by early November 1998. These observations can be explained as follows: cerambycid larvae and eggs are found in the trunk or crowns of large trees, away from the heat of the fires, and there were abundant food logs and food trees weakened by the drought and fires. Furthermore, the numbers of the beetles' natural enemies, such as ants, might have been reduced during the period. The cerambycids in the upper layers of the forest recovered more quickly than those in the lower layers.

## 14.3.4 Platypodidae and Scolytidae (Wood Borers)

A total of 1310 individuals (541 Platypodidae and 769 Scolytidae) representing 87 species (50 Platypodidae and 37 Scolytidae) were caught in the nine Malaise traps between January and October 1998. Platypodids and scolytids are the well-known bark and ambrosia beetles, which often cause economic damage to logs and timber

(Kikata et al. 1994). The larvae always feed within plant tissue. Some groups feed on ambrosia fungi (Lawrence and Britton 1991). *Hypothenemus* sp. 1 (Scolytidae, Fig. 14.1i), *Genyocerus abdominalis* (Schedl) (Platypodidae, Fig. 14.1g), and *Platypus parallelus* (Fabricius) (Platypodidae, Fig. 14.1h) were predominant.

The number of species collected in the four ground traps on the forest floor peaked before and after the first fire, but remained consistently low after the second (Fig. 14.7a). This pattern occurred because the three predominant species were collected only before the second fire. The trend in the number of individuals was similar to that in the number of species, but the peaks were more distinct in the former and the fluctuations depended on the abundance of the three above-mentioned species. *Genyocerus abdominalis* was abundant before the fires (January), while *Hypothenemus* sp. 1 was largely captured just before and after the first fire. These two species disappeared after the second fire. In contrast, *Platypus parallelus* was captured quite constantly, except for the first 3 months after the fires (Fig. 14.7b).

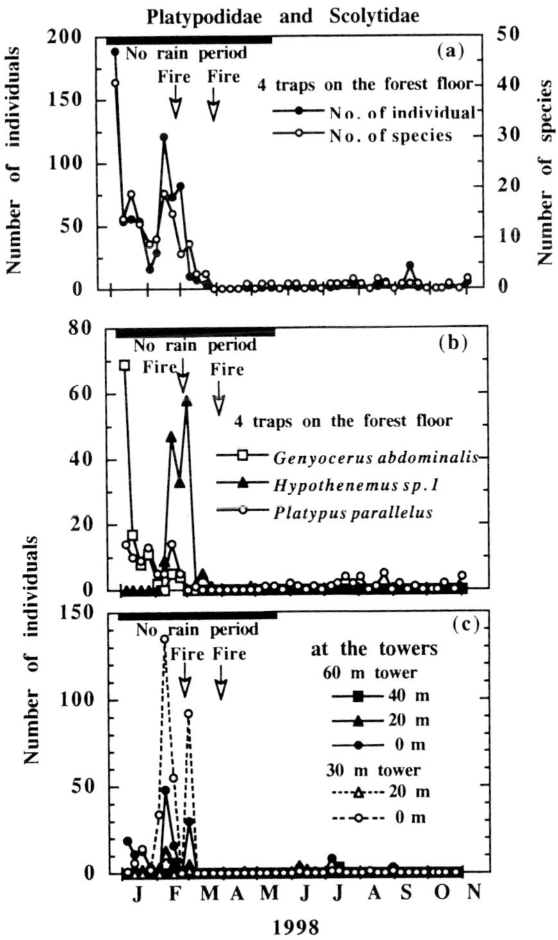


**Fig. 14.7a–c.** Changes in **(a)** the number of species and individuals of Platypodidae and Scolytidae and **(b)** the number of the three predominant species caught in the Malaise traps set on the forest floor, and **(c)** the number of individuals of Platypodidae and Scolytidae caught in the Malaise traps set on the towers

At the 30 m tower, the number of individuals collected in the ground-level trap increased just before and after the first fire because of a marked increase of *Hypothenemus* sp. 1. Very few species or individuals were collected at a height of 20 m on the 30 m tower. At the 60 m tower, the trend at ground level was similar to that of the four traps on the forest floor. The numbers of species and individuals collected at 20 and 40 m on the 60 m tower remained very low. The numbers of species and individuals in this group decreased with an increase in the height of the traps (Fig. 14.7c).

*Hypothenemus* beetles prefer small twigs and the pith of dead branches. It appears that during a fire, individuals of this species are smoked out and disperse, which might be why many of these beetles were caught during the fires. *Genyocerus abdominalis* and *Platypus parallelus* belong to the Platypodidae, and the galleries formed by their colonies reach sapwood. Therefore, they can survive within logs even during the fires. However, *Genyocerus abdominalis* may not be resistant to dry conditions, since the abundance of this species decreased as the forest dried and their numbers did not recover until November. In contrast, *Platypus parallelus* appears to be more tolerant of dryness, since individuals were caught several months after the fires, although the levels were low.

## 14.4 Conclusion

The extremely high diversity of coleopteran insects makes it difficult to understand their behaviors; the habits of many are not known and some species have not even been given a scientific name. The drought and fires that attacked the humid tropical forest in East Kalimantan in 1998 had a significant impact on the forest ecosystem, and allowed us to clearly observe the unusual behaviors of coleopteran insects.

Of the families surveyed, the Bostrychidae, which can attack dry wood, was the least affected by the burning. The next most tolerant family was the Cerambycidae, which recovered to about half the level of early January. The recovery in the high canopy layer was more rapid than that on the forest floor, perhaps a result of increased survival in the high canopy because of the abundant food and habitat in newly dead trees and branches after the fires. In contrast, the fires seriously affected the Carabidae, Platypodidae, and Scolytidae and their numbers had still not recovered six months after the fires.

Some species showed remarkable behaviors during the fires: *Celanephes parallelus* (Carabidae) was predominately collected immediately after a fire. Many *Xylotrechus australis* (Cerambycidae) and *Hypothenemus* sp.1 (Scolytidae) were caught just before and after the first fire. It is highly unlikely that there was an outbreak of these species during the fires, and it therefore seems that they were captured while escaping from the burned area surrounding the traps. These two species are widely distributed and are probably able to fly long distances. In the tall tree stand, the number of cerambycid species in the upper parts of trees increased after the fires, and at the same time the number of individual Bostrychidae also increased. This trend was more distinct at the 60 m tower, which was surrounded by dipterocarp trees about 40 m tall, than at the 30 m tower, where *Macaranga* trees about 20 m high predominated and 90% of the trees were killed by the droughts and fires. These results suggest that the wood-boring beetle population is capable of early recovery in tall tree stands after a ground fire. On the other hand, insects that inhabit the forest floor suffered crushing damage as a result of the ground fires.

In the post-fire period, the insect infestations were cyclic. In the Cerambycidae, there was a very clear small population peak in June and July, and a marked peak in October and November (Fig. 14.6b). The cyclic peaks accord with their life history. The unbalanced species composition in the post-fire period, as shown by the extraordinary outbreak of Cerambycidae in October and November, has just started. We do not know when the normal balance will be restored or when the numbers of species and individuals of Carabidae, Platypodidae, and Scolytidae will return to normal, or even whether some species will return. The unbalanced beetle populations may affect other organisms through the network of interactions between living things. However, we know too little about the beetles' behaviors to describe these interesting inter-relationships. We should continue long-term observation of the insect world in the tropics.

## References

Holliday NJ (1984) Carabid beetles (Coleoptera: Carabidae) from a burned spruce forest (*Picea* spp.). Can Entomol 116:919–922

Holliday NJ (1992) Carabid beetles (Coleoptera: Carabidae) during post fire regeneration of boreal forest: properties and dynamics of species assemblages. Can J Zool 70:440–452

Kikata Y, Hijii N, Kajimura H (1994) Monograph of the new species of bark beetles and ambrosia beetles (Coleoptera: Scolytidae and Platipodidae) from some tropical countries, intercepted at Japanese port. Shizuo Ohno collection. Bull Nagoya Univ Furukawa Museum Special Report 3:1–188

Lawrence JF, Britton EB (1991) Coleoptera. In: I.D. Naumann, P.B. Carne (eds) The insects of Australia II, 2nd edn. pp 543–683

Linsley EG, Chemsak JA (1997) The Cerambycidae of North America, Part III: Bibliography, index, and host plant index. Univ Calif Public Entomol 117:1–534

Makihara H (1999) Atlas of longicorn beetles in Bukit Soeharto Education Forest, Mulawarman University, East Kalimantan, Indonesia. PUSREHUT Special Publication No. 7, Samarinda, Indonesia

McCullough DG, Werner RA, Neumann D (1998) Fire and insects in northern and boreal forest ecosystems of North America. Ann Rev Entomol 43:107–127

Ohkura M (1985) Carabidae (Perigoninae-Dryptinae) (in Japanese). In: Ueno S-I, Kurosawa Y, Sato M (eds) The Coleoptera of Japan in color 2:163–179

Schulte A (1996) Dipterocarp forest ecosystem theory based on matter balance and biodiversity. In: Schulte A, Schöne D (eds) Dipterocarp forest ecosystems: Towards sustainable management. World Scientific Publishing, Singapore, pp 3–28

Werner RA (1983) Biomass, density and nutrient content of plant arthropods in the taiga of Alaska. Can J For Res 13:729–730

Whitmore TC (1984) Tropical rain forests of the Far East, 2nd edn. Clarendon, Oxford

# Part 4

# Human Impacts on Forest Ecosystems

# 15 Mechanism of Changes in the Kenyah's Swidden System: Explanation in Terms of Agricultural Intensification Theory

MAKOTO INOUE

## 15.1 Introduction

### 15.1.1 Views on Swidden Agriculture

According to the FAO (1995, 1997), 12 594 000 ha of tropical forests diminished annually in the 1990s. The main factors of tropical deforestation are slash and burn agriculture, reclamation of farmland, conversion to pasture, industrial wood felling, firewood collecting, and illegal logging. Moreover, the land tenure system, economic growth, and population increase have been indicated as underlying causes of tropical deforestation. In this manner, most of the causes of deforestation seem to have been made clear (Brown and Pearce 1994, Sponsel et al. 1996), even though discussion on swidden agriculture or shifting cultivation still remains controversial.

The term "shifting cultivation" is often taken to mean that the people are themselves "shifting" or semi-nomadic, even though most of them live in relatively permanent settlements. In fact people use fire and fallow their fields after harvest. The field, once set on fire, is called "swidden". Therefore I think "swidden agriculture" is preferable to the commonly used term "shifting cultivation" (Dove 1981). In any case, swidden agriculture can be regarded as one of the most important local resource management systems in the tropics.

Development and forest economists, working for the FAO, World Bank, or bilateral aid agencies, depicted local inhabitants as threats to the forests in the 1950s (Peluso et al. 1995). In this context, swidden agriculture was considered to be "encroachment" on the public domain. Such recognition continues to influence most of the traditional foresters who call it "slash and burn agriculture". Mr. Katsura Watanabe, a Japanese ex-officer of the FAO, called traditional foresters' inclination to love straight trees and to hate people the "foresters' syndrome". New foresters who move away from the foresters' syndrome may be called "social foresters" (Gelder and O'Keefe 1995).

On the other hand, anthropologists and human geographers carried out fieldwork to dispel many detrimental myths about swidden agriculture in the 1950s (Conklin 1957) and 1960s (Spencer 1967). Since then many researchers, through field-work, have tried to refute the detraction of swidden agriculture as being the

Ecological Studies, Vol. 140
Guhardja et al. (eds.) Rainforest Ecosystems of East Kalimantan: El Niño, Drought, Fire and Human Impacts

most significant cause of forest degradation (Peluso et al. 1995). In this context, swidden agriculture is considered to be rather sustainable as long as population density is low enough to retain "recurrent cultivation" with a relatively long fallow period.

Basically, the object of interest between traditional foresters and anthropologists seems to be different. The attention of traditional foresters is concentrated on the land utilization practiced by the newcomers; anthropologists observe the land utilization practiced by indigenous people. As a result, anthropologists do not regard "slash and burn agriculture" practiced by the newcomers as "swidden agriculture". Actually, even recurrent types of swidden agriculture performed by indigenous people are undergoing drastic changes.

### 15.1.2 A Simple Category of Swidden Agriculture

How can we comprehensively understand the state of swidden agriculture ? To tackle this problem, I have introduced a simple category of swidden agriculture in terms of sustainability as shown below (Inoue 1988), even though several geographers (Conklin 1957, Greenland 1974) had classified swidden agriculture previously.

1) Traditional swidden agriculture: sustainable swidden agriculture practiced by indigenous people. Most are recurrent types with a long fallow period, but pioneer types are also included as long as they are sustainable.

2) Transitional (quasi-traditional) swidden agriculture: swidden agriculture by indigenous people, originating from both recurrent and pioneer types, but not so sustainable because of a short fallow period. Most existing swidden agriculture by indigenous people may be included here.

3) Non-traditional swidden agriculture: unsustainable slash and burn agriculture often with cash crops, usually practiced by newcomers but sometimes by indigenous people who once practiced traditional swidden agriculture.

Non-traditional swidden agriculture is practiced mostly by landless farmers who have been pushed out of the lowland rural areas. Research in agricultural economics and developmental economics has revealed the mechanism of increase in the number of the landless farmers (Watanabe 1986, Kano 1982, Takigawa and Saito 1968). Focusing on swidden agriculture, social conditions have recently been forcing an increase in the number of non-traditional swiddenners. On the other hand, there has been little discussion about the mechanism of changes in the form of swidden agriculture practiced, from traditional, by way of transitional (quasi-traditional), to non-traditional.

### 15.1.3 Purpose of This Chapter

In this chapter, dynamic changes in swidden agriculture practiced by the Kenyah people will be discussed. Specifically, the shortening of the fallow period will be explained theoretically.

## 15.2 Changes in Swidden Agriculture by the Kenyah Dayak

### 15.2.1 Review of Studies on Swidden Agriculture

Previous studies on swidden agriculture can be summarized into three groups: microscopic "ethnological studies", macroscopic "comparative geographic studies" (Sasaki 1970), and "natural scientific and agronomic studies".

There are two methods of empirical and positive studies on the dynamics of swidden systems. One is the method of observing the same village for 10 or 20 years. The other is the cross-sectional comparative study of the villages which are resided in by the same ethnic group and distributed geographically. The former is included in ethnological studies and the latter in comparative geographical studies.

However, as mentioned by Conklin (1961), there are no detailed monographs based on field surveys over 1–2 years on the ethnography on the numerous peoples conducting swidden agriculture. According to an evaluation by Sasaki (1970), comparative studies on swidden agriculture have been merely simple descriptions of each swidden agricultural technique in various areas. That is, only a simple comparison has been made on the complex of swidden agriculture techniques established in different cultural backgrounds and different natural environments, and no criteria or standards have been established for the comparison. Excellent studies such as the comparative studies on the Karen and Lua tribes in Thailand (Kunstadter 1978) are no exception to these shortcomings.

### 15.2.2 Characteristics of the Study

From the results of field studies over almost three years in the late 1980s, I verified the reality of changes in the swidden system by the Kenyah Dayak people, typical indigenous people of Borneo island. A characteristic feature of the research was to ascertain the consecutive changes in swidden agriculture practiced by the Kenyah (Inoue and Lahjie 1990), unlike other excellent comparative studies among a few different ethnic groups (Colfer 1983, Hadi et al. 1985, Kartawinata et al. 1984, Vayda 1981, Vayda et al. 1980).

In order to describe these consecutive changes, the infiltration of a monetary economy, more specifically "an average annual cash expenditure per household", was applied as an indicator of socio-economic conditions. As an indicator of the impact of swidden agriculture on their environment, I applied a "degree of recovery of secondary vegetation during the fallow period".

In conclusion, it was clarified that the Kenyah's swidden agriculture was quite reasonable from an ecological point of view, but at the same time it was so fragile that it had changed drastically from traditional to non-traditional swidden agriculture.

### 15.2.3 The Kenyah and Study Sites

Proto-malayan ethnic groups in Borneo island are called "Dayak" generically. Dayak people are classified into five groups linguistically, one of which is the "Kayan-Kenyah group" including the Kenyah people. The native land of the Kenyah people distributed in East Kalimantan is called "Apo Kayan", located in the central plateau of Borneo island. In the beginning of the 1950s they began to move out from Apo Kayan to the Mahakam and the Kayan basins. The Apo Kayan region was then depopulated until 1990.

The inhabitants of most of the Kenyah villages, both in Apo Kayan and in other locations, are self-sufficient in staple foodstuffs, but trade with other regions is vital for them to obtain the other necessities of life. Six villages were selected as research sites: two villages (village A and B) in the Apo Kayan, three (villages C, D, E) in the middle reaches of the Mahakam river, and one (village F) in the vicinity of the provincial capital, Samarinda.

### 15.2.4 Recurrent System of Swiddens and Its Change

As the Kenyah people migrated downstream, their livelihoods were incorporated into the market economy, which led to changes in their lifestyles and social structure, such as work organization, customary land tenure systems, and mutual aid systems in daily life (Inoue and Lahjie 1990).

In the course of a series of such changes, their sustainable swidden agriculture system was also changed to a less sustainable one. Only the rotation system is summarized here. Almost all the swiddens are shifted every year and the main annual crop is upland rice, often mixed with maize, cucumber, and eggplant. Traditionally, the Kenyah people classify the secondary growth after harvesting rice as a succession stage. The period of each category is not fixed because the recovery rate of the vegetation depends on site conditions such as soil fertility. For example, the vegetation becomes "Jekau", or secondary forest with few weeds, within seven years on some plots but only after ten years on other plots. The point is that they do not fell the forest again until secondary growth, known as "Jekau Lataq" or old secondary forest, with tree diameters of as large as the human waist, is established, as in the most isolated village (village A) in Apo Kayan.

However in the villages in the middle reaches of the Mahakam river, people make swiddens again even if the secondary vegetation is still as small as "Jue Dumit", or shrub size. Therefore, the rotation system in these villages may have a negative influence on the environment in the near future. Moreover in the village in the vicinity of Samarinda (village F) the people have introduced commercial pepper production on former swiddens where upland rice was harvested.

In the following section, the mechanism of changes in the swidden system will be discussed, based on the comparative studies summarized above.

## 15.3 New Concepts for Study of the Shortening of Fallow Period

### 15.3.1 Brief Review of Agricultural Intensification Theories

Population density and shortening of the fallow period have been discussed in agricultural intensification theories, which might become the key to elucidate the changes in the swidden system.

Boserup (1965) argued that intensive productive technology is stimulated by the increase in population. That is, population density is an independent variable, and the intensity of productive technology, or intensity of land use, are dependent variables. The intensity of land use here means frequency of cultivation, that is the ratio of cultivation period to fallow period. In primitive swidden agriculture, intensification is done by the shortening of the fallow period. As a result of shortening the fallow period, it becomes necessary to introduce a method for the maintenance and recovery of soil fertility, such as green manure, compost, etc. In the course of shortening of the fallow period, weeding labor must increase. Terracing techniques and irrigation facilities may also be introduced.

Thus, the intensification of swidden agriculture needs more intensive labor input, which causes a decline of labor productivity. Furthermore, when the fallow period is shortened further, the surrounding vegetation deteriorates, the grassland area increases and animal labor for plowing is introduced. The reduction in labor productivity is then stopped.

This theory, the so-called "Boserup theory", cannot be overlooked when discussing agricultural intensification, especially the intensification of swidden agriculture.

On the other hand, Malthus once argued that the population depended on the given resources and technological standards. Geertz (1963) argued that wet rice productivity was maintained at the minimal survival level while more and more labor was put into wet rice cultivation in the course of subdivision of agricultural land as a result of population increase. This is the so-called "involution theory". Brookfield (1972) then integrated the Malthusian theory, the Boserup theory and the involution theory.

Vasey (1979) has developed a theory concerning the environment of the system, which Boserup neglected and Brookfield could not discuss sufficiently. He concludes that the degree of agricultural intensification is affected by ecological adaptation at a certain population density. However, if so, the land cannot be degraded.

### 15.3.2 Defects of Previous Agricultural Intensification Theories

It is clear that agricultural intensification theories have directly discussed the relationship between population density and fallow period. However, the defects of the previous intensification theories become clear if the example of the Kenyah is considered just briefly.

Firstly, Vasey (1979) developed an intensification theory considering the relationship with the environment. However, the environment he considered was limited to the natural environment, and did not include the socio-economic environment. The example of the Kenyah showed that the socio-economic environment, such as the infiltration of a monetary economy, land tenure system, national forest policy, has a strong impact on the changes in the swidden system.

Secondly, in agricultural intensification theories including that of Boserup, the swidden system of the village, where the villagers are cutting down virgin forest and expanding the swiddens, has not been considered. This means that regions where many new swidden villages had been established by migration to virgin forested areas and the logged over forest areas, such as in East Kalimantan, would be excluded from the intensification theory of swidden agriculture.

To overcome these problems, the stages of the swidden village with migration is first discussed, then new concepts are proposed such as "potential population density" and "intensity of land use frequency", and lastly the shortening of the fallow period is explained theoretically.

### 15.3.3 Stages of a Swidden Village

Generally, how does the swidden village develop? Provided a population is constant, the starting point is where a new village is formed due to migration by a part of the population (Fig. 15.1).

People practice swidden agriculture by cutting down virgin forest and logged over forests cut by timber companies, and by burning the forests (Stage I). At this stage the area of forest that can be utilized as swidden is rapidly expanded. After a certain number of years, some of the people begin to carry out swidden agriculture after cutting and burning the secondary vegetation of former swiddens (Stage II). However, there are still many swiddens produced from virgin forests, and the rotation system of the former swiddens has not been established. Furthermore, after several years, the village area, or the area of forest that can be utilized as swidden, reaches an almost established state (Stage III). In most of these swiddens, a clear rotation system is developed, and new swiddens are made on the secondary forests of the former swiddens. Provided the population is constant, the village will be maintained stably in this state for long periods as long as the economic environment does not change.

Next, we will consider the case where the population is increasing. This is a more realistic model. The increased food demand due to the population increase is

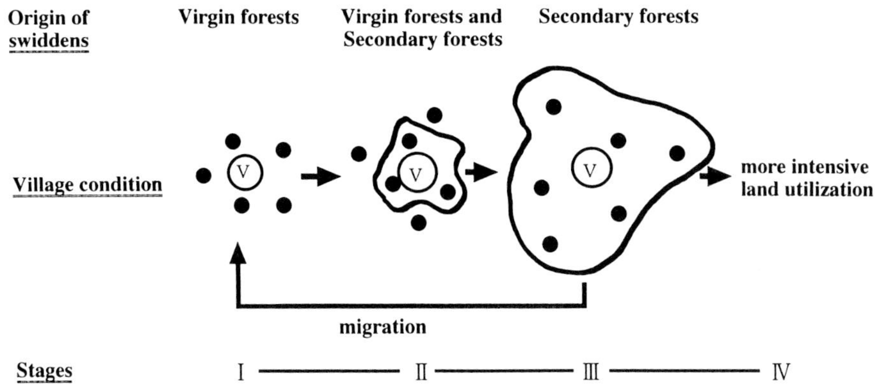


**Fig. 15.1.** Stages of swidden village. In Stage I, residents in a village make swiddens (*black circles* ) from virgin forests because the place of their residence (*V*) is surrounded by them. In Stage II, some of the swiddens are made from secondary forests or former swiddens (enclosed by *solid line*). In Stage III, most of the swiddens are made from secondary forests. Some of the residents migrate to establish a new hamlet of Stage I. In Stage IV, more intensive land utilization may be developed

met by the expansion of cultivated land and by the additional input of excess labor. That is, development up to stage III can be seen in common for both when population is constant and when it increases. However, when the population has increased above a certain level, the village can no longer remain at Stage III. The fallow period is shortened, and the frequency of land use increases (Stage IV). Boserup called this process, from Stage III to Stage IV, "agricultural intensification". It is clear that previous intensification theory considered only the process from Stage III to Stage IV.

With the increase in population, some people migrate from the village to a place with a lower population density. This is the process from Stage III to Stage I. When the territory of a new village is limited because of the existence of other villages in the vicinity, the transition time from Stage I to Stage III must be shortened.

### 15.3.4 New Indicators for Analysis

Concerning the central dogma of agricultural intensification theory, stages I and II had not been considered previously. Here, we will establish a concept that incorporates stages I and II and allows discussion at the same level as stages III and IV.

First we assume two villages, such as village X in stage II and village Y in stage III, within the same forest area that is much larger than the village area (see Fig. 15.2), provided the population density of the two villages is the same statistically. When discussing the intensification of swidden agriculture, would it be correct to consider the population density of X and Y to be the same? The people in village X practice swidden agriculture by cutting the virgin forest, and the area of the village is expanding. On the other hand, in village Y, the people are crowded in a certain area. The population density that the people actually feel might be different.

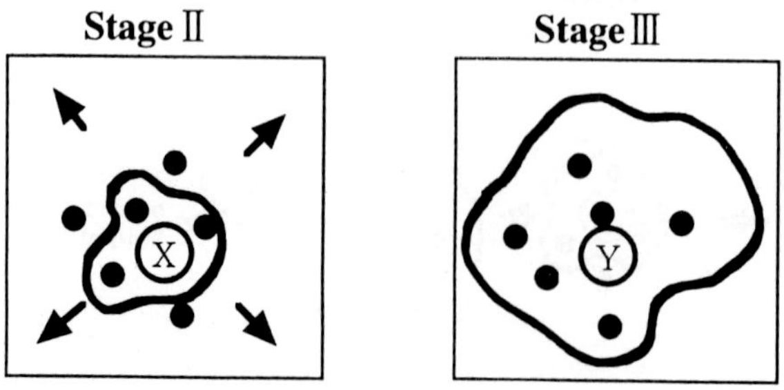


**Fig. 15.2.** Basic concept of potential population density (PPD). Statistically, village *X* of Stage II and *Y* of Stage III have the same population density

A new indicator of "potential population density (PPD)" is then defined as the following equation:

$$PPD = P / S \qquad (1)$$

where P indicates population and S indicates swidden area which consists of cultivated land and fallow.

In the case of village Y, the PPD can be calculated simply by applying the following formula:

$$\text{PPD at Stage III and IV} = P / [MS \times H \times (F+C) / C] \qquad (2)$$

where F, C, MS, and H indicate fallow period, cultivation period, mean area of swidden under cultivation per household, and number of households respectively.

However, in the case of village X, the calculation is slightly more difficult, because the swidden land the people need can be classified into two kinds according to their origin. One is land that is cultivated recurrently, and the other is virgin forest and logged over forests. The PPD of the former can be obtained by Eq. 2. But we cannot obtain an actual number of the PPD of the latter, because cultivation period and fallow period are indefinite.

For comparing the actual data of 6 villages, we then have no choice but to estimate a relative size of the PPD in each village from the "population density per unit cultivated land" and the "percentage of swiddens made from secondary forests or former swiddens".

The PPD can also express how crowded the inhabitants feel. That is, they do not feel crowded when the swiddens are spreading toward the surrounding tropical forest area. However, even at the same population density statistically, when the rotation system of swidden has been established, or swidden area has already been fixed, the inhabitants feel as if they are crowded into a certain area and the population density they feel becomes high. In this sense, the PPD can be regarded as a psychological population density.

Thus as Boserup mentioned, labor will be introduced more intensively in stages III and IV, and consequently, labor productivity will be lowered. However, the intensity of land use frequency will be improved with the increase in the potential population density.

Next, "intensity of land use frequency" is defined. Based on Boserup's idea,

Ruthenberg (1976) defined the intensity of land use frequency with the following formula:

$$R = 100 \times C / (C+F) \qquad (3)$$

where R indicates the intensity of the land use frequency proposed by Ruthenberg. The larger the value of R is, the higher the intensity is. Under this definition, however, both the single cropping system of wet rice cultivation and the double cropping system have the same intensity, which seems to be contradictory to common sense.

We would like to define the intensity generally as follows:

$$IN = 100 \times T / (L+F) \qquad (4)$$

where IN, T, and L indicate the intensity of the land use frequency proposed by Inoue, the number of times of harvests during the land use period, and land use period, respectively.

Thus, in a 10-year period, the intensity for double cropping would be 100 × 20 / (10+ 0) = 200, and that for single cropping is 100 × 10 / (10 +0)= 100. Double cropping has twice the intensity of single cropping. In the case of plantation forestry, where the main felling is assumed to be at 40 years old with thinning performed 5 times before then, the intensity would be 100 × 6 / (40 + 0) = 15. The intensity would be much lower than that of single cropping wet rice, which is in agreement with general understanding.

However, here, the origin of the swidden becomes the question. The IN must be calculated according to the percentage of swidden made from secondary forests and virgin forests. After abandoning the swiddens made from virgin forests, the swiddens must be reused rotationally. However, the duration of the rotation remains unclear at present. At the stage of reutilization of the former swiddens made from virgin forest, the percentage of swiddens made from secondary forests increases, and the fallow period for the swidden made from secondary forests is considered to change. The future duration of the fallow period can not be forecast and a discussion based on the actual values is not possible. Thus for the time being, emphasis is put on the current fallow period and the future is disregarded. The virgin forest is thus regarded as a secondary forest with an unlimited fallow period. As a result, the intensity of land use frequency when felling virgin forest can be considered to approach zero. We thus get Eq. 5 as an equation for actual calculation, developed from Eq. 4.

$$IN = PS \times 100 \times T / (L+F) \qquad (5)$$

where PS indicates the percentage of swiddens made from secondary forest.

When pepper cultivation is introduced as a cash crop, as in village F, the intensity for the pepper field must also be obtained. If the pepper fields are utilized permanently, the land use period will be 16 years including upland rice production, the fallow period will be 0 years, and the number of times harvests are made during the land use period will be 14 because pepper production starts in the 3rd year after planting. However, if the pepper field is abandoned due to deterioration of soil fertility, the fallow period becomes unknown and the intensity of land use frequency becomes considerably lower.

## 15.4 Intensification Process of the Kenyah's Swidden System

### 15.4.1 Indicators of Swidden Intensification of the Kenyah

Table 15.1 shows indices of swidden intensification of a Kenyah village. "Labor productivity in the year under review" means rice production per hectare harvested just before the survey conducted by the author. Villages from A to F are listed in order of increasing infiltration of a monetary economy (Inoue, 1990).

In fact, except for village B in which the swiddens were completely made from secondary forests, it was impossible to obtain the actual value of the PPD. However, the potential population density of each village could be compared relatively. The higher the population density was per unit of cultivated land in the year under review and the higher the percentage of swidden made from secondary forests, the higher the potential population density or psychological population density was considered to be.

In Fig. 15.3, the PPD is higher towards the upper right and lower towards the lower left. The PPD was the highest in the Apo Kayan region (villages A and B), in the remote area of east Kalimantan, and the lowest in the suburbs of Samarinda city (village F) and intermediate in the Mahakam basin (villages C, D and E).

From the 1950s to the present, more than a half of the population has been moving from the Apo Kayan region, mainly for economic benefit. This population decline is assumed to be due to the high PPD. Actually the PPD of village B was 41 persons km$^{-2}$, exceeding the carrying capacity for swidden agriculture in Southeast

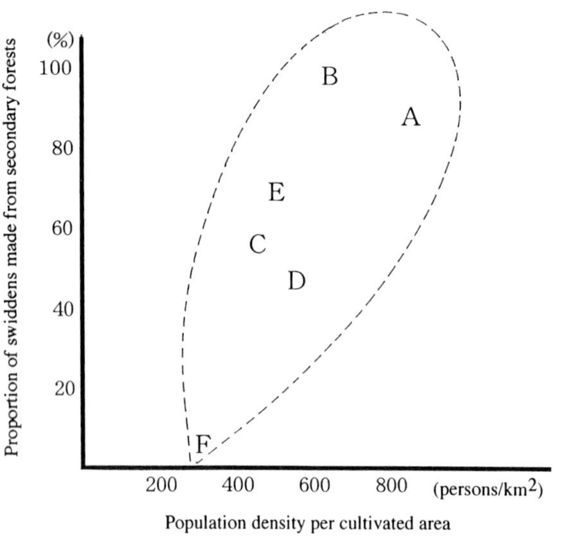


**Fig. 15.3.** Relative PPD of the villages. The more upper right a village is plotted, the higher PPD the village has. Thus the PPD of village *A* and *B* is higher than that of village *E*; the PPD of village *E* is higher than that of village *C* and *D*; the PPD of village *F* is the lowest

**Table 15.1.** Indices of swidden intensification of the Kenyah village

| Villages | PPD[a] (persons/km$^2$) | Population density per unit cultivated are (persons/km$^2$) | IN | Labor input per hectare (man-day/ha) | Labor productivity (kg/man-day) | Land productivity in the year under review (kg/ha) | Proportion of swiddens made from secondary forests[c] (%) |
|---|---|---|---|---|---|---|---|
| A | - | 833 | 3.0 | 726 | 3.9 | 2313 | 90 |
| B | 41.4 | 607 | 9.1 | 501 | 5.0 | 2099 | 100 |
| C | - | 460 | 9.7 | 231 | 7.8 | 1692 | 58 |
| D | - | 523 | 12.5 | 292 | 8.8 | 1970 | 50 |
| E | - | 507 | 20.0 | 305 | 16.8[b] | 4704[b] | 70 |
| F | - | 305 | 2.0 (paddy)<br>87.5 (pepper) | 172 | 11.4 | 1632 | 2 |

PPD, potential population density; IN, intensity of land use frequency.
All the numerical values are based on data obtained by the author in the late 1980s.
Data on "labor input per hectare" involve only upland rice production, and data on "labor productivity" and "land productivity in the year under review" show the productivity of the upland rice only.
Upland rice production is calculated in weight of dried unhusked rice.
Applying the method of analysis of variance, the null hypothesis of H : mean values in all the villages are the same was rejected at the 5% significance level, in relation to the values "labor input per hectare", "labor productivity" and "land productivity in the year under review".
[a] Only in the case of village B can we calculate the value of PPD because all the swiddens were made from secondary forests.
[b] Cultivating soybean on swiddens seems to affect the rise of productivity of upland in the village E.
[c] Swiddens, except for ones made from secondary forests, are made from virgin forests or logged over forests which have not been felled by swidden cultivators.

Asia of 25–30 persons km$^{-2}$ (Sasaki, 1970). The carrying capacity reported by Sasaki was obtained by dividing the population by the total area of swidden which is used in the rotation system, and the value can be compared with the PPD of village B.

Thus, the further downstream the people migrated, the greater the infiltration of a monetary economy into the village could be seen, but the potential population density decreased. At the same time, the intensity of land use frequency increased, or the fallow period was shortened. However, in the case of village F, the intensity of swidden use decreased since there were almost no swiddens made from secondary forests. For pepper fields, if the soil remains fertile and the land is used permanently, the intensity will show a high value of 87.5. However, since soil fertility could not be maintained, and pepper fields will probably be abandoned and idle for a long period, the intensity must show a lower value.

## 15.4.2 Comparison of the Kenyah's Case with the General Rule

Table 15.2 shows these results in comparison with the general rule of intensification theory for swidden. The direction of the arrow shows the time elapsed for the general rule and the direction of migration for the case of the Kenyah .

A marked difference between the general rule and the Kenyah's case can be seen in the intensity of land use frequency (IN). For the general rule, the mechanism

**Table 15.2.** Comparison between the general rule and the case of the Kenyah villages

| Indices | | Stages of villages | | |
|---|---|---|---|---|
| | | Stage I | | Stage III |
| PPD | General | low | ⟶ | high |
| | The Kenyah | low | ⟵ | high |
| Labor intensity | General | saving | ⟶ | intensive |
| | The Kenyah | saving | ⟵ | intensive |
| Labor productivity | General | high | ⟶ | low |
| | The Kenyah | high | ⟵ | low |
| Land productivity | General | low | ⟶ | high |
| | The Kenyah[a] | low | ⟵ | high |
| IN | General | low | ⟶ | high |
| | The Kenyah | low | ← high ← | low |
| Social organization | General | underdeveloped | ⟶ | developed |
| | The Kenyah | partially revived | ←vanished← | developed |
| Infiltration of monetary economy | The Kenyah | much | ⟵ | little |
| Location | The Kenyah | suburban | ⟵ | interior |
| Swidden system | The Kenyah | unsustainable | ⟵ | sustainable |

[a]In the case of the Kenyah, land productivity means "land productivity in the year under review". Long-term land productivity of the villages of stage I and III are reversed, as shown later.

shows a direction of "increase in potential population density" to "intensification of land use frequency" to "labor intensification or more labor input" to "decrease in labor productivity, increase in land productivity in the year under review, and increase in long-term land productivity". However, in the Kenyah's case, it becomes "decrease in potential population density" to "intensification of land use frequency" to "labor dis-intensification or saving of labor" to "increase in labor productivity and decrease in land productivity in the year under review ".

Among these changes, the process of "decrease in potential population density" to "saving of labor" to "increase in labor productivity and decrease in land productivity in the year under review" is in accordance with the general rule. That is, in the downstream area with a low potential population density, the swidden practice becomes labor saving and the labor productivity increases, but since the absolute amount of labor input per land acreage is low, the land productivity in the year under review is decreased. In other words in the villages in remote areas, the PPD is high and the inhabitants are in a crowded state in certain areas of secondary forest and the swidden technique becomes labor intensive. As a result of the large amount of labor input, the labor productivity is lowered.

What is the theoretical basis for the realization of the apparently contradictory process of "decrease in potential population density " to "intensification of land use frequency" to "saving of labor" ?

Before giving this explanation, I would like to clarify the differences between the Kenyah's case and the general rule. Figure 15.4 shows the relationship between the PPD and the IN. A transition in the direction from upper left to lower right shows the direction of the transition with the passage of time for a certain village, which can be said to be the "general rule". On the other hand, in the Kenyah's case, a transition from the upper right to lower left (thick arrows) can be seen as a result of migration. Thus, in the Kenyah's case as well, unless migration occurs, the transition should be considered to occur with the passage of time (upper left to lower right) in each village following the general rule.

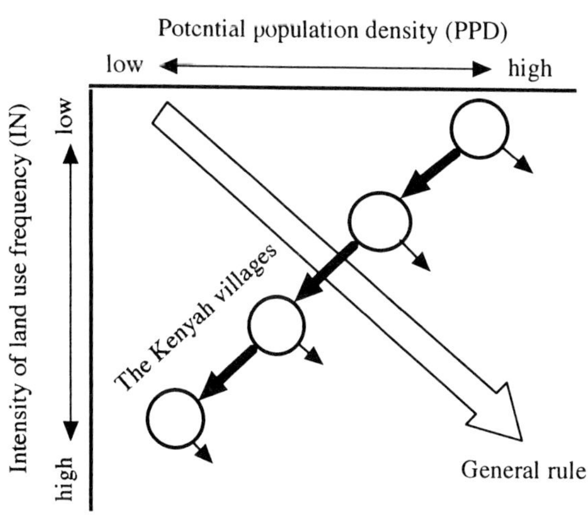


**Fig. 15.4.** Relationship between PPD and intensity of land use frequency (IN). The Kenyah villages (*open circles*) are connected each other through migration (*thick arrows*). The economic condition or infiltration of a monetary economy at the site of migration determines the vertical position of a village. Beginning at this position, the general rule comes into play, and the fallow period is shortened by the increase in population density as shown by the *narrower arrows*

On the other hand, the direction of the arrow in the general rule can be reversed. That is, when migrating to a lower population density area, land utilization becomes extensive or dis-intensified (Brookfield 1972, FAO Forestry Department 1985). This is because the high labor productivity of the extensive swidden agriculture becomes an incentive.

### 15.4.3 Why Does the IN Increase When the PPD Decreases?

As shown above, it is clear that the case of the migrating Kenyah does not follow the general rule. In spite of the decrease in PPD by migration, the fallow period is shortened. This is a point that cannot be explained by previous intensification theories. As mentioned already, this is considered to be due to the environment surrounding the swidden system.

First, the natural environment must be examined. The largest difference in the environment surrounding the Kenyah villages is site condition: the highland area or large river basin area. Another important difference is vegetation cover: the area is dominated by virgin forest, logged over forest, or secondary forest including forest fallow.

The most important factor that can be considered to cause such differences in the intensity of land use frequency would be soil fertility. Intensive land use is possible where the soil is fertile, but impossible where it is infertile. The soil fertility is high in the large river basin and low in the highland areas. Furthermore, soil fertility is highest in the virgin forest, followed by logged over forests and secondary forests of swidden land (Jordan, 1985). Thus, the same order applies to the intensity of land use frequency. If this explanation is correct, immediately after migration to a virgin forest area in the Mahakam basin, which has relatively fertile soil, the area of a village is not expanded so much, and swidden agriculture is repeated with short fallow periods. Then with the passage of years, due to a decrease in soil fertility, the fallow period is extended and agriculture becomes extensive. As long as they use a traditional fallow period, or as long as we hypothesize that the people will take steps toward the protection of environment, this theory may be correct. Since the vegetation recovers rapidly in a fertile area, the fallow period is short, but when the soil fertility is lowered, the recovery of vegetation is delayed, and the fallow period must be extended.

Thus, if their customary practice remains unchanged, it may be concluded that the reversal of the relationship between PPD and the IN would be due to the difference in the natural environment. However, it has already been reported that their customary practice has changed (Inoue, 1990). The further downstream the villagers migrate, the higher the frequency of land use as swidden is. Thus, soil fertility continues to decrease. Furthermore, although the soil fertility should be higher the further downstream, the single-year land productivity of upland rice is low.

In conclusion, the shortening of the fallow period can be considered to be due to the collapse of customary practice rather than to the site condition. Thus, it is clear that the difference in the natural environmental conditions cannot be the decisive factor responsible for the shortening of the fallow period accompanying migration.

Next, the socio-economic environment should be examined. The infiltration rate of a monetary economy expresses clearly the difference in the economic environment of each village. As already mentioned the infiltration of a monetary economy increases the further downstream the people migrate.

Furthermore, logging companies have concessions in areas to develop the forest region formally, and it is difficult in reality for the villagers to use a certain piece of land exclusively. That is, the use of the forest by the Kenyah overlaps with the legal rights of the logging companies, and there is no guarantee that they will be able to use the area permanently.

These conditions are considered to have motivated the Kenyah people to change their traditional swidden system, which had conserved the forest. This is why the IN increased (or shortening of fallow period) in spite of the decrease in the PPD.

### 15.4.4 Why is Labor Saved When Land Use Frequency is Intensified?

The final point to be examined is the process of "intensification of land use frequency" to "saving of labor". Here, the presumption proposed by Brookfield (1972) that the people prefer to achieve maximum income with minimum effort is applicable. First, the intensification of land use frequency clearly increases the long-term production output per unit acreage (Table 15.3). Here, "long-term land productivity (LLP) = land productivity in the year under review (LPY) / (fallow period + cultivation period) ". When actually calculated, it becomes the following equation, because the fallow period is unlimited in the case of swidden made from virgin forests:

$$LLP = PS \times LPY / (F+L) \quad (6)$$

The LLP of upland rice increases with the shortening in the fallow period further downstream. Furthermore, if cash crops such as pepper are introduced, the average

**Table 15.3.** Long-term land productivity

| Villages | Main crops | Average value of output per year over long term (Rupiah/hectare) |
|---|---|---|
| A | Upland rice | 27,840 |
| B | Upland rice | 76,320 |
| C | Upland rice | 65,440 |
| D | Upland rice | 98,520 |
| E | Upland rice + Soybean | 188,160 + 44,186 = 232,346 |
| F | Upland rice + Pepper[a] | 13,040 + 2,587,800 = 2,600,540 |

Annual average value of output per hectare of upland rice is calculated for the price at village F in January 1989. Therefore the figure can indicate the height of the long term land productivity among the villages.

[a]In the case of village F, pepper gardens are assumed to be utilized permanently.

annual output for the long term will be increased greatly. However, since this calculation is based on the presumption that soil fertility will be maintained for a long period, the actual value will be lower than this.

Thus productivity was increased by the shortening of the fallow period. But if the productivity increases, according to Brookfield's presumption, the people will try to lower the input of labor. This is the process of "intensification of land use frequency" to "saving of labor".

### 15.4.5 Summary of the Section

As shown in Fig. 15.5, the process of intensification of the swidden system of the Kenyah can be regarded as a combination of two processes. In process I, the land productivity of upland rice decreases even when the soil is fertile downstream. However, as shown in Process II, the long-term land productivity increases when the land use frequency is high even if the single-year land productivity is low and when some cash crops are introduced. However, at the same time, process II is directly connected to deforestation and forest degradation.

## 15.5 Conclusion

By applying the agricultural intensification theory to these studies on the shortening of the fallow period , the changes in the swidden system in the process of migration

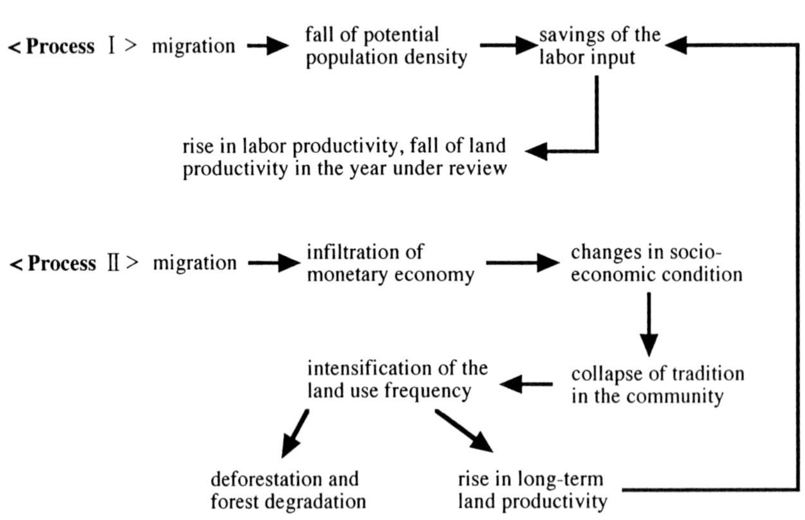


**Fig. 15.5.** Process of the intensification of a swidden system

can be explained not only as a correlation of geographical comparisons but also as a socio-economic causal relationship.

The Kenyah people, released from the potential population pressure as a result of migration from the Apo Kayan region, choose to reduce input of labor and increase labor productivity. In a society of labor extensive production, the need for developed labor organizations is decreased, and the organization develops toward the independence of each household. On the other hand, in the migrated areas, they confront the difference in the socio-economic environmental conditions, especially the high infiltration of a monetary economy. The change in the economy causes a change in the social structure such as the mutual support system. Furthermore, because the land can not be used exclusively due to the presence of logging companies, the customary land tenure system and rotation system of the swiddens changes.

The survival of the Kenyah was originally dependent on the stability of the surrounding resources. The Kenyah, however, are in the process of losing their character as forest conservators, even if the excessive harvest from nature which causes reduction and depletion of the resources must be avoided. In other words, this is a process of the development of individualism and also a process of the transition to a modern land tenure system. This is the motivation for the process of the "infiltration of monetary economy" to "intensification of land use". This is the force that overrules the usual process of "decrease in population pressure " to "extensive land use".

However, the intensification of land use is acceptable when the people show consideration toward conservation of soil fertility, but will be directly linked to the destruction of the environment if the customary practice is abandoned and measures are not taken to maintain soil fertility.

At last we should notice that the changes in the swidden system of the Kenyah are strongly affected by external factors such as large scale economic activities supported by development policy which could not be discussed here.

# References

Boserup E (1965) The conditions of agricultural growth. George Allen & Unwin Ltd, London

Brookfield HC (1972) Intensification and disintensification in Pacific agriculture. Pacific Viewpoint 15:30–48

Brown K, Pearce DW (eds) (1994) The causes of tropical deforestation — the economic and statistical analysis of factors giving rise to the loss of the tropical forests. UCL Press, London

Colfer CJP (1983) Change and indigenous agroforestry in East Kalimantan. Borneo Res Bull 15(1):3–21 and 15(2):69–87

Conklin H (1957) Hanunoo agriculture. FAO, Rome

Conklin H (1961) The study of shifting cultivation. Curr Anthropol 2(1):27–61

Dove M (1981) Swidden systems and their potential role in agricultural development — a case-study from Kalimantan. PRISMA, 21:81–100

FAO (1995) Forest resources assessment 1990 — Global synthesis (FAO Forestry Paper No.124). FAO, Rome

FAO (1997) State of the World's Forest 1997. FAO, Rome

FAO Forestry Department (1985) Changes in shifting cultivation in Africa. Unasylva 30 (150):40–50
Geertz C (1963) Agricultural involution. University of California Press, Berkeley
Gelder BV, O'Keefe P (1995) The new forester. Intermediate Technology Publications, London
Greenland DJ (1974) Evolution and development of different types of shifting cultivation (FAO Soils Bulletin No.24). FAO, Rome, pp 5–13
Hadi S, Hadi S, Hidayat R (1985) Indonesia. In: Swidden cultivation in Asia, Volume Three. UNESCO regional office for education in Asia and Pacific, Bankok, pp 74–149
Inoue M (1988) Socio-economic mechanism of deforestation in the tropics (in Japanese). Forest Econ 41(10):9–21
Inoue M (1990) Changes in the swidden system practiced by the Kenyah Dayak (in Japanese with English summary). J Southeast Asian Studies 28 (2):222–255
Inoue M, Lahjie AM (1990) Dynamics of swidden agriculture in East Kalimantan. Agroforestry Syst 12:269–284
Jordan CF (1985) Nutrient cycling in tropical forest ecosystems. Wiley, Chichester, pp 88–146
Kano H (1982) Landless households in Indonesia. In: Takigawa T (ed) Low income classes of the rural area in Southeast Asia (in Japanese). Institute for Asian Economic Development, Tokyo, pp 77–114
Kartawinata K, Jessup T, Vayda AP, Colfer CJP (1984) The impact of development on interactions between people and forests in East Kalimantan — a comparison of two areas of Kenyah Dayak settlement. Environmentalist 4 (Suppl 7):87–95
Kunstadter P (1978) Subsistence agricultural cconomies of Lua' and Karen hill farmers, Mae Saring district, Northern Thailand. In: Kunstadter P et al. (ed) Farmers in the forest. East-West Center, Hawaii, pp 74–133
Peluso NL, Vandergeest P, Potter L (1995) Social aspects of forestry in Southeast Asia — a review of postwar trends in the scholarly literature. J Southeast Asia Studies 26(1):196–218
Ruthenberg H (1976) Farming systems in the tropics. Clarendon, Oxford, pp 1–18
Sasaki T (1970) Swidden agriculture in the tropics (in Japanese). Kokonshoin, Tokyo
Spencer JS (1967) Shifting cultivation in Southeastern Asia. University of California Press, Berkeley
Sponsel LE, Headland TN, Bailey RC (eds) (1996) Tropical deforestation — the human dimension. Columbia University Press, New York
Takigawa T, Saito H (eds) (1968) Land tenure and rural social structure in Asia. Institute for Asian Economic Development, Tokyo
Vasey DE (1979) Population and agricultural intensity in the humid tropics. Hum Ecol 7 (3):269–283
Vayda AP (1981) Research in East Kalimantan on interaction between people and forests — a preliminary report. Borneo Res Bull 13(1):3–15
Vayda AP, Pierce CJ, Brotokusumo M (1980) Interactions between people and forest in East Kalimantan. Impact of science on society 30(3):179–190
Watanabe T (1986) Development economics (in Japanese). Nihonhyouronsha, Tokyo

# 16 Tropical Secondary Forest and Its Succession Following Traditional Slash-and-Burn Agriculture in Mencimai, East Kalimantan

Yasuyuki Okimori and Paulus Matius

## 16.1 Introduction

Secondary forests in Southeast Asian regions have been extending rapidly to constitute various types of forest. Recently, the principal disturbances due to human impact, such as mechanized commercial logging, slash-and-burn agriculture, and the development of large-scale estates of cash crops, have accelerated the formation of secondary forests in the tropics (Kartawinata et al. 1989; Chapter 1, this volume). The structures of these secondary forests are dependent on the frequency and intensity with which disturbances occurred in previous forests.

We focused on the recovery potential of tropical rainforest following human disturbances, such as logging impact (Chapter 18, this volume) and slash-and-burn agriculture (this chapter), to make use of practical rehabilitation techniques for degraded forests. There are two major systems of slash-and-burn agriculture in East Kalimantan (see Chapter 15, this volume). One is traditionally carried out by the local people who have been living their life in a close relationship with the forests. The other is carried out intensively by people who have immigrated from outside the area since the 1960s to cultivate cash crops (Chapters 1 and 17, this volume). It is important to distinguish the two systems, i.e. traditional and recent, in order to understand the processes of vegetational changes caused by slash-and-burn agriculture. Recently, *Imperata cylindrica* grasslands have spread widely after the recent slash-and-burn agricultural activities of newly immigrated people in East Kalimantan (Chapter 17, this volume), whereas the traditional slash-and-burn cultivation system carried out by the local people can be effective in sustaining the recovery potential of secondary forest (Kartawinata et al. 1989, Inoue and Lahjie 1990).

Brown and Lugo (1990) reviewed previous publications on tropical secondary forests from the viewpoints of land-use values, species richness, vegetation structure, biomass, nutrient cycling and relations with people. However, few investigations have been performed on secondary forests of different ages to predict the process of succession. This chapter reveals the recovery processes from the initial succession stage to near-primary forests in fallow lands after traditional slash-and-burn agriculture, by measuring the forest structures in a series of secondary forests with

Ecological Studies, Vol. 140
Guhardja et al. (eds.) Rainforest Ecosystems of East Kalimantan: El Niño, Drought, Fire and Human Impacts

different ages in Mencimai village in East Kalimantan in 1990. The successional process after intensive slash-and-burn agriculture by newly immigrated people is discussed in Chapter 17 in this volume.

## 16.2 Secondary Forests in Mencimai Village

### 16.2.1 Mencimai Village

Mencimai village lies in Barong Tongkok, Kutai regency, East Kalimantan (see Fig. 1.2 in Chapter 1, this volume). The village is about 230 km upstream on the Mahakam River from Samarinda, the capital of East Kalimantan Province. Although the major geology of East Kalimantan is formed of Tertiary sedimentary rocks, the geology of Barong Tongkok region is formed of Neogene volcanic rock, of which distribution is limited to this region in Kalimantan (Chapter 8, this volume). Sixty families of the Benuaq people, a tribe of Dayak, lived at Mencimai village, and were still performing traditional slash-and-burn agriculture to produce upland rice in 1990.

The forests around the village had been converted rapidly into paddy and other crop fields, because Javanese people had transmigrated since the 1960s and established a new village bordering the Mencimai village. Benuaq villagers had also been converting the fallow fields to plantations of rubber, one of the important cash crops. The primary mixed dipterocarp forests and old secondary forests remained in a limited area in this village when we observed them in 1990. Thus, the forest area in this village consisted of patches of various vegetation types in a mosaic, which involved primary forests, secondary forests with different canopy heights, shrubs, rubber plantations and fruit garden forests.

### 16.2.2 Ages of Secondary Forests and Analysis of Forest Structure

It is important to know the successional age of the fallow fields in order to ascertain vegetation dynamics. The Benuaq people know how the slash-and-burn events influence the time scale of the subsequent development of each area of fallow land and they use this knowledge to plan their next slash-and-burn event. They classify the fallow lands into the following categories according to differences in the successional stages: (1) the initial stage of secondary growth (1–3 years) is called *urat* and involves shrubs and herbs; (2) the young secondary forest which follows *urat* is called *baling batakng* (5–15 years) and consists of pioneer tree species; (3) the old secondary forest, aged 20–30 years, is called *batakng*; and (4) the older stage forest which is called *bengkal tua*, particularly when over 45 years old. *Asli bengkal* is the name given to primary forest. We repeatedly interviewed particular farmers who had good knowledge of the history of each secondary forest.

We set up five study plots involving different ages and developments of stand structures: (1) 2-year-old shrub (termed the 2-year-old shrub stand); (2) a young secondary forest stand (10–15-year-old, termed the 15-year-old stand): (3) an old secondary forest stand (25–30-year-old, termed the 30-year-old stand); (4) another old secondary forest stand (35–40-year-old, termed the 40-year-old stand); and (5) the oldest secondary forest stand (termed the 70-year-old stand, the age of which was given by the farmers). All tall trees over 10 cm in diameter at breast height (DBH) were measured in each plot, and described as tree stage (DBH ≥ 10 cm), pole stage (10 cm > DBH ≥ 2 cm), sapling (DBH < 2 cm or the top canopy height >l.5 m), and seedling (the top canopy height ≤1.5 m).

These five stands were situated along a stream within a 650-m length range and there were no marked differences in topography, although there were slight variations in soil texture, based on clay content, between the plots (Chapter 7, this volume). Soil conditions differed with the development of secondary forests from fallow. Nutrient content in the soils was decreased by slash-and-burn agriculture, and afterwards it gradually increased with the development of vegetation from the fallow lands. However, the recovery speed of soil nutrients with time might differ between the plots, because of the differences in soil textures (Chapter 7, this volume). Light intensity at 2.0 m above the ground surface relative to full sunlight at the open site (relative light intensity, RI) was about 55%, 9%, 3%, 2%, and 1% in the 2-year-, 15-year-, 30-year-, 40-year-, and 70-year-old stands, respectively.

The community structure of woody vegetation in each secondary forest after fallow was described using an importance value (IV) (Curtis and Macintosh 1951) with modification (IV*). The IV* is derived by the following equation:

$$IV^*=(RF+RD+RC)/3 \quad (1)$$

where RF, RD, and RC are the relative frequency, the relative density, and relative coverage of a species to the total of all species, respectively. RFs represent the number of quadrats where a species was present. RCs was determined based on basal area for tree and pole stages, and on crown area for sapling and seedling stages.

## 16.3 Changes in Forest Structure and Species Richness

The frequency distribution of the DBH of trees in each stand is shown in Fig. 16.1. In the 15-year-old stand, most of the trees were concentrated within the diameters of 10–15 cm. The frequency of large trees increased with age in the older secondary forests. Basal areas of the 15-, 30-, 40- and 70-year-old stands were 11.4, 18.0, 17.0 and 33.3 $m^2ha^{-1}$, respectively.

Regarding the variation in species richness (Fig. 16.2), the number of species at the tree and pole stages increased with the age of the forest stands. For example, there were 35 species at the tree stage in the 15-year-old stand and 95 species in the 70-year-old stand. Similarly, 27 species were at the pole stage in the 2-year-old shrubs, which increased to 175 species in the 70-year-old stand. The 70-year-old stand had the highest level of species richness of all the tree stages.

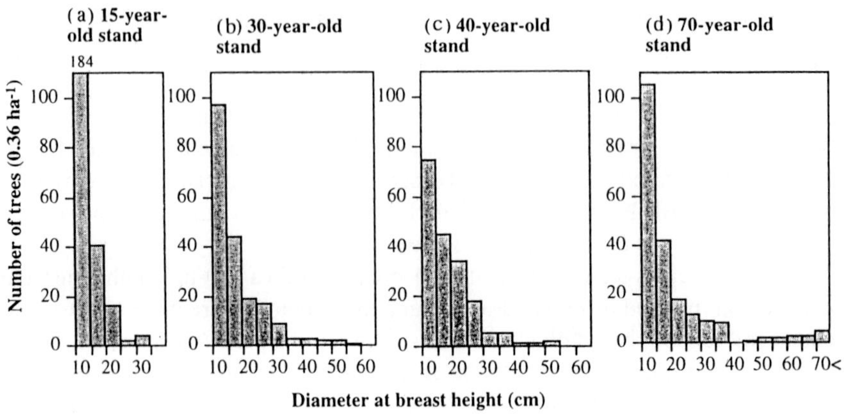


**Fig. 16.1.** Diameter distributions of trees (DBH ≥ 10 cm) in secondary forests of different ages following slash-and-burn agriculture at Mencimai, East Kalimantan

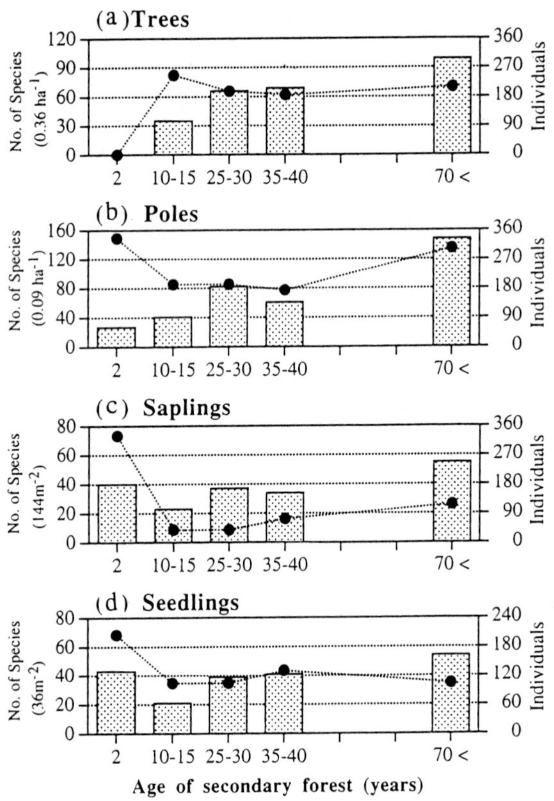


**Fig. 16.2a–d.** Number of species (*bars*) and densities (*filled circles*) of trees, poles, saplings and seedlings in secondary shrubs and forests of different ages following slash-and-burn agriculture at Mencimai, East Kalimantan. Tree stages are classified into four categories: **(a)** trees (DBH ≥10 cm), **(b)** poles (2 cm ≤ DBH < 10 cm), **(c)** saplings (DBH < 2 cm and H ≥ 1.5 m), **(d)** seedlings (H < 1.5 m)

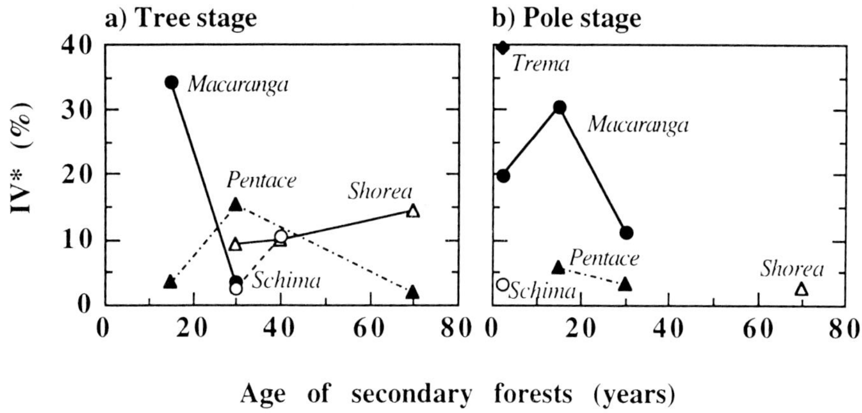


**Fig. 16.3a,b.** Changes in dominant species in terms of IV* (importance value with modification) following slash-and-burn agriculture at Mencimai, East Kalimantan. **a)** tree (DBH ≥ 10 cm) stages, **b)** pole (2 cm ≤ DBH < 10 cm) stages. The change in IV* of the most dominant species at each age of the stands at tree stage are shown. The species are as follows: *Macaranga*: *Macaranga gigantea*; *Pentace*:*Pentace laxiflora*; *Schima*: *Schima wallichii*; *Shorea*: *Shorea parvifolia*; *Trema*: *Trema orientalis*

Patterns in secondary succession are complex, because they are influenced by many factors, such as soil conditions, intensity and frequency of slash-and-burn agriculture, cultivation periods and conditions of the surrounding forests. One of the typical patterns in mixed dipterocarp forests was observed in Mencimai forest. Figure 16.3 shows the changes in the dominant species in terms of IV* at the tree and pole stages. Pioneer shrub trees such as *Trema orientalis* (Ulmaceae) and *Melastoma malabathricum* (Melastomataceae) grew gregariously about 2 years after cessation of cultivation. The appearance of some climax tree species and the deaths of pioneer trees were found at about 30 years after slash-and-burn agriculture. The IV* of Euphorbiaceae decreased 15 years after fallow, while that of Dipterocarpaceae began to increase afterwards. The 30- to 40-year-old secondary forest consisted of primary forest species in a canopy layer, such as *Shorea parvifolia* (Dipterocarpaceae) and *Pentace laxiflora* (Tiliaceae), mixed with the remainders of the pioneer tree species, thus constituting a more diversely structured forest. The oldest secondary forest, at 70 years, had already developed to a near-primary forest with increased species richness and a well-stratified canopy.

## 16.3.1 Two-Year-Old Shrubs

The two-year-old shrubs stand following slash-and-burn agriculture was dominated by pioneer plant species, including trees, herbs, ferns and vines, with short and dense canopies (Fig. 16.4). The woody plants already covered herbs and ferns. The most predominant species among the woody plants at the pole stage was *Trema orientalis* with 39.2% of IV*, followed by *Macaranga gigantea* (Euphorbiaceae) with 19.9%.

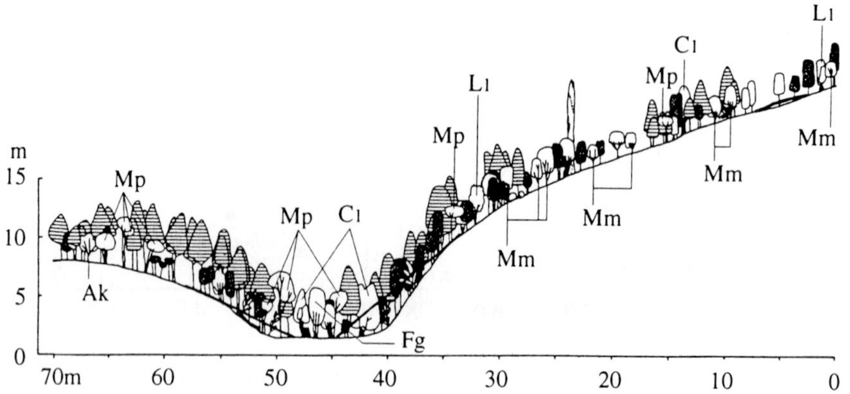


**Fig. 16.4.** Profile diagram of the two-year-old shrub stand following slash-and-burn agriculture at Mencimai. Plot area 70 m × 5 m; all trees over 5 m in height are shown. The major species are *Trema orientalis* (*horizontally hatched*) and *Macaranga*, including *M. gigantea*, *M. triloba,* and *M. trichocarpa* (*gray*). Abbreviations for other important species are listed in appendix 16.1

These species are very common after slash-and-burn events. The life-span of the *Trema orientalis* shrub is only a few years, and shrubs consisting of the same age trees would die simultaneously (Okimori, personal observation in Jambi, Sumatra island). The canopy initially occupied by *T. orientalis* would be replaced by some tall pioneer species such as *Macaranga* spp, as follows.

## 16.3.2 Fifteen-Year-Old Stand

Pioneer species, *Macaranga gigantea* and *M. triloba,* were predominant in the 15-year-old stand, with high IV*'s of 34.2% and 24.9% at the tree stage, respectively. The structure of this forest was so simple that the canopy was a 10–20-m tall monolayer (Fig. 16.5). The pole stage also consisted mainly of *Macaranga gigantea* (30.5% of IV*) and *M. triloba* (16.3% of IV*). However, these two pioneer tree species did not appear in the sapling and seedling stages, suggesting their poor recruitment faculties. *Pentace laxiflora* appeared in the tree, pole and sapling stages. This species is common in mixed dipterocarp forests, especially on clay-rich soils (Ashton 1988). Thus, climax tree species were already found in the tree stage in the 15-year-old stand.

## 16.3.3 Thirty-Year-Old Stand

The 30-year-old stand was characterized by no markedly predominant species, by an increase in the number of species (Fig. 16.2) and by the development of a stratified canopy (Fig. 16.6). Both climax and pioneer species were found in the tree and pole stages. Two climax species, *Pentace laxiflora* and *Shorea parvifolia*, were predomi-

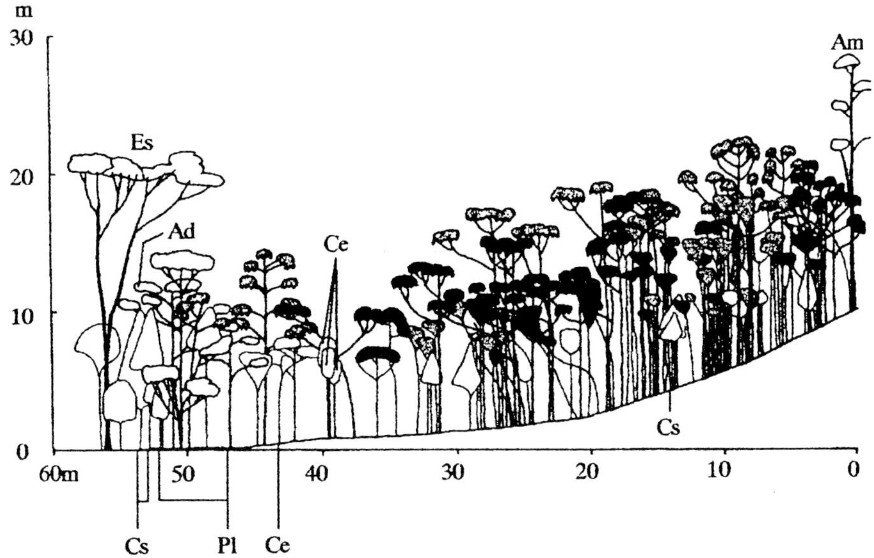


**Fig. 16.5.** Profile diagram of 10- to 15-year-old secondary forest following slash-and-burn agriculture at Mencimai. Plot area 60 m × 10 m; all trees over 5 m in height are shown. The major species are *Macaranga triloba* (*stippled*) and *M. gigantea* (*filled*). Abbreviations for other important species are listed in appendix 16.1

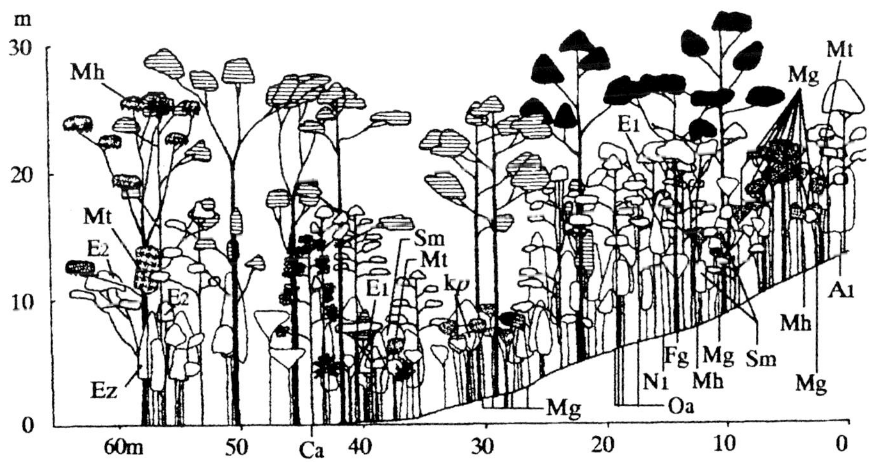


**Fig. 16.6.** Profile diagram of 25- to 30-year-old secondary forest following slash-and-burn agriculture at Mencimai. Plot area 60 m × 10 m; all trees over 5 m in height are shown. The major species are *Macaranga* (*stippled*), *Pentace laxiflora* (*horizontally hatched*), and *Shorea parvifolia* (*filled*). Abbreviations for other important species are listed in appendix 16.1

nant with 15.5% and 9.3% of IV* at the tree stage, respectively, and they appeared in the top canopy layer (Fig. 16.5). Some pioneer species had high priority IV* at the tree stage, 14.1% of IV* in total for the three *Macaranga* species and 2.7% of IV* for *Campnosperma auriculatum* (Anacardiaceae). The climax species, *Pentace laxiflora* and *Shorea parvifolia*, seemed to have replaced the pioneer *Macaranga* spp. as the major species of the canopy layer.

Many seedlings of *Pentace laxiflora* were recruited to the larger stages, with 32.0% of IV* at the seedling stage. Good regeneration of *Pentace* spp. as dense carpets of seedlings had been reported in Sarawak, North Borneo (Whitmore 1984). Saplings and seedlings of *Macaranga* spp. were almost absent. In Lempake near Samarinda, *Macaranga gigantea, M. pruinosa, Artocarpus tamaran* (Moraceae), and some other late secondary species prevailed in 35-year-old secondary forest following pepper plantation (Riswan and Abdulhadi 1992). However, in their study site, 60% of the total tree species were already primary species, and almost all of the saplings and seedlings were primary species. Thus, secondary tree species are gradually replaced by primary species after 25–30 years of fallow.

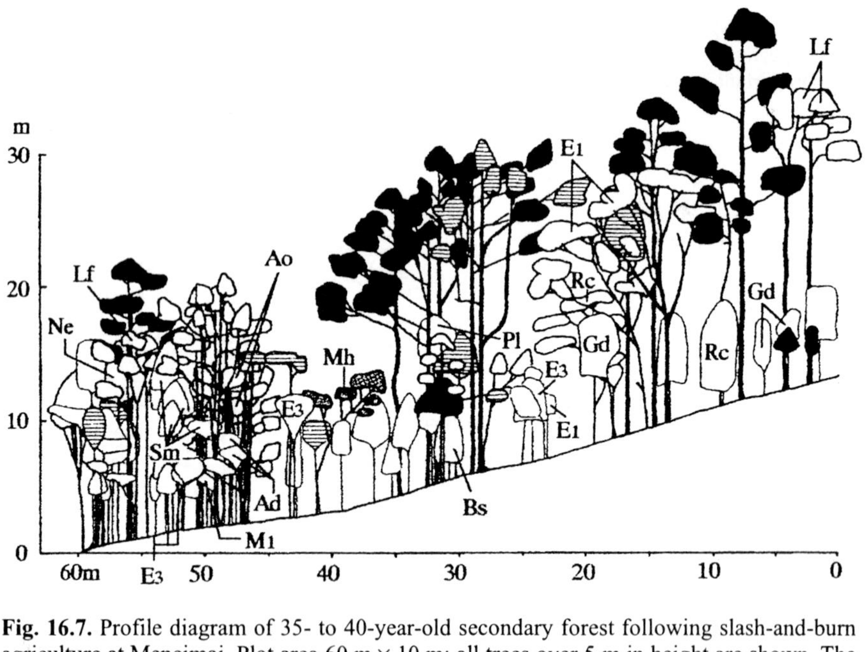


**Fig. 16.7.** Profile diagram of 35- to 40-year-old secondary forest following slash-and-burn agriculture at Mencimai. Plot area 60 m × 10 m; all trees over 5 m in height are shown. The major species are *Shorea parvifolia* (*filled*) and *Schima wallichii* (*horizontally hatched*). Abbreviations for other important species are listed in appendix 16.1

### 16.3.4 Forty-Year-Old Stand

Stand structure, and the number of species and stems in this stand were similar to those of the 30-year-old stand, but a slight difference was found in species composition (Fig. 16.7). The main canopy tree species were *Schima wallichii* (Theaceae), *Shorea parvifolia* and *Nephelium eriopetalum* (Sapindaceae) with 10.5%, 9.8% and 5.3% of IV* in the 40-year-old stand, respectively. The pioneer species of *Macaranga* spp. had retreated in this stand. *Eugenia* sp.1 (Myrtaceae) and *Millettia sericea* (Fabaceae) appeared frequently, rather than *Schima wallichii* and *Shorea parvifolia*, at the pole, sapling, and seedling stages.

### 16.3.5 Seventy-Year-Old Stand

The largest trees in the 70-year-old stand reached 40 m high, and the canopy layer was divided into three strata (Fig. 16.8). The heights of the first, second and third strata were 25–40 m, 10–20 m and less than 10 m, respectively. There were no emer-

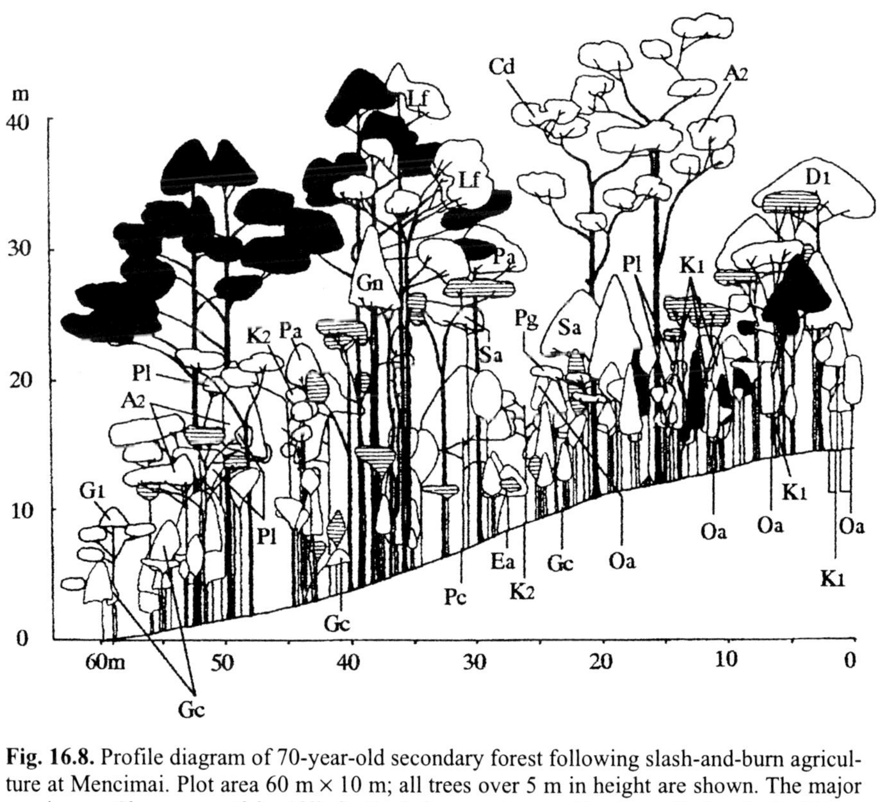


**Fig. 16.8.** Profile diagram of 70-year-old secondary forest following slash-and-burn agriculture at Mencimai. Plot area 60 m × 10 m; all trees over 5 m in height are shown. The major species are *Shorea parvifolia* (*filled*), *Nephelium encinatum* (*horizontally hatched*). Abbreviations for other important species are listed in appendix 16.1

gent trees in this stand. The remarkable characteristic of this stand was the predominance of dipterocarp species. The values of IV* for *Shorea parvifolia* and *S. leprosula* at tree stage were 14.3% and 8.4%, respectively, and these two species occupied 49.3% of the relative basal area. The number of species was highest among the studied stands (Fig. 16.2). This forest stand had surely advanced to a mature phase of the succession process, though there were still no emergent trees.

## 16.4 Dynamics of Pioneer and Sprouting Trees

Herbs, ferns and vines start to grow immediately after cessation of cultivation. Mackie et al. (1986) described the rapid re-establishment of woody plants in Apo Kayan, East Kalimantan. Herbs colonized 1-year-old fallow with 67% of coverage, though woody vegetation grew rapidly and dominated the field by the 4th year. In Mencimai, pioneer trees had overwhelmed herbaceous plants 2 years after the cessation of cultivation.

Whitmore (1984) divided the pioneer trees into two groups, short-lived pioneers (about 15–30 years) and long-lived pioneers (more than 80 years). The genus *Campnosperma* is listed as a long-lived pioneer. *Campnosperma auriculatum* found as the largest tree in 30-year-old or 40-year-old stands was a typical long-lived pioneer tree species. On the other hand, most of the *Trema* and *Macaranga* species, some *Mallotus* species, and *Commersonia bartramia* are listed as short-lived pioneer species. A large *Macaranga hypoleuca* (37.9 cm in DBH) was found in the 30-year-old stand though there were no large ones in the 40-year-old stand. Although the flowering and fruiting of *M. gigantea* and *M. triloba* were observed at the 15-year-old stand in May 1990, there were no seedlings of *Macaranga* species in this stand, probably due to the low light intensities at the forest floor (9.1% RI in the understorey). Thus, *Macaranga* species could have a life-span up to about 30 years. This characteristic resulted in the simple vertical structure of the forest, with a high density at the tree stage, a low density of juveniles, and low richness of species (Fig. 16.2).

Sprouting from cut stumps was often encountered in the regeneration process of the fallow land. The sprouting stems accounted for 3.8% of trees in the 70-year-old stand and 31.7% in the 40-year-old stand (Table 16.1). Mackie et al. (1986) reported

**Table 16.1.** Number of sprouting stems in the secondary forests following slash-and-burn agriculture in Mencimai, East Kalimantan

| | Age of forest stands (years) | | | | |
|---|---|---|---|---|---|
| | 2[a] | 15[b] | 30[b] | 40[b] | 70[b] |
| Total number of stems | 335 | 247 | 197 | 186 | 210 |
| Number of sprouting stems | 59 | 21 | 33 | 59 | 8 |
| % of sprouting stems | 18 | 9 | 17 | 32 | 4 |

[a]For poles ($2 \leq$ DBH < 10 cm) in 0.09 ha plot.
[b]For trees (DBH $\geq$ 10 cm) in 0.36 ha plot.

that 43 tree species could sprout from stumps and they accounted for 20% of total basal area in a 17-year-old secondary forest following slash-and-burn agriculture in East Kalimantan. Some tree species, *Artocarpus* (Moraceae), *Nephelium* (Sapindaceae) and *Schima* (Theaceae), could sprout from cut and burnt stumps. *Schima wallichii* has a thick, fire-resistant bark. It easily sprouts to survive successfully and gradually extends to make a fire-climax formation (Symington 1943). The faculty of sprouting would contribute to the appearance of this species for a long time from 2-year-old to 40-year-old stands as shown in Fig. 16.3.

## 16.5 Regeneration Characteristics of *Shorea parvifolia*

It is interesting that *Shorea parvifolia*, a climax dipterocarp, occurred frequently and grew gregariously in the older stands more than 30 years after fallow. This species belonging to the light-red meranti group is one of the canopy tree species in primary forests in Malaysia, Sumatra, Borneo and the intervening islands (Symington 1943). Although it is commonly considered that dipterocarp species do not easily regenerate in degraded forests, *S. parvifolia can* easily regenerate through the successional process after fallow.

The first reason for this is the existence of mother trees in and around the fallow lands. Since the traditional farmers usually clear a narrow forest stand covering only a few hectares, many mother trees succeeded in surviving. Actually, there were some large trees of *Shorea parvifolia* in residual forests around the 40-year-old and 70-year-old plots. Also, the irregular shape of the cultivated land provides a longer periphery than regular-shaped land with the same area, which may increase the possibility of seed dispersal into the cultivated lands (Kartawinata et al. 1989). Furthermore, *Shorea parvifolia* were not harvested by local sawmillers because the wood was too light to use. The second reason is that this species is a light-demanding and fast growing dipterocarp species (Symington 1943). Many seedlings and saplings of *Shorea parvifolia* were found in sites opened by logging operations.

The dipterocarp trees come into flower simultaneously in an irregular cycle lasting 3 to 7 years and bear mass fruits (Burgess 1972, Cockburn 1975). We observed gregarious flowering in the forests in Mencimai in 1990. If a great number of seeds are provided at the right time in the initial stage of secondary succession, a light-demanding species such as *Shorea parvifolia* gets the opportunity to establish with the pioneer species. Ashton (1964) also mentioned the good regeneration of *Shorea parvifolia* in Brunei on old fallow fields where seeds were provided from the edge of the forests.

## 16.6 Conclusion

In the fallow of traditional slash-and-burn agriculture, pioneer species such as *Macaranga* spp. were predominant during at least 15 years following cessation of cultivation. However, primary species, such as *Shorea parvifolia* and *Pentace laxiflora*, might replace *Macaranga* spp. in the 30-year-old forests. High species richness and development of multi-strata canopy were characterized in the secondary forests more than 30 years after fallow. Both *Shorea parvifolia* and *Pentace laxiflora* were found in the secondary forests after cessation of slash-and-burn agriculture in Mencimai forests and *Macaranga* spp. clearly retreated in the 40-year-old forest. After 70 years of fallow the dipterocarp species became predominant, however, there were still no emergent trees.

For natural recovery to proceed after slash-and-burn agriculture, at least two

**Appendix 16.1.** Abbreviation list for Figures16.4–8.

| Abbrev. | Species Name | Abbrev. | Species Name |
|---|---|---|---|
| Am | *Adina minutiflora* | Gc | *Gymnacranthera contracta* |
| A2 | *Alseodaphne* sp. | K2 | *Knema* sp. |
| Ad | *Artocarpus dadah* | K1 | *Koompassia malaccensis* |
| Ak | *Artocarpus kemando* | kp | kiaq pangir (local name) |
| Ao | *Artocarpus odoratissimus* | Lf | *Litsea firma* |
| A1 | *Artocarpus* sp. | L1 | *Litsea* sp. |
| Bs | *Baccaurea stipulata* | Mg | *Macaranga gigantea* |
| Ca | *Campnosperma auriculatum* | Mh | *Macaranga hypoleuca* |
| Cd | *Canarium dichotomum* | Mt | *Macaranga triloba* |
| C1 | *Castanopsis* sp. | Mp | *Mallotus paniculatus* |
| Ce | *Clausena excavata* | Mm | *Melastoma malabathricum* |
| Cs | *Cratoxylum sumatranum* | M1 | *Millettia* sp. |
| D1 | *Dehaasia* sp. | Ne | *Nephelium eriopetalum* |
| Es | *Elaeocarpus stipularis* | N1 | *Nephelium* sp. |
| E1 | *Elaeocarpus* sp. 1 | Oa | *Ochanostachys amentacea* |
| E2 | *Elaeocarpus* sp. 2 | Pg | *Palaquium gutta* |
| Ea | *Eugenia* cf. *argyrocalyx* | Pl | *Pentace laxiflora* |
| E3 | *Eugenia* sp. | Pc | *Polyalthia curtisii* |
| Ez | *Eusideroxylon zwageri* | Pa | *Prunus arborea* |
| Fg | *Ficus geocarpa* | Rc | *Rhodamnia cinerea* |
| G1 | *Garcinia* sp. | Sa | *Santiria apiculata* |
| Gn | *Gironniera nervosa* | Sm | *Scaphium macropodum* |
| Gd | *Guioa diplopetala* | | |

essential criteria must be met. Firstly, the cultivated area should be limited to a small area of a few hectares, because of the importance of the presence of mother (seed-bearing) trees in the neighboring old-secondary or primary forests. Secondly, the fallow period should be extended to about 30–40 years, which may allow the development of a forest structure with high species richness and a multi-strata canopy. If we always follow these essential criteria in slash-and-burn agriculture processes, natural regeneration with dipterocarp species would progress smoothly in the tropics of Southeast Asia.

## References


Ashton PS (1964) A manual of the dipterocarp trees of Brunei State. Oxford University Press, pp 206–207

Ashton PS (1988) Manual of the non-dipterocarp trees of Sarawak, Vol.2. Dewan Bahasa dan Perpustaka, Kuala Lumpur, pp 449–454

Brown S, Lugo AE (1990) Tropical secondary forests. J Trop Ecol 6:1–32

Burgess PF (1972) The phenology of dipterocarps, Malay For 35:103–123

Cockburn PF (1975) Phenology of dipterocarps in Sabah. Malay For 38:160–170

Curtis JT, Macintosh RP (1951) An upland forest continuum in the prairie-forest border region of Wisconsin. Ecol 32:476–496

Inoue M, Lahjie AM (1990) Dynamics of swidden agriculture in East Kalimantan. Agroforestry Systems 12:269–284

Kartawinata K, Jessup TC, Vayda AP (1989) Exploitation in Southeast Asia. In: Lieth H, Werger MJA (eds) Tropical rainforest ecosystems. Elsevier , Amsterdam, pp 591–610

Mackie C, Jessup TC, Vayda AP, Kartawinata K (1986) Shifting cultivation and patch dynamics in an upland forest in East Kalimantan. In: UNESCO-MAB Regional workshop on impact of man's activities on upland forest ecosystem. Serdang, Malaysia, February, 1986, pp 486–518

Riswan S, Abdulhadi R (1992) Succession after disturbance of lowland mixed dipterocarp forest by shifting agriculture in East Kalimantan, Indonesia. In: Goldammer JG (ed) Tropical forests in transition. Birkhauser, Basel, pp 77–84

Symington CF (1943) Forester's manual of dipterocarps. Malayan Forest Records 16, Penerbit University Malaya, Kuala Lumpur

Whitmore TC (1984) Tropical rainforests of the Far East (2nd edn). Clarendon, Oxford, pp 83–91

# 17 The Role of Slash-and-Burn Agriculture in Transforming Dipterocarp Forest into *Imperata* Grassland

Yoshiyuki Kiyono and Hastaniah

## 17.1 Introduction

Slash-and burn agriculture is one of the major causes of forest degradation in the tropics and is widespread in Kalimantan (Spencer 1966, Kartawinata et al. 1981). In the interior of East Kalimantan, some secondary forests have returned to conditions similar to primary dipterocarp forest after 70 years in fallow (Chapter 16, this volume). In the same region, however, fallow forests of predominately pyrophytic tree species have developed simultaneously, as a result of repeated slash-and-burn agriculture. This is swidden agriculture in the narrow sense (Whitten et al. 1987). As long as the area remains fallow, this type of forest regenerates continuously (Kiyono and Hastaniah 1996, 1997). In contrast, logged-over dipterocarp forests degenerate to different communities after slash-and-burn use. The vegetation sometimes contains *Imperata cylindrica* grass. *Imperata* grassland is a degenerate plant community found worldwide in the tropics; large areas are found alongside the road between Samarinda and Balikpapan. It is widely believed that repeated slash-and-burn agriculture converts forest into *Imperata* grassland. According to our observations, however, slash-and-burn agriculture itself does not seem to allow *Imperata* to form grassland, even when it is practiced with a short fallow period (about five years). Why then do grasslands form in slash-and-burn areas?

Many studies on farming systems and the social aspects of slash-and-burn agriculture in Kalimantan have been published (e.g., Jessup 1981, Dove 1983, Vayda and Sahur 1985, 1996, Fraser 1998). However, little attention has been given to secondary plant succession in slash-and-burn fallow land. The function of slash-and-burn agriculture in the transformation of forest to *Imperata* grassland must be clarified. In this chapter, we will discuss the factors that seem to govern secondary plant succession after the slash-and-burn use of logged-over dipterocarp forest, and try to clarify the processes of subsequent grassland formation and plant succession. The discussion is based on a vegetation survey and on information obtained by interviewing farmers in and around Bukit Soeharto, which is near the mid-point of the Balikpapan-Samarinda road, in East Kalimantan.

Ecological Studies, Vol. 140
Guhardja et al. (eds.) Rainforest Ecosystems of East Kalimantan: El Niño, Drought, Fire and Human Impacts

## 17.2 Transformation of Dipterocarp Forests into *Imperata* Grasslands

In lowland Kalimantan, commercial logging has harvested mainly red *meranti*, one of the *Shorea* group, using a selective felling system in the primary dipterocarp-*Eusideroxylon zwageri* forest, since the late 1960's (Kartawinata et al. 1981). In Bukit Soeharto and its vicinities, intensive commercial logging was practiced in the 1970's. After the commercial logging, farmers migrated into the logged-over forest from other islands, including Sulawesi and Java. The Banjar, Kutai, and Dayaks, indigenous peoples of Kalimantan, also migrated from the south and interior of the island. They started slash-and-burn farming alongside the newly established roadways (Kiyono and Hastaniah 1996). The completion of the highway between Samarinda and Balikpapan in 1976 attracted settlers who planned to develop plantations of perennial crops after harvesting upland rice (Vayda and Sahur 1985). The farmers usually fell the forests in May–June and burn them in July–September. Upland rice is sown in October and harvested the following February. The migrant communities usually cultivate upland rice for a one- or two-year period. The scrub that grows on the fallowland is often converted into plantations of pepper (*Piper nigrum*), rambutan (*Nephelium lappaceum*), cloves (*Syzygium aromaticum*), and other perennial cash crops.

Forest fire is another major agent that brings about the destruction and degeneration of forest. Frequent wildfires resulted from the increased farming activity. Drought and fires in 1982–83, 1987, 1991, 1994, 1997–98 and other years had great impact on the forest in the region (Chapter 3, this volume).

As a result of the combination of these human and natural factors, the thick primary forests that once covered the area have been degraded and destroyed at a rapid rate since the late 1960's, to be replaced by a vast mosaic of grassland and degraded secondary forest.

### 17.2.1 Secondary Plant Succession after Slash-and-Burn Agriculture

A vegetation survey was taken on 33 stands consisting of logged-over and burned forests and secondary vegetation in and around Bukit Soeharto, in 1993. The results revealed that the structure and composition of the plant communities changed drastically with the influence of fires and slash-and-burn practices. In the stands that were affected by fire in 1982–83, but which had never been subject to slash-and-burn, the predominant tree species was *Macaranga gigantea*, a pioneer tree up to 30 m tall (top left in Fig. 17.1). These stands also contained dipterocarp trees that survived the forest fires. Fourteen out of 33 forest stands were first cleared for cultivation 4 to 7 years after the 1982–83 fires. Of the 14 stands, 9 were cleared for a second cultivation after a 2 to 6 year fallow period. *Macaranga gigantea* rarely regenerates, because it is very vulnerable to felling and burning (Table 17.1). After the first cropping period, instead of *Macaranga gigantea*, small-tree pioneer species, such as *Macaranga trichocarpa*, *Macaranga tanarius*, *Mallotus paniculatus*, *Trema orientalis*, and *Homalanthus populneus* germinate and establish, and can grow to a height of about 10 m (Fig. 17.2, middle left in Fig. 17.1). Since there are few, if any, adults of these species in either the dipterocarp or *Macaranga gigantea* forests, the trees must grow from seeds buried in the soil. However, after the second cropping period, these small-tree species also fail to regenerate because of their high vulnerability to felling and burning (Table 17.1). Instead, exotic shrub species, such as *Austroeupatorium inulifolium* and *Piper aduncum*, become dominant (bottom left in Fig. 17.1). The seeds of the former species are dispersed by the wind and those of the latter are dispersed by animals.

It appears that the more often slash-and-burn practice is repeated, the lower the succeeding stand height becomes. After 5 years, *Macaranga gigantea* forests are about 10 m high, while small-tree forests are 6 m high, and shrub communities are only 2 m high. The base nutrient status of soils under degraded ecosystems is improved, rather than worsened. However, there is less available N and P for plants, and this might be associated with the decreased tree height in the fallow after repeated slash-and-burn (Chapter 4, this volume). Inherent species characteristics might also be associated with the change.

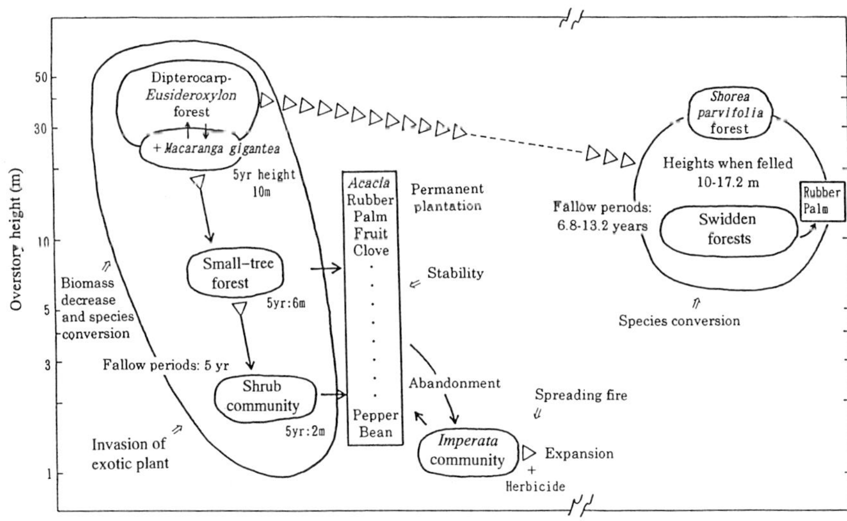


**Fig. 17.1.** The sequence of plant communities in lowland and lower montane East Kalimantan (modified from Kiyono and Hastaniah 1997). The *vertical axis* represents overstory height. The *horizontal axis* represents the history of past slash-and-burn use, ranging from 10 to 20 years to more than 300 years

**Table 17.1.** Interspecific variations of vulnerability to felling and/or fire events. (modified from Kiyono and Hastaniah 1997)

| Species | DBH (cm) | Number of individuals | % sprouted | Treatment[a] | Vulnerability to felling and burning |
|---|---|---|---|---|---|
| Tree species | | | | | |
| *Cotylelobium lanceolatum* | 30–60 | 5 | 100 | F & B | Low |
| *Shorea balangeran* | 10–25 | 12 | 92 | F & B | Low |
| *Dipterocarpus cornutus* | 30–60 | 3 | 0 | F | High |
| *Dipterocarpus humeratus* | 30–40 | 3 | 0 | F | High |
| *Dipterocarpus oblongifolius* | 50–80 | 21 | 0 | F & B | High |
| *Dipterocarpus tempehes* | 20–40 | 4 | 0 | F & B | High |
| *Shorea johorensis* | 30–100 | 3 | 0 | F | High |
| *Shorea laevis* | 20–75 | 3 | 0 | F | High |
| *Shorea leprosula* and *S. parvifolia* | 25–60 | 21 | 0 | F & B | High |
| *Shorea ovalis* | 20 | 1 | 0 | F | High |
| *Shorea smithiana* | 25 | 1 | 0 | F | High |
| *Schima wallichii* | 8–20 | 28 | 100 | F & B | Low |
| *Vitex pinnata* | 5–25 | 137 | 100 | F & B | Low |
| *Peronema canescens* | 11–21 | 46 | 100 | F & B | Low |
| *Nauclea* sp. | 7–16 | 13 | 100 | F & B | Low |
| *Artocarpus anisophyllus* | 6–13 | 23 | 100 | F & B | Low |
| *Vernonia arborea* | 7–17 | 42 | 98 | F & B | Low |
| *Eusideroxylon zwageri* | 44–92 | 26 | 92 | F & B | Low |
| *Macaranga gigantea* | 11–22 | 90 | 0 | F & B | High |
| *Macaranga triloba* | 10–20 | 14 | 0 | F & B | High |
| Small-tree species | | | | | |
| *Trema orientalis* | 5–10 | 14 | 0 | F & B | High |
| *Macaranga tanarius* | 5–13 | 28 | 7 | F & B | High |
| *Macaranga trichocarpa* | 2–4 | 112 | 1 | F & B | High |
| *Macaranga denticulata* | 6–9 | 38 | 0 | F & B | High |
| *Homalanthus populneus* | 4–8 | 14 | 0 | F & B | High |
| *Mallotus paniculatus* | 4–6 | 59 | 2 | F & B | High |
| *Fagraea racemosa* | 5–14 | 24 | 100 | B | Low |
| Shrub species | | | | | |
| *Piper aduncum* | 3–5 | 91 | 81 | B | Low |
| *Austroeupatorium inulifolium* | 1–2 | 129 | 98 | B | Low |
| *Melastoma malabathricum* | 1–2 | 62 | 90 | B | Low |
| *Millettia sericea* | 1–7 | 48 | 100 | B | Low |

[a]F: felling, B: burning.

**Fig. 17.2.** A small-tree forest in September 1993. Tall trees were burned by the forest fires in 1982–83

### 17.2.2 Biological Invasion of *Austroeupatorium inulifolium*

Shrub species that are resistant to felling and fires increase in fallow land with repetition of slash-and-burn practice. Among fire-tolerant species, *Austroeupatorium inulifolium* is a unique plant (Fig. 17.3), which is native to America. It was probably first planted in the Botanical Garden in Bogor, west Java in the late 19$^{th}$ century, and spread from there (Stoutjesdijk 1935, unpublished). In Kalimantan, the species was first recorded in the 1940's (Monk et al. 1997), and local people named it *rumput jepang* (Japanese weed) (Kiyono and Hastaniah 1996). It became more common near Bukit Soeharto after the fires in 1982–83. Seeds or other migrules of this shrub species may have been brought to new regions with seedlings for industrial plantations (HTI) and agriculture. This species is very fire-resistant (Table 17.1) and survives repeated slash-and-burn (Stoutjesdijk, unpublished data). Once this species becomes predominant, small-tree species can rarely dominate the fallow. The increasing numbers of cattle and water buffalo in the area do not feed on *Austroeupatorium inulifolium*. In addition, the species is less sensitive to the herbicide, a glyphosate: isopropylamine salt of *N*-(phosphonomethyl) glycine, that has been in common use in agriculture and forestry since the 1980's in East Kalimantan (Kiyono and Hastaniah 1997). The combination of repeated slash-and-burn practice, increased numbers of cattle and water buffalo, and the widespread use of the herbi-

**Fig. 17.3.** An *Austroeupatorium inulifolium* shrub community

cide appears to have accelerated the invasion and expansion of *Austroeupatorium inulifolium* in East Kalimantan, and may have inhibited the regeneration of the forests after the fires in 1982–83.

### 17.2.3 Formation of *Imperata* grassland

Small-tree forests and shrub communities are often converted into orchards, plantations of pepper (Fig. 17.4) or other perennial crops, or vegetable farmlands. Some plantations have been abandoned and reclaimed by *Imperata cylindrica* grassland (Fig. 17.5, bottom middle in Fig. 17.1). Since *Imperata* requires strong light, it migrates into plantations that are weeded intensively. According to our 1994 survey in and around Bukit Soeharto (Kiyono and Hastaniah 1997), 16 out of 18 grasslands developed in abandoned plantations of perennial crops. The fruit and clove tree plantations were abandoned after fire damage, most of which occurred in 1982–83, while pepper cultivation was abandoned as a result of fluctuations in its price in the late 1980's (around Bukit Soeharto, the market price in 1989 was only 13% of that in 1988). Some people believe that pepper fields were abandoned because of the decreased soil productivity (e.g., AICAF 1992). However, all the farmers we interviewed gave up pepper farming because of the drop in the market price, not because of a decrease in yield. In fact, 17 out of 18 grasslands that were established in 1982–1991 formed in plantations that were abandoned because of the drop in pepper price or the effects of wildfire. These observations lead us to conclude that the formation of *Imperata* grassland is closely related to the abandonment of perennial cash crop

**Fig. 17.4.** A pepper plantation. Support stakes for pepper vines are usually ulin (*Eusideroxylon zwageri*) wood

**Fig. 17.5.** *Imperata cylindrica* grasslands. The land was once a pepper field, and many support stakes were for pepper vines

plantations. In one location, *Imperata* grassland was established in a field where tillage was practiced. This suggests that intensive weeding and surface soil disturbance by tillage remove young shoots from tree stumps or the seeds of former vegetation species, and eventually leave only crops and migrant plants in the plantations.

Recent migrants from Sulawesi, Java, and South Kalimantan established perennial crop plantations along roads in and around Bukit Soeharto and then abandoned them, mostly between 1970 and the early 1990's. The conversion of logged-over forest into *Imperata* grassland has also occurred in other regions of East Kalimantan (Kiyono and Hastaniah 1997), and the main cause of grassland formation has been similar to that in and around Bukit Soeharto. Logged-over forests were converted into *Imperata* grassland in as little as 10 to 20 years after settlement.

## 17.3 Succession in *Imperata* Grassland

### 17.3.1 Biomass and Net Primary Productivity in *Imperata* Grassland

The net primary productivity (NPP) in *Imperata* grassland (Table 17.2) is comparable to values reported for primary and secondary forests in the tropics: 25.8 t $ha^{-1}$ $year^{-1}$, *Schima wallichii* secondary forest (Kiyono and Hastaniah 1997); 26.9 t $ha^{-1}$ $year^{-1}$, mature dipterocarp forest (Kira 1978); and 14.7–17.5 t $ha^{-1}$ $year^{-1}$ (above ground), disturbed dipterocarp forest (Chapter 10, this volume). The biomass of the grassland (ca. 7.5 t $ha^{-1}$) is far smaller than that of the forest. The high NPP of *Imperata* grassland is due to high net photosynthetic rates, because of the $C_4$ photosynthetic reactions of the species. Tanaka et al. (Chapter 5, this volume) found that *Imperata*

**Table 17.2.** Biomass, net primary productivity, necromass, and seed production of *Imperata cylindrica* grassland. (modified from Kiyono and Hastaniah 1997)

| No. | Topography and soil texture type | Biomass (ton $ha^{-1}$) | Net primary productivity (ton $ha^{-1}$ $year^{-1}$) | Necromass[a] (ton $ha^{-1}$) | Seed production after burning (million $ha^{-1}$) |
|---|---|---|---|---|---|
| 1 | Convex-coarse | 7.1 | 17.7 | 4.3 | 1045 |
| 2 | Concave-medium | 5.6 | 16.7 | 3.7 | 904 |
| 3 | Convex-fine | 7.2 | 18.3 | 3.4 | 405 |
| 4 | Concave-coarse | 9.1 | 25.4 | 7.2 | 355 |
| 5 | Concave-coarse | 6.9 | 20.5 | 3.5 | 914 |
| 6 | Convex-medium | 7.6 | 27.8 | 9.6 | 1448 |
| 7 | Convex-very coarse | 8.6 | 29.0 | 13.5 | 14 |
| 8 | Concave-very coarse | 8.2 | 24.8 | 11.4 | 1000 |
| Mean | | 7.5±2.0 | 22.5±4.8 | 7.1±4.6 | 861±133 |

[a]Aboveground. 7-8 months after burning in nos. 1–5 and for nos. 6–8, 19 months.

grassland might have well-developed nutrient conservation mechanisms, comparable to those of natural forest. This might also be related to high NPP in the grassland. However, the high NPP does not result in biomass accumulation of the grassland. More organic matter is produced and transferred to the soil than stored in aboveground vegetation.

### 17.3.2 Recovery from *Imperata* Grassland Inhibited by Repeated Fires

Even on *Imperata* grassland, agriculture is sometimes practiced using herbicides. This practice became common near Bukit Soeharto around 1980. Wildfires spread from the slash-and-burn fields and other accidental fires disturb the normal succession (orthosere) in grasslands. One hectare of *Imperata* grassland contains 107 plant species, including pyrophytic trees such as *Vernonia arborea*, *Cratoxylum formosum*, and *Vitex pinnata* (Kiyono and Hastaniah 1997). Consequently, the grasslands may succeed to secondary forest if they do not catch fire for several years. However, since *Imperata* catches fire easily and recovers immediately, frequent wildfires maintain *Imperata* grasslands. Around Bukit Soeharto, 39% of the land covered with savanna-like communities (containing dense *Imperata* and scattered overstory trees such as *Vernonia arborea* and *Fagraea racemosa*) was replaced by *Imperata* grassland after the fires in 1991–1992 (Kiyono and Hastaniah 1997).

Prolonged drought, a characteristic of the climate in East Kalimantan, occurs at intervals of several years, and is linked to severe El NiÒo-Southern Oscillation (ENSO) events (Chapter 2, this volume). Increased human activity, including slash-and-burn agriculture, causes frequent wildfires during droughts (Chapter 3, this volume). The repeated wildfires linked to droughts and human factors will maintain *Imperata* grasslands in East Kalimantan.

## 17.4 Conclusion

Humans have played the most important role in forest conversion in East Kalimantan. Logged-over dipterocarp forest has been transformed into *Imperata cylindrica* grassland within 10 to 20 years of settlement. Slash-and-burn agriculture, as practiced by recent migrants in East Kalimantan, destroys the sensitive vegetation of the original dipterocarp forests and replaces it with degenerated vegetation, namely, small-tree forests and shrub communities. Signs of degeneration include decreased biomass, species conversion, and the invasion of exotic plants. Although the degenerated vegetation is often converted into plantations, subsequent plantation abandonment triggers the establishment of *Imperata* grassland. *Imperata* grasslands are thus established because of the abandonment of perennial crop plantations, and not because of shortened fallow periods with repeated slash-and-burn agriculture. In this area, the combined effect of wildfires and the slump in white pepper prices were the main causes for plantation abandonment. Slash-and-burn agriculture itself rarely causes grassland formation in fallowland, but frequent wildfires from slash-and-burn fields disturb normal succession (orthosere) and serve to expand grassland areas.

## References

AICAF (1992) The final report on the study on the integrated agriculture and forestry development project for tropical forest conservation (for the East Kalimantan Province in Indonesia) (in Japanese). AICAF, Tokyo

Dove MR (1983) Theories of swidden agriculture, and the political economy of ignorance. Agroforest Syst 1 IS-2. 2:85–99

Fraser AI (1998) Social, economic and political aspects of forest clearance and land-use planning in Indonesia. In: Maloney BK (ed) Human activities and the tropical rainforest. Kluwer, Amsterdam, pp 133–150

Jessup TC (1981) Why do Apo Kayan shifting cultivators move? Borneo Res Bull, 13 (1):16–32

Kartawinata K, Adisoemarto S, Riswan S, Vayda AP (1981) The impact of man on a tropical forest in Indonesia. Ambio 10:115–119

Kira T (1978) Primary productivity of Pasoh Forest — a synthesis. Malay Nat J 30:291–297

Kiyono Y, Hastaniah (1996) Forest conversion and the role of slash-and-burn agriculture in lowland Kalimantan. Borneo Res Bull 27:70–76

Kiyono Y, Hastaniah (1997) Slash-and-burn agriculture and succeeding vegetation in East Kalimantan. PUSREHUT Special Publication 6, Univ Mulawarman, Samarinda

Monk KA, Fretes Y de, Reksodiharjo-Lilley G (1997) The ecology of Nusa Tenggara and Maluku. The ecology of Indonesia, series V, Periplus, Singapore

Spencer JE (1966) Shifting cultivation in southeastern Asia. Univ California Press, San Francisco

Vayda AP, Sahur A (1985) Forest clearing and pepper farming by Bugis migrants in East Kalimantan: Antecedents and impact. Indonesia 39:93–110

Vayda AP, Sahur A (1996) Bugis settlers in East Kalimantan's Kutai National Park. Their past and present and some possibilities for their future. CIFOR, Jakarta

Whitten AJ, Mustafa M, Henderson GS (1987) The ecology of Sulawesi. Gadjah Mada UP, Yogyakarta

# 18 Impact of Different Intensities of Selective Logging on a Low-Hill Dipterocarp Forest in Pasir, East Kalimantan

Yasuyuki Okimori and Paulus Matius

## 18.1 Introduction

Selective logging systems are commonly employed in natural forests in tropical countries. Under such systems, the removal of trees imitates the natural phenomena of tree fall, because only some harvestable and commercial trees are removed. However, the rate of timber harvest commonly exceeds the natural rate of tree fall. Therefore, these systems lead to a decline in forest resources and the subsequent extension of degraded forests. The precise impact of selective logging on the residual tropical forests needs to be examined.

In recent years, scientists have studied this situation, since selective logging is purported to be an important technique for sustainable forest management. Much research into assessing the impact of logging has been conducted in Southeast Asian countries, including Indonesia (Tinal 1978, Abdulhadi et al. 1981, Cannon et al. 1994, Bertault and Sist 1997), the Philippines (Weidelt and Banaag 1982), and Malaysia (Nicholson 1958, Fox 1968, Burgess 1971). Intensive studies have also been carried out in the tropics in the Brazilian Amazon (Uhl and Vieira 1989) and Costa Rica (Webb 1997).

The main activities of selective logging operations involve felling big trees with chainsaws, and constructing roads to pull out the tree trunks with heavy machines. These activities have different effects on the remaining trees, soil, microenvironment, and other organisms, and depend on the logging intensity, forest structure, topography, etc. This chapter aims to clarify the direct impacts of different logging intensities on the aboveground part of a low-hill dipterocarp forest.

## 18.2 Research Site, Methods, and Trees Harvested

This study was conducted at 400–500 m above sea level, in the northern Meratus Mountains, Pasir regency, East Kalimantan (see Fig 1.2 in Chapter 1, this volume). The forest at the study site was typical of the low-hill dipterocarp forests found in

Ecological Studies, Vol. 140
Guhardja et al. (eds.) Rainforest Ecosystems of East Kalimantan: El Niño, Drought, Fire and Human Impacts

East Kalimantan. The Indonesian Ministry of Forestry established the TPTI guidelines (the standard Indonesian selective logging and planting system) for selective logging operations, in 1989. According to the guidelines, trees over 50 cm in diameter at breast height (DBH) can be cut and extracted for commercial use. The contractor surveys the trees in a certain area of the concession forest and marks the harvestable trees, which are then cut with chainsaws. Tractors enter the forest plot to pull out the logs, which are yarded at a landing.

Two one-hectare plots (100 m × 100 m) were established in an area that had been harvested the previous year. One was logged lightly and the other heavily. Two large dipterocarp trees, *Dryobalanops beccarii* and *Shorea johorensis*, were harvested in the lightly logged plot, while 16 trees from 7 dipterocarp species (*Dipterocarpus* sp., *Dryobalanops beccarii*, *Shorea johorensis*, *S. laevis*, *S. leprosula, S. parvifolia,* and *Shorea* sp.) were harvested in the heavily logged plot.

## 18.3 Effects of Logging at Different Intensities

### 18.3.1 Canopy Opening

We examined the effects of selective logging and tractor yarding with secondary and tertiary roads. The mechanized harvesting systems employed in Southeast Asian countries include truck, tractor and high-lead yarding (Weidelt and Banaag 1982). Tractor yarding is more common in Malaysia and Indonesia. Logging roads are classified into three categories (Uhl and Vieira 1989): 1) primary road: main roads to which all boles are dragged; 2) secondary road: spur roads connected to primary roads; 3) tertiary road: flange roads that are only used once to pull out a single tree bole.

The tractor path was usually about 5 m wide, but sometimes exceeded 10 m where the tractor changed direction or the tree fell. The areas of the tractor paths in the two plots were almost the same, despite the difference in the number of harvested trees (Fig 18.1), 18% and 19% in the lightly and heavily logged plots, respectively. In other studies, the area of tractor paths varied widely, from 10% to 40%. Examples include 26% in a mixed dipterocarp forest in Brunei (Kobayashi 1993); 12% in Sabah, of which 6% was log-landings and dumps and a further 6% was soil tipped down hillsides (Burgess 1971); and 14% on flat to gently undulating topography (Nicholson 1958). Fox (1968) reported figures of over 40% for the tractor-paths and set-up area for felling trees. Abdulhadi et al. (1981) found that 30% of the ground was covered with tractor paths and skidding trails in Lempake, East Kalimantan. These different figures depend on the topography, type of logging road, and clearance for felling logs.

The canopy gaps created by the logging operation were larger in the heavily logged plot. The total gap area was 23% in the lightly logged plot, and a continuous upper canopy layer still covered the majority of the tractor paths (Fig. 18.1a). The gap area was 41% in the heavily logged plot; stumps were spread along the tractor paths, and a long narrow gap extended above the paths (Fig. 18.1b). In Paragoninas, in the Brazilian Amazon, a harvest of 8 trees per hectare reduced the number of crowns by 12%, which reduced the total canopy cover from 80% in unlogged forest to 43%, with considerable accompanying damage (Uhl and Vieira 1989). A significant relationship between post-logging canopy cover and extraction intensity was found in a selectively logged forest in northern Costa Rica (Webb 1997). Therefore, intensive extraction extends the canopy openings more in the residual stands.

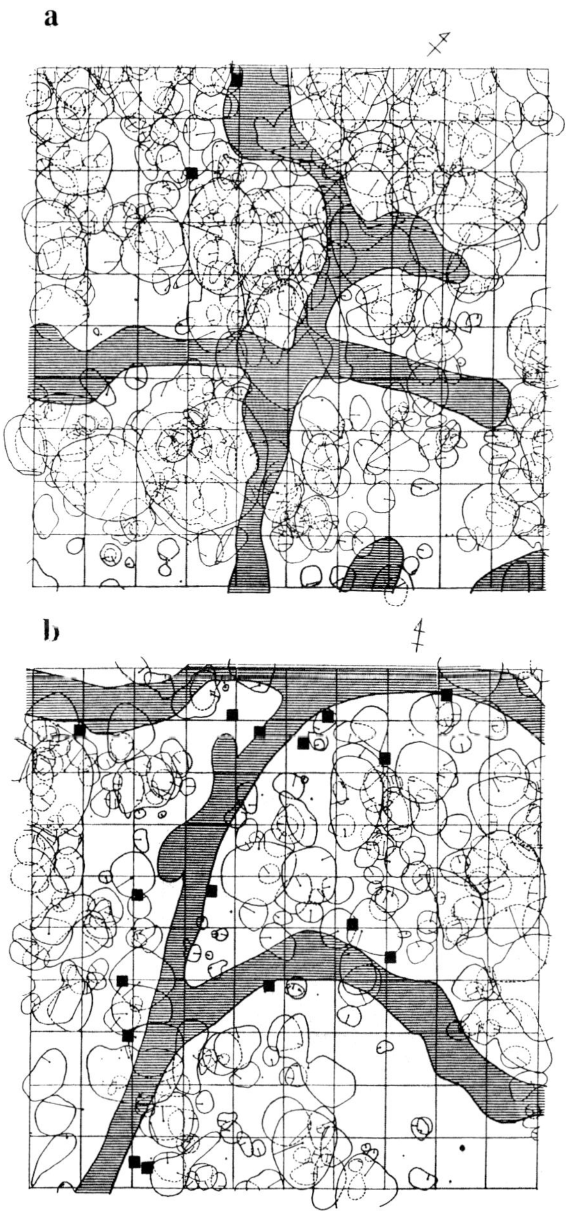

**Fig. 18.1a,b.** Crown projection of trees (DBH ≥ 10 cm) and tractor path distribution in the (**a**) lightly and (**b**) heavily logged plots of 1 ha (100 × 100 m). The *thin and broken lines* represent crowns projecting from the exposed upper canopy and crowns under the exposed canopy, respectively. *Squares, hatched, and gray areas* represent stumps, tractor paths, and canopy gaps, respectively

## 18.3.2 Damage to Trees

In Indonesia, less than 10% of the total number of trees is cut under the conventional harvesting system, yet the basal area of the trees remaining is equal to half or less of the total basal area before-logging. As in primary dipterocarp forests, the ratios of the number and basal area of dipterocarps to the total number of trees were larger in the larger diameter class above 60 cm (Fig. 18.2). After logging, the proportions of undamaged residual trees were 63% and 68% in the lightly and heavily logged plots, respectively (Fig. 18.3). In terms of basal area, the proportion of undamaged re-

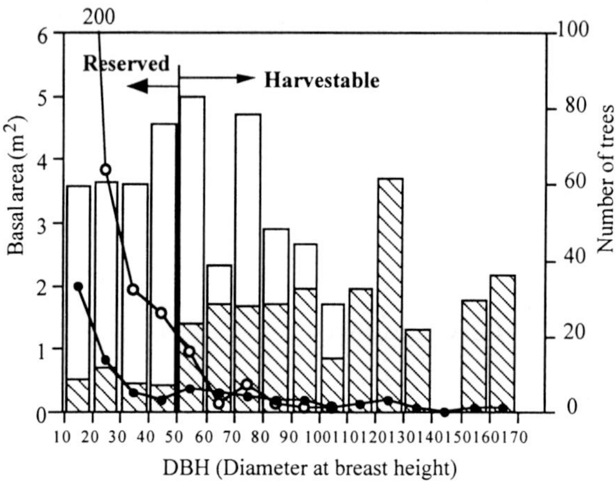


**Fig. 18.2** Distribution of trees (*circles*) and basal area (*columns*) according to DBH class. The *solid circles* and *shaded parts of columns* represent the values for dipterocarp trees. The *open circles* and the *blank part of columns* represent the values for non-dipterocarp trees

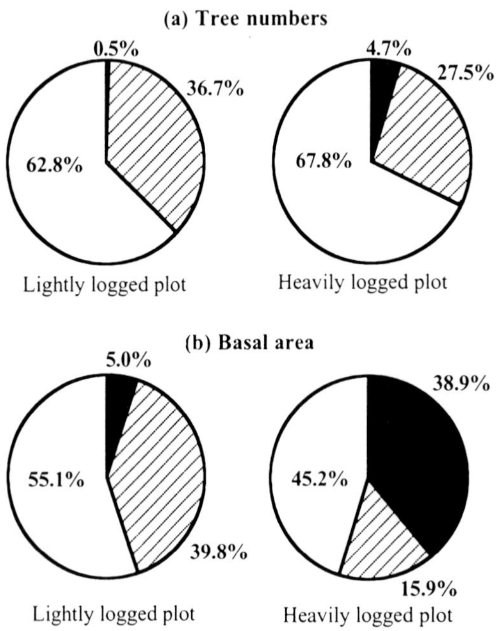


**Fig. 18.3a,b.** Effect of selective logging on **(a)** the number of trees and **(b)** basal area. *Solid, stippled*, and *blank areas* represent felled, damaged, and undamaged trees, respectively

sidual trees was 55% and 45% in the lightly and heavily logged plots, respectively. Even the removal of less than 5% of the total number of trees resulted in the loss of about 40% of the basal area in a lightly logged plot. The greatest logging damage was to diameter classes smaller than 60 cm (Fig. 18.4). Fallen trees and broken trunks were the most common fatal damage, and represented about two thirds of the total damage in terms of the number of trees in either plot (Fig. 18.5). In Trengganu, West Malaysia, a selective logging operation felled 10% of the timber by basal area in a 40 ha sample plot. This destroyed 55% of the residual trees (DBH > 10 cm) and left 35% of the trees in the plot undamaged (Burgess 1971). In Paragoninas, in the Brazilian Amazon, harvesting less than 2% of the total number of trees with 16% of the total basal area (on average 8 trees per ha were harvested), killed or damaged 26% and 44% of the remaining trees by number and basal area, respectively (Uhl and Vieira 1989). These values are similar to those observed in this study (Fig. 18.5).

The intensity of logging damage was related to the size of the trees harvested and the distribution of tractor paths. There was not a remarkable difference in the num-

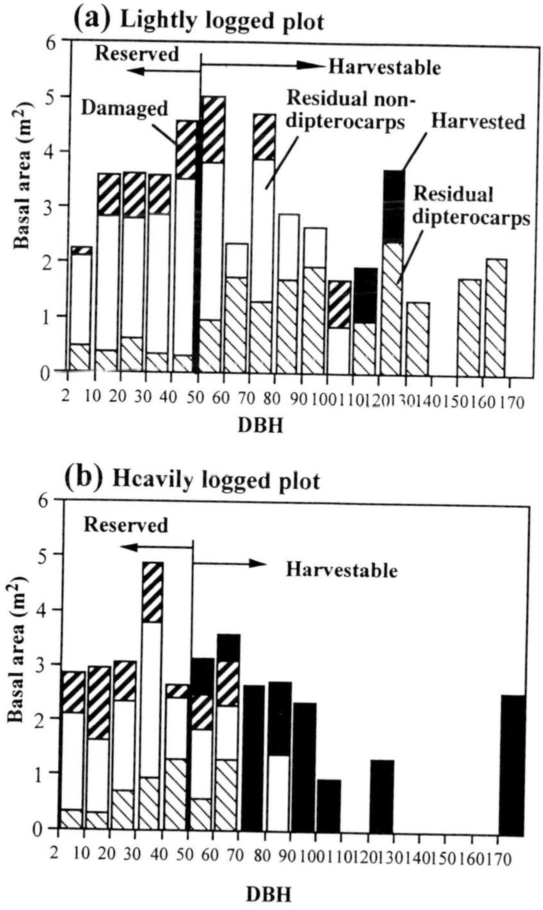


**Fig. 18.4a,b.** Effect of selective logging on the distribution of basal area according to diameter class in (**a**) lightly and (**b**) heavily logged plots. *Solid, shaded, thick hatched*, and *blank areas* represent harvested, residual dipterocarp, damaged, and residual non-dipterocarp trees, respectively

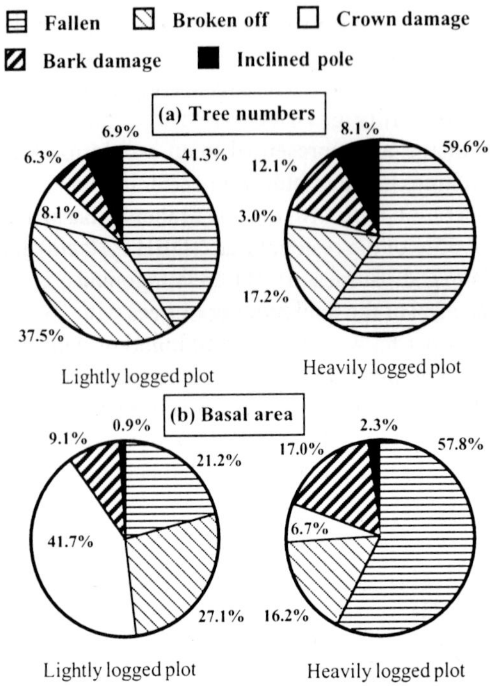


**Fig. 18.5a,b.** Percentage of damaged trees among the residual trees in terms of **(a)** tree numbers and **(b)** basal area

ber of undamaged residual trees in the lightly and heavily logged plots. However, many researchers have reported that logging impact is related to felling intensity and that there is considerable variation according to the amount of timber extracted. For example, Nicholson (1958) found that there was a fairly good positive correlation between the amount of damage (total damage, broken and fallen trees, and damaged bark and crowns) and the logging intensity (according to the basal area of the harvested timber). A significant relationship was also found between the percentage of damaged residual trunks and the extraction intensity (killed and damaged trunks) in 7 plots in a swamp forest, in Costa Rica (Webb 1997). Webb's study suggested that the percentage of damaged trunks to the total number of trunks before logging might increase exponentially with the number of extracted trunks. The discrepancy between our study and others might be derived from differences in the amount of damage and the size of trees extracted (112 and 129 cm DBH in the lightly logged plot and an average of 95 cm in the heavily logged plot). The discrepancy also resulted from the fact that the area of the tractor path was almost the same in both plots.

### 18.3.3 Effect of Logging on the Growth of the Remaining Small Trees

The dynamics of the survival, growth, and recruitment of the residual trees should be used to assess the recovery of the logged-over forests. Every woody species, even non-dipterocarp trees, should be included in the successors, because soil erosion is reduced and biodiversity maintained with a covering of vegetation.

The differences in the canopy cover after a logging operation resulted in different light conditions on the forest floor in the lightly and heavily logged plots. The average relative light intensity (RI) of 100 subplots (10 m × 10 m) was 12% and 54% in the lightly and heavily logged plots, respectively. In the lightly logged plot, 84% of the subplots had a RI lower than 20% and 68% had a RI below 10%. The RI along the gaps was over 20% and these bright places covered almost half the area of the heavily logged plot. As described below, pioneer species readily invade the bright areas.

About one year after the logging, the densities of both saplings and seedlings were higher in the heavily logged stand. Kuusipalo et al. (1996) reported a similar result: seedling density tends to be higher in a logged-over stand than in an unlogged stand, due to the higher light intensity in the logged-over stand. The density of saplings of Euphorbiaceae was higher in the heavily logged stand than in the lightly logged stand (Table 18.1). This was because the juveniles of pioneer species of this family such as *Macaranga triloba*, *M. gigantea*, *M. trichocarpa*, and *Endospermum malaccense*, immediately colonized the gap corridors of the tractor paths. Seedlings of another pioneer species, *Anthocephalus chinensis* (Rubiaceae), were frequently observed along the tractor paths.

In dipterocarps, the frequency of saplings and seedlings relative to total saplings and seedlings were 32% and 19%, respectively, on average for both the heavily and lightly logged stands (Table 18.1). They probably existed in both stands prior to logging. The seedlings of *Shorea leprosula* and *S. laevis*, which are canopy tree species, develop epinastic growth in their terminal shoots under gap conditions. If a certain number of undamaged dipterocarp trees remain after logging, these trees may join the mass of recruits.

**Table18.1.** Sapling and seedling densities in the lightly logged plot (LLP) and heavily logged plot (HLP) at one year after the logging operation in a low-hill dipterocarp forest in Pasir, East Kalimantan

| | Saplings[a] | | Seedlings[b] | |
|---|---|---|---|---|
| | LLP (No. ha$^{-1}$) | HLP (No. ha$^{-1}$) | LLP (No. ha$^{-1}$) | HLP (No. ha$^{-1}$) |
| Dipterocarpace | 850 | 1025 | 3500 | 4000 |
| Euphorbiaceae | 375 | 1375 | 4500 | 5000 |
| Rubiaceae | 225 | 125 | 1000 | 1000 |
| Others | 1100 | 925 | 8200 | 13400 |
| Total | 2550 | 3450 | 17200 | 23400 |

[a]Taller than 150 cm in height and less than 2 cm in diameter at breast height.
[b]Less than 150 cm in height.
Sampling area is 400 and 100 m$^2$ for saplings and seedlings, respectively.

## 18.4 Countermeasures for Reducing the Damage and Promoting Sustainable Management

Guidelines to reduce the impact of logging should be established. The most important aspect of this study was to determine how to use the results to establish practical guidelines to reduce the amount of tree damage, while maintaining the level of harvest. Putz and Pinard (1993) suggested the importance of rearranging the entire logging operation, including cutting vines (Fox 1968). They proposed 'reduced-impact logging' guidelines, aimed at reducing logging damage, and reconsidered all of the field operation procedures. Their guidelines consider a harvest plan (a stock map, involving all harvest areas), pre-felling vine cutting, skid trail planning (located on a map and checked in the field), tree felling direction (to avoid damage and for workers safety), and other arrangements. 'Reduced-impact logging' was tested in logging concessions in Malaysia (Pinard and Putz 1996) and Indonesia (Bertault and Sist 1997). They reported that logging damage was substantially reduced following their guidelines, as compared with conventional logging.

A mechanized harvesting system with controlled selection, similar to the one used in the trials in Malaysia and Indonesia, and the use of small bulldozers for skidding can limit the residual damage during extraction (Webb 1997). Johns (1997) was concerned with the effects of logging on biodiversity, and stressed field procedures with no or minimal cost that aimed specifically at increasing the biological value of the plot. Measures included inspecting all the logging procedures and adding post-felling procedures (clean-up procedures, water run-off control, post-felling plot inventory, post-felling biodiversity inventory).

Harvest systems with reduced-impact logging are sorely needed to maintain the residual timber stock at proper levels. These systems should be combined with proper silvicultural treatments to ensure the survival, growth, and recruitment of the residual trees. For proper forest management, we believe that information is urgently needed concerning 1) the number of trees that can be safely cut, 2) accurate growth and mortality rates for the mid-sized residual trees, 3) measures to promote the regeneration of the commercial species, and 4) rehabilitation and reforestation techniques.

In conclusion, harvest systems with reduced-impact logging are needed to maintain the residual timber stock at a proper level. The mid-sized trees that remained after logging could become the timber stock for the next harvesting. The reduced-impact systems should then be followed by silviculture treatments to ensure the survival and growth of the residual commercial trees. There is also a keen need to promote the regeneration of commercial species even in properly treated forests.

## References

Abdulhadi R, Kartawinata K, Sukardjo S (1981) Effects of mechanized logging in the lowland dipterocarp forest at Lempake, East Kalimantan. Malay For 44:407–418

Bertault J-G, Sist P (1997) An experimental comparison of different harvesting intensities with reduced-impact and conventional logging in East Kalimantan, Indonesia. For Ecol Manage 94:209–218
Burgess PF (1971) Effect of logging on hill dipterocarp forests. Malay Nat J 24:231–237
Cannon CH, Perart DR, Leighton M., Kartawinata K (1994) The structure of lowland rainforest after selective logging in West Kalimantan, Indonesia. For Ecol Manage 67:49–68
Fox JED (1968) Logging damage and the influence of climber cutting prior to logging in the lowland dipterocarp forest of Sabah. Malay For 31:326–347
Johns AG (1997) Timber production and biodiversity conservation in tropical rainforests. Cambridge University Press, Cambridge
Kobayashi S (1993) Harvesting impact and rehabilitation in tropical rainforests. J Agr Met 48:663–666
Kuusipalo J, Jafarsidik Y, Ådjers G, Toumela K (1996) Population dynamics of tree seedlings in a mixed dipterocarp rainforest before and after logging and crown liberation. For Ecol Manage 81:85–94
Ministry of forestry (1989) Guideline for TPTI (Indonesian selective cutting system), Ministerial ordinance No.564/KPTS /IV– BPHH/1989 (in Indonesian), Jakarta
Nicholson DI (1958) An analysis of logging damage in tropical rainforest, North Borneo. Malay For 21:235–245
Pinard MA, Putz FE (1996) Retaining forest biomass by reducing logging damage. Biotropica 28:278–295
Putz FE, Pinard MA (1993) Reduced-impact logging as a carbon-offset method. Cons Biol 7:755–757
Tinal UK (1978) Mechanical logging damage after selective cutting in the lowland dipterocarp forest at Beloro, East Kalimantan. In: Suparto RS, Soerianegara I, Hamzah Z, Haeruman H, Hadi S, Manan S, Basjarudin H, Sukotjo W (eds) Proceeding of symposium on the long-term effects of logging in Southeast Asia. BIOTROP Special Publication No. 3, BIOTROP, Bogor, pp 91–96
Uhl C, Vieira ICG (1989) Ecological impacts of selective logging in the Brazilian Amazon: A case study from the Paragoninas region of the State of Para. Biotropica 21:98–106
Webb EL (1997) Canopy removal and residual plot damage during controlled selective logging in lowland swamp forest of northeast Costa Rica. For Ecol Manage 95:117–129
Weidelt HJ, Banaag VS (1982) Aspects of management and silviculture of the Philippine dipterocarp forests. Schriftenreihe der GTZ No. 132, Eschborn

# 19 A Case Study of Second Felling in a Logged-Over Dipterocarp Forest

Yosep Ruslim, Paulus Matius, and Maman Sutisna

## 19.1 Introduction

Dipterocarp forest covers most of the lowlands and hills of East Kalimantan, and 90% of the dipterocarp forest has been granted to forest concessions. Most of the lowland forests have been selectively logged once by the concessions according to Indonesian government regulations. Extensive logging operations have continued since the early 1970s, and as a result a large logged-over forest remains in the lowlands of East Kalimantan. The large wildfires in 1982–83 accelerated the degradation of the forests in East Kalimantan (e.g., Goldammer and Seibert 1990). The fire damage was especially severe in the lowland forests of the Kutai region, in East Kalimantan, which was one of the main log production areas in the 1970s.

The government's seventh Development Plan, for the years 2000 to 2004, estimates that log production in this period will increase to 1.5 times the present rate. In the plan, about 1/3 of the total amount (57 million $m^3 year^{-1}$ in Indonesia) is expected to be produced from natural forest. Presently, however, the primary dipterocarp forest is restricted to hinterlands far from roads and rivers. This means that in the near future log production from the natural forests must partly depend on the logged-over dipterocarp forest that has already been selectively logged once. In the regulations of the Indonesian Selective Logging System, one felling cycle is between 25 and 40 years, depending on the conditions for logging in the forest (Weidelt 1996). It has been more than 25 years since parts of the forest were logged selectively in the initial exploitation period in the 1970s.

This chapter discusses the effects of second felling in a logged-over dipterocarp forest. There are many studies on logging methods and the damage caused by felling and skidding operations in primary forests in the tropics (e.g., Nicholson 1979, Abdulhadi et al. 1981, Weidelt and Banaag 1982, Jonkers 1987, Ahrenholz 1991, Kobayashi 1994, Bertault and Sist 1997, and Chapter 18, this volume), but few studies have examined second felling (Weidelt 1996). It is important to develop an integrated, effective felling and skidding system for the second felling, both to reduce the damage from logging and to obtain better growth and regeneration of succeeding trees. This study examined logging conducted according to the Indonesian Selective Cutting and Planting System (TPTI), in order to determine how much and what kind

Ecological Studies, Vol. 140
Guhardja et al. (eds.) Rainforest Ecosystems of East Kalimantan: El Niño, Drought, Fire and Human Impacts

of damage occurs, and how the forest structure changes with conventional (low felling: LF) and heavy felling (HF) operations in the second felling. It is also valuable to know how the forest structure changes after logging. This study is part of a cooperative research program between the Faculty of Forestry, Mulawarman University, Center of International Forestry Research (CIFOR), and Japan International Cooperation Agency (JICA).

## 19.2 Study Site and Forest Structure Before Logging

The study site was established in Bukit Soeharto Educational Forest (BSEF) of Mulawarman University in 1997. The site is a lowland dipterocarp forest, which was logged selectively in the early 1970s and disturbed once in the 1982–83 wildfire. Matius et al. discuss the species composition and structure of the site before the second felling in Chapter 9 in this volume. The study site was divided into 9 one-hectare plots (Fig. 19.1). Trees were felled in 6 of the 9 plots with two different felling intensities and 3 plots were left untouched as controls. To outline the forest structure in the study site, the mean tree density and basal area are shown in Table 19.1, together with data from the primary dipterocarp forest in a commercial logging area (INHUTANI I) in Berau region (Sist and Saridan 1997), which is about 300 km north of BSEF. There were fewer small trees (10–20 cm in diameter at breast height [DBH]) in the primary dipterocarp forest in East Kalimantan (Sukardjo et al. 1990, Sist and Saridan 1997), but similar numbers of larger trees. Dipterocarps made up about 12% of the total number of trees and 39% of the basal area (Table 19.1). The basal area of dipterocarps in primary forests usually ranges between 43% and 55% (see Chapter 3, this volume). The decreased number of small trees and the lower percentage of basal area in the study site reflect the previous selective logging operation and the 1982–83 wildfire. Nevertheless, the number of harvestable trees (above 50 cm in diameter) was similar to that of the primary forest in Berau. These figures represent the typical forest structure in BSEF.

## 19.3 Felling and Skidding Methods

Two felling treatments were used in the study plots. Light felling (LF) is the conventional method used in the forest concessions, in which trees with a diameter of at least 50 cm are harvested for merchantable timber. In heavy felling (HF), trees with a diameter of at least 30 cm are harvested. There were three replications of each felling treatment (Fig. 19.1).

Dipterocarp species are preferentially selected for felling because of their high commercial value. The species in this family are the dominant trees, with heights of 40 to 50 m, and their crowns cover large areas above the species in the middle canopy (30 to 40 m high). On a slope, the natural felling direction of a cut tree is downhill,

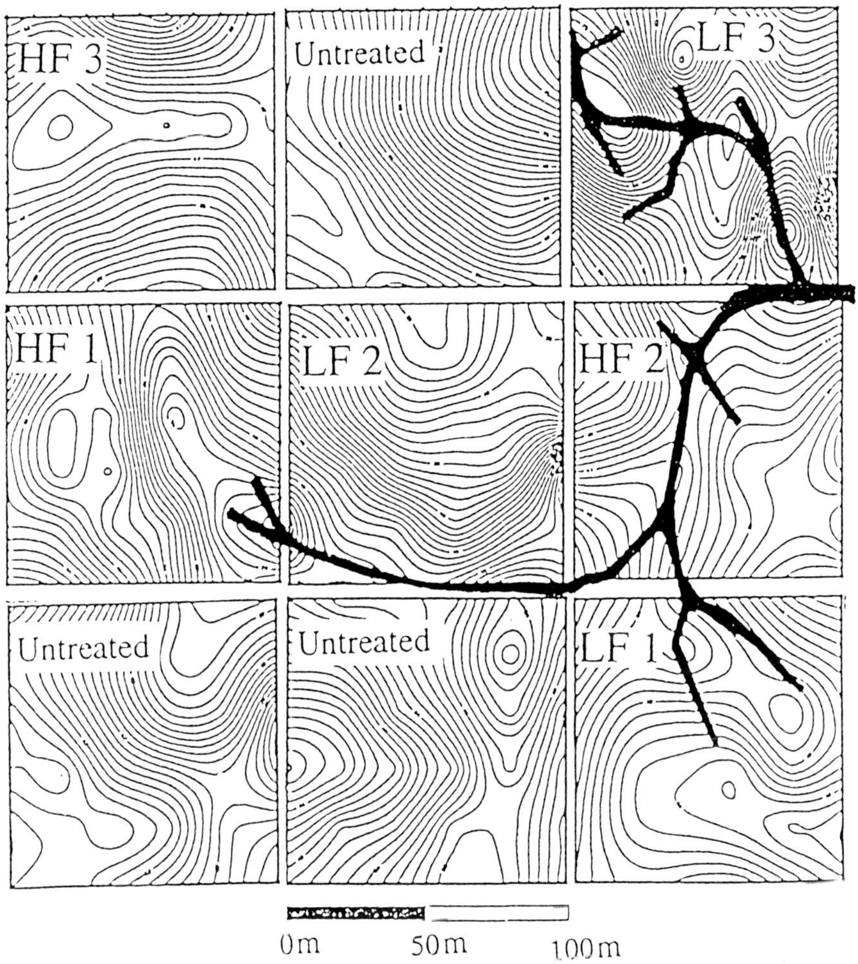


**Fig. 19.1.** Location of plots, skidding roads and topography in one plot. The benchmark (50m) contour line was at the southeast corner of each plot. *LF*, light (conventional) felling, *HF*, heavy felling. The numbers after LF and HF are replicates

**Table 19.1.** Comparison of structure between primary dipterocarp forest in Berau[a] and forest in this study site (BSEF)

| | DBH class | >10 cm | | >30 cm | | >50 cm | |
|---|---|---|---|---|---|---|---|
| Species | Site | BSEF | Berau | BSEF | Berau | BSEF | Berau |
| All | TD(trees $ha^{-1}$) | 429 | 530 | 81 | 99 | 30 | 34 |
| | BA($m^2$ $ha^{-1}$) | 27.2 | 31.5 | 18.5 | 22.3 | 12.8 | 14.8 |
| Dipt. | TD(trees $ha^{-1}$) | 52 | 135 | 29 | 46 | 19 | 24 |
| | BA($m^2$ $ha^{-1}$) | 10.7 | 15.9 | 10.2 | 13.1 | 8.9 | 10.4 |

[a]Berau: Study site of STREK project (from Sist and Saridan 1998).
BSEF, Bukit Soeharto Education Forest; DBH,diameter at breast height; Dipt., Dipterocarps; TD, tree density; BA, basal area.

and this causes extensive damage, both to the surrounding trees and to the felled tree itself. Therefore, before cutting, the felling direction was planned (1) to minimize the risk of destroying remaining trees, (2) so that the tree fell uphill or parallel to the slope, and (3) so that the log would lie in a position that was favorable for skidding to the timber yard. The felling was directed using an undercut and a back cut. If there was only one tree of a species in a plot, it was not felled.

After cutting the trunk, skidding was carried out in the conventional way with a bulldozer in 3 plots: LF1, LF3, and HF2. In the other plots, the felled logs were left on the forest floor.

# 19.4 Effects of Logging on the Forest Stand

## 19.4.1 Timber Volume Harvested

The number of trees and volume of timber harvested from each plot are shown in Table 19.2. The average volumes of the harvested trees in the LF and HF plots were 54.7 $m^3 ha^{-1}$ (12 tree $ha^{-1}$) and 85.5 $m^3 ha^{-1}$ (27 tree $ha^{-1}$), respectively. The proportion of felled trees to harvestable trees (diameter classes are ≥50 cm in LF and ≥30 cm in HF) was about 40% of the basal area in both treatments. In existing commercial selective logging in this state, the harvested timber volume is generally about 50–100 $m^3 ha^{-1}$ (about 10 trees $ha^{-1}$) (Bertault and Sist 1997). In comparison, many more trees were logged in the HF treatment, but the timber volume was in the same range as commercial logging. The harvested species were mainly dipterocarps: about 80% in number and 93% in basal area. The other species logged were *Dehaasia* spp., *Litsea* spp., *Palaquim* spp., and *Ganua motleyana*. These species are all saleable (Kessler 1996).

**Table 19.2.** Forest structure before logging and harvested trees

| | Before Logging | | Harvestable trees | | Harvested trees | | |
|---|---|---|---|---|---|---|---|
| | Numbers | Basal area | Numbers | Basal area | Numbers | Basal area | Volume[a] |
| Plot | (trees $ha^{-1}$) | ($m^2 ha^{-1}$) | (trees $ha^{-1}$) | ($m^2 ha^{-1}$) | (trees $ha^{-1}$) | ($m^2 ha^{-1}$) | ($m^3 ha^{-1}$) |
| LF1 | 402 (48) | 23.8 ( 6.7) | 23 (10) | 9.0 ( 4.9) | 6 ( 6) | 3.4 (3.4) | 38.6 |
| LF2 | 363 (40) | 24.5 ( 9.3) | 29 (15) | 12.2 ( 7.7) | 11 ( 9) | 5.1 (4.6) | 52.1 |
| LF3 | 408 (43) | 28.4 (11.6) | 34 (25) | 13.4 (10.5) | 18 (16) | 6.3 (6.1) | 73.3 |
| Average | 391(44) | 25.7 ( 9.2) | 29 (18) | 11.9 ( 7.7) | 12 (10) | 4.9 (4.7) | 54.7 |
| HF1 | 394 (56) | 27.6 (14.5) | 88 (47) | 20.0 (14.2) | 37 (31) | 9.5 (8.8) | 103.4 |
| HF2 | 402 (27) | 26.6 ( 6.5) | 80 (13) | 17.7 ( 6.2) | 15 (10) | 5.1 (4.4) | 63.3 |
| HF3 | 524 (64) | 26.8 (10.0) | 78 (29) | 17.3 ( 9.4) | 29 (22) | 8.3 (7.4) | 90.0 |
| Average | 440 (49) | 27.3 (10.3) | 82 (30) | 18.3 ( 9.2) | 27 (21) | 7.6 (6.9) | 85.5 |

LF, light felling (≥50 cm in DBH); HF, heavy felling (≥30 cm in DBH)
Dipterocarps are shown in parentheses.
[a]From Rudjehan (1997, unpublished)

### 19.4.2 Tree Damage Caused by Felling and Skidding

All trees with a diameter over 10 cm at breast-height were measured and counted in May 1997, before felling treatments. After logging, a second inventory was conducted in November 1997 to record the damage to trees using the following criteria: (a) killed (dead trees), (b) injured (surviving trees with injuries), (c) withered (dead trees caused by the severe drought between July and October 1997), and (d) healthy.

The number of trees damaged (killed or injured) by the felling and skidding was 85±10 and 109±30 (mean ±1SD) trees $ha^{-1}$ in the LF and HF plots, respectively (Table 19.3). The percentage of trees standing before logging that were damaged was 21.7% in LF and 24.8% in HF, although the difference was not significant. Slightly more dipterocarp trees were killed in the HF plots ($P<0.05$). Higher rates of killed trees occurred in the small diameter classes (Fig. 19.2), because small trees more often suffered fatal damage when trunks were broken by falling trees or when uprooted by skidding.

The percentage of damaged trees in each plot ranged from 17% to 28% of the trees present before logging (Table 19.3). In East Kalimantan, this percentage is reported to vary from 15% to 50% (Tinal and Palanewen 1978, Abdulhadi et al. 1981, Bertault and Sist 1997, Chapter 18, this volume). In East Mindanao, Philippines, more than 50% of the remaining trees were damaged (Weidelt and Banaag 1982, Jonkers 1987). Although the impact of logging on a forest varies according to the logging and transportation methods, forest structure, species composition, and topography, the rate of damage in our study area was within the range reported for logging operations in primary dipterocarp forest.

**Table 19.3.** Forest structure after logging and damaged trees

| Plot | Remaining trees[a] | | Killed trees | | Injured trees | | % of damaged trees (Damaged Total)[b] | |
|---|---|---|---|---|---|---|---|---|
| | Numbers (trees $ha^{-1}$) | Basal area ($m^2$ $ha^{-1}$) | Numbers (trees $ha^{-1}$) | Basal area ($m^2$ $ha^{-1}$) | Numbers (trees $ha^{-1}$) | Basal area ($m^2$ $ha^{-1}$) | Killed | Injured |
| LF1 | 320 (42) | 17.7 (3.4) | 41(0) | 1.3 (0.0) | 21 (3) | 1.6 (0.9) | 10.2 | 5.2 |
| LF2 | 275 (29) | 16.2 (4.6) | 64 (2) | 3.0 (0.1) | 29 (5) | 1.6 (0.2) | 17.6 | 8.0 |
| LF3 | 244 (25) | 16.4 (5.5) | 85 (2) | 3.0 (0.0) | 17 (0) | 1.3 (0.0) | 20.8 | 4.2 |
| Average | 280 (32) | 16.8 (4.5) | 63 (1) | 2.4 (0.0) | 22 (3) | 1.5 (0.4) | 16.1 | 5.6 |
| HF1 | 234 (18) | 14.3 (4.8) | 78 ( 6) | 2.6 (0.8) | 24 (0) | 1.0 (0.0) | 19.8 | 6.1 |
| HF2 | 257 (12) | 16.8 (1.6) | 68 ( 5) | 2.4 (0.5) | 12 (0) | 0.7 (0.0) | 16.9 | 3.0 |
| HF3 | 352 (31) | 14.8 (2.5) | 115 (11) | 3.4 (0.2) | 30 (2) | 1.1 (0.0) | 21.9 | 5.7 |
| Average | 281 (20) | 15.3 (3.0) | 87 ( 7) | 2.8 (0.5) | 22 (1) | 0.9 (0.0) | 19.8 | 5.0 |

Dipterocarps are shown in parentheses.
[a]Remaining trees are trees present before logging minus trees logged, killed and withered (see text)
[b]Proportion(%) of no. of damaged trees/no. of trees present before logging.

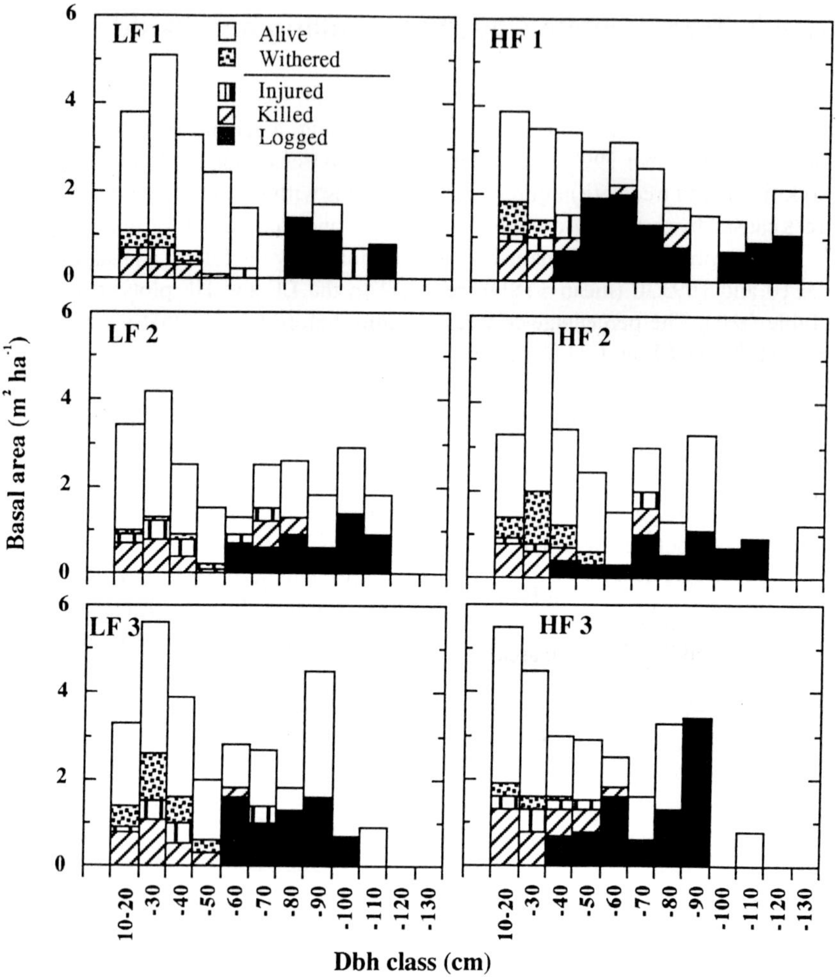


**Fig. 19.2.** Basal area distribution according to diameter class of the trees logged, killed, injured, withered, and remaining after light (LF) and heavy (HF) felling treatments. The figures following LF and HF are replicate numbers

## 19.4.3 Traffic Area and Soil Damage

Skidding traffic intensity was quantified by measuring the length and width of the skidding tracks. The traffic intensity is expressed as the percentage of the plot area covered by skidding tracks that were driven on more than once. The soil damage was graded into four levels (Table 19.4). Skidding affected from 3% to 11% of the logged area (Fig. 19.1, Table 19.4). The number of felled trees and the direction in which the felled timbers are skidded determines the traffic intensity and the soil damage, i.e., the extent of the traffic area and the grade of soil damage are proportional to the distance traveled and the number of bulldozer operations (Table 19.4).

**Table 19.4.** Traffic area and soil damage in LF1, LF3 and HF2

| Plot | Logged tree (trees ha$^{-1}$) | Traffic area (%) | Grade of soil damage (%) | | | |
|---|---|---|---|---|---|---|
| | | | Level 1 | Level 2 | Level 3 | Level 4 |
| LF1 | 6 | 3.3 | 1.0 | - | - | 2.3 |
| HF2 | 15 | 8.1 | 0.4 | 1.3 | 4.1 | 2.3 |
| LF3 | 17 | 17.1 | 1.5 | 4.1 | 9.1 | 2.4 |

Level 1: ground vegetation was existing on the tracks; Level 2: mineral soil was exposed; Level 3: distinct soil compaction was observed on the tracks; Level 4: soil structure was destroyed by earth-works.

Logging with heavy machines not only causes severe damage to the remaining trees, but also to the forest soil (Kobayashi 1994). Heavy machinery churns up the ground, and denudes and compacts the soil on the hauling paths and roads. Soil compaction reduces water infiltration and increases run-off, which leads to soil erosion and prevents the regeneration of forest trees. In East Kalimantan, Abdulhadi et al. (1981) found that after removing only 11 trees per hectare, 30% of the forest area had been devastated by the bulldozers. Similarly, Ahrenholz (1991) reported that hauling ruined 20.7% of the area.

## 19.5 Remaining Trees and Predicting the Next Felling

The tree density and basal area in the HF plots was markedly reduced after logging, although the felling rate according to basal area and the damage caused by felling and skidding were almost the same in the two felling treatments, as mentioned above. The difference in tree number and basal area before (Table 19.2) and after (Table 19.3) logging is the sum of the number of logged, killed, and withered trees. Most of the withered trees were *Macaranga* spp., and no dead dipterocarps were observed. The average number of withered trees was 33.3±23.6 and 42.7±17.0 trees ha$^{-1}$ in the LF and HF plots, respectively. After logging, about 72% of the trees remained in the LF plots and 64% in the HF plots, and the respective basal area remaining was about 65% and 56%, respectively.

In number and basal area, the percentages of dipterocarps remaining were 11.5% and 26.3% in LF, and 6.8% and 16.1% in HF, respectively. The reduction in basal area was extremely large compared with previously reported values of about 50% in primary dipterocarp forests (Kartawinata et al. 1981, Sukardjo et al. 1990, Sist and Saridan 1997) and 37–40% in partially disturbed forest in BSEF (see Chapter 10, this volume). Therefore, there was a marked reduction of large dipterocarp trees after the logging.

Vegetation recovery by pioneer species such as *Trema, Macaranga, Mallotus,* and *Homalanthus* began one or two months after the logging operation, as previ-

**Table 19.5.** Number of remaining dipterocarp trees in logged area

| | LF1 | LF2 | LF3 | Average | HF1 | HF2 | HF3 | Average |
|---|---|---|---|---|---|---|---|---|
| Reserved | 15 | 9 | 12 | 12 | 6 | 4 | 7 | 5.6 |
| Harvestable | 2 | 5 | 8 | 5 | 8 | 1 | 4 | 4.3 |

Reserved:20 to 49 cm in DBH. Harvestable: more than 50 cm in DBH.

ously reported (Fox 1972). However, regeneration was halted by the severe drought and large wildfire that occurred in early 1998 (see Chapter 3, this volume). Vegetation recovery restarted after normal rainfall began in the middle of 1998. However, no dipterocarp seedlings have appeared because their seeds have not yet been produced until March 1999.

The diameter class between 20 and 49 cm are the trees reserved for the next harvest. There were very few reserved dipterocarps left in the logged-over area (Table 19.5). According to estimates of tree growth in a partially degraded dipterocarp forest in BSEF (by T. Toma, unpublished data), the average growth rate of the reserved trees is 0.44±0.13 cm $year^{-1}$ for all species, and 0.92±0.33 cm $year^{-1}$ for dipterocarps. These growth rates are similar to previously reported values (Weidelt 1996). Therefore, it will take about 70 years for the smallest reserved trees (20 cm) to exceed 50 cm in diameter. Considering the dipterocarps, which generally have high commercial value, the smallest reserved tree should reach the harvestable diameter class (over 50 cm) in about 35 years. After 35 years, the number of harvestable dipterocarps is expected to be 14 to 20 tree $ha^{-1}$ (average 17) in LF plots and 5 to 14 tree $ha^{-1}$ (average 10) in HF plots, if mortality is zero. Using conventional felling methods (LF) and assuming that the felling rate of dipterocarps is the same as in this study (55% for harvestable trees), there should be 8 to 11 harvestable trees in LF areas and 3 to 8 in HF areas. This number would also include some trees with no commercial value, such as those that are injured, crooked, or have rotten hearts.

This prediction suggests that LF felling is the maximum allowable. The minimum harvestable diameter must be at least 50 cm, and the harvested timber volume must be less than 50 $m^3$ $ha^{-1}$ in the logged-over dipterocarp forest. Otherwise, it will be very difficult to conduct a third felling cycle within 40 years. Furthermore, the recruitment of dipterocarp seedlings must be considered, since tree fruiting is closely related to tree diameter, and in some species trees less than 50 cm in diameter do not fruit (see Chapter 11, this volume).

Consequently, this study indicates that (1) harvesting good quality timber from the logged-over forest economically requires a felling cycle that is longer than 30 years. (2) If a third felling is to be considered, a larger minimum harvestable diameter (≥50 cm) and a longer felling cycle (≥40 year) are desirable ecologically and economically. (3) As long as conventional felling methods continue, the dipterocarp forest will deteriorate with every felling cycle, with an increase in the number of small trees. (4) Enrichment planting of favorable tree species is essential in degraded forest areas that have been burned by wildfires, because, other than pioneer species, small, succeeding trees are very scarce.

The 1998 wildfire affected the study site. In the control plots, the number of species and individuals was reduced by 60% and 30%, respectively, while in the felled plots these was reduced by 40% and 20%, respectively. Wildfires caused by human activity must be actively prevented.

# References

Abdulhadi R, Kartawinata K, Sukardjo S (1981) Effects of mechanized logging in the lowland dipterocarp forest at Lempake, East Kalimantan. Malay For 2 & 3:407–418

Ahrenholz T (1991) Die Erscliessungssituationen bei der Nutzung tropischer imergüner Feucthwälder in Ostkalimantan, Indonesia. eine Fallstudie an einem Beispielbetrieb. Diplomarbeit der Forstw Fäkultat der Georg-August Univ, Göttingen

Bertault J-G, Sist P (1997) An experimental comparison of different harvesting intensities with reduced-impact and conventional logging in East Kalimantan, Indonesia. For Ecol Man 94:209–218

Fox JED (1972) Logging damage and the influence of climber cutting prior to logging in the lowland dipterocarp forest of Sabah. Malay For 41:326–347

Goldammer JG, Seibert B (1990) The impact of droughts and forest fires on tropical lowland rain forest of East Kalimantan. In: Goldammer JG (ed) Fire in the tropical biota. Ecological Studies 84. Springer, Berlin, pp 11–31

Jonkers WBJ (1987) Vegetation structure, logging damages in a tropical rain forest in Suriname. Agric Univ, Wageningen

Kartawinata K, Abdulhadi R, Partomihardjo J (1981) Composition and structure of a lowland dipterocarp forest at Wanariset, East Kalimantan, Malay For 44:397–406

Kessler PJA (1996) Not only dipterocarps: an overview of tree species diversity in dipterocarp forest ecosystem of Borneo. In: Schulte A, Schöne D (eds) Dipterocarp forest ecosystems. World Scientific, Singapore, pp74–101

Kobayashi S (1994) Effects of harvesting impacts and rehabilitation of tropical forest. J Plant Res 107:99–106

Nicholson DI (1979) The effects of logging and treatment on the mixed dipterocarp forest of South East Asia. FAO, Rome

Sist P, Saridan A (1998) Description of the primary lowland forest of Berau. In: Bertault J-G, Kadir K (eds). Silvicultural research in a lowland mixed dipterocarp forest of East Kalimantan. CIRAD-foret, FORDA, INHUTANI, Indonesia, pp 51–73

Sukardjo S, Hagihara A, Yamakura T, Ogawa H (1990) Floristic composition of a tropical rainforest in Indonesian Borneo. Bull Nagoya Univ For 10:1–43

Tinal U, Palanewen JL (1978) Mechanical logging damage after selective cutting in the lowland dipterocarp forest at Beloro, East Kalimantan. Biotrop Spec Publ 3:91–96

Weidelt HJ (1996) Sustainable management of dipterocarp forests — opportunities and constraints. In: Schulte A, Schöne D (eds) Dipterocarp forest ecosystems. World Scientific, Singapore, pp249–273

Weidelt HJ, Banaag VS (1982) Aspects of management and silviculture of Philippine dipterocarp forest. GTZ, Eschborn

# 20 Effects of Forest Fragmentation on the Behavior of Bornean Gibbons

Teruki Oka, Ecep Iskandar, and Dadang Iman Ghozali

## 20.1 Introduction

According to a report compiled by the World Conservation Monitoring Centre (WCMC, 1996), the remaining closed-forest area (that is, the area having a tree-crown cover of over 30%) in Indonesia is about 117.9 million ha, and is being reduced at a rate of 1.2 million ha per year, due to forest fires and clearing for agriculture and plantations. However, the WCMC report does not mean that a vast forest has been gradually reduced from its edges inward. In other words, we must remember that deterioration of forests means not only gradual reductions in area but also division into smaller, isolated fragments (Harris 1980, 1984, Wilcove et al. 1986, Shafer 1990).

The ability of animals to tolerate and adapt to fragmentation of habitats should vary from species to species since it depends on species-specific morphological, physiological, and behavioral characteristics (Chivers 1991). A number of studies have focused on the effect of fragmented habitat on animal populations, resulting in considerable ecological insights into this subject (Saunders et al. 1991, Redpath 1995, Abensperg-Traun et al. 1996, Bellamy et al. 1996, Forys and Humphrey 1996, Smith et al. 1996). Fragmentation has caused a massive decline in population size in some species (Saunders et al. 1991, Fahrig and Merriam 1994), or a division into many small, isolated sub-populations in others (Pimm 1991).

The Bornean gibbon, *Hylobates muelleri*, is found in the lowland tropical rainforest in Kalimantan. An adult weighs about 5.0–6.5 kg (e.g., Leighton 1987) and shows no distinct sexual dimorphism. *Hylobates* is frugivorous, with a diet consisting mainly of juicy, ripe, and sugar-rich fruits (Rodman 1978, Gittins and Raemaekers 1980, Chivers and Raemaekers 1986). Some studies on the feeding behavior of gibbons suggested that they have a direct effect on the regeneration of these fruiting trees (Whitington and Treesucon 1991).

*Hylobates* is generally monogamous (reviewed by Dunbar 1988, see also Brockelman et al. 1998) and territorial (e.g., Gittins 1980). Females first menstruate at around 8 years of age, and bear their first young at 9 years of age (Carpenter 1940). Subsequently, they reproduce every 2 or more years (Carpenter 1940, Tilson 1981) throughout a reproductive life spanning 10–20 years (e.g., Ellefson 1974, Mitani

Ecological Studies, Vol. 140
Guhardja et al. (eds.) Rainforest Ecosystems of East Kalimantan: El Niño, Drought, Fire and Human Impacts

1990). Sexually mature young males and females emigrate from the natal groups at 8–10 years of age to establish a new group (e.g., MacKinnon and MacKinnon 1977). The territory is routinely maintained by singing performances of mated pairs at dawn (e.g., Mitani 1984). Their monogamy and territoriality account for the stability of size and structure of gibbon groups (e.g., Mitani 1990).

Gibbon activity usually starts at dawn and stops well before sunset. They are almost completely arboreal, only rarely venturing onto the ground. Normally they brachiate using their long arms, hands and flexible forelimb joints, so that they move more easily through the forest canopy than sympatric orangutans and macaques (Temerin and Cant 1983, Grand 1984, Leighton 1987). Recently, however, this evolutionary adaptive locomotion has become less adaptive in the deteriorated forests because the discontinuity of the forest canopy does not permit them to move from one remnant to another. In this chapter, we will describe how gibbons are influenced by the fragmentation and isolation of the forest and discuss their present status from the viewpoint of conservation.

## 20.2 Forests Inhabited by Gibbons

The Jalan Raya highway in East Kalimantan, Indonesia, is 120 km in length and connects Samarinda with Balikpapan. Since the 1960s, people from Sulawesi, Java, South Kalimantan and the Madura Islands have immigrated along this road, and have converted a vast forest into rice or pepper fields, retaining large stands of fragmented forest. In 1978, the Indonesian Government acted to prevent the destruction of these forests by granting protected forest status to a large portion of this region. Part of the semi-natural forest area within the protected forest has been under the management of Mulawarman University as the Bukit Soeharto Education Forest (BSEF) since 1979. In addition, a large tract of land that includes some of this forest area (710 $km^2$ in 1996) has been maintained as the recreational forest Taman Hutan Raya (Tahura) Bukit Soeharto (see Chapter 1, this volume) since 1982.

The field investigation of the present study was conducted in the BSEF and around the Tahura Bukit Soeharto from October 1995 to September 1997. The BSEF consists of semi-natural dipterocarp forest from which most of the available trees were selectively removed in the 1970s, together with degraded forests which suffered from large-scale forest fires in 1982–83. The flora of the BSEF is described in detail in chapter 9 in this volume. There are main roads (Jl. Ibu-1, Jl. Ibu-2 and Jl. Enggang B) and tracks (Jl. Landak, Jl. Enggang A and Jl. Banir) for research and educational activities (see Fig. 20.1). The main roads are 4–6 m wide, in some places more than 10 m, with no canopy covering in most places. In contrast, the narrow (2–3 m) and steep (>30°) tracks are covered with branches which prevent the passage of any vehicles.

There are numerous isolated forests around the Tahura Bukit Soeharto, and seven forests in which a gibbon family was found were selected for the field survey. Each fragmented forest was 16–25 ha in area and was surrounded by rice fields, pepper fields and grasslands which were dominated by alang-alang, *Imperata cylindrica*.

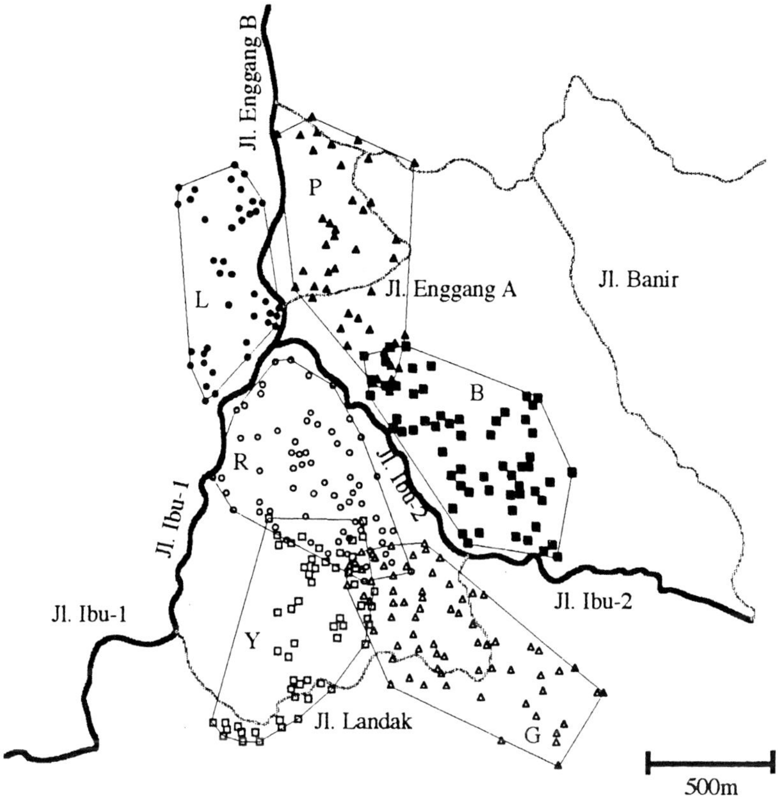


**Fig. 20.1.** Home ranges of six gibbon families in the BSEF. The main roads *Jl. Ibu-1*, *Ibu-2*, and *Enggang B* are open to all vehicles (*thick lines*) although the others, the tracks, are not (*dotted lines*). Data from December 1995 to November 1996 were combined. Each *symbol* shows the location of the tree that was used by the member of a given family

The trees, most about 15 m in height, were sparsely distributed as a result of the repeated burnings (e.g., there were two or three additional burnings after the forest fires of 1982–83) by immigrants practicing slash-and-burn agriculture (see Chapter 17, this volume). Trees more than 40 m in height, such as dipterocarps and fig trees, remained even after slash-and-burn agriculture due to their sacred status. The floristic composition of these secondary forests varied with the severity of the practice of slash-and-burn agriculture (see Kiyono and Hastaniah 1997).

## 20.3 Gibbon Families in the BSEF

In the BSEF, nine groups of gibbons were identified. In the core area of this forest (about 2.5 km$^2$), there were 20 individuals from six different groups (Table 20.1) and one solitary male, giving an average group size of 3.3 individuals. Each group was a nuclear family based on a mated pair. In family B, a young female was markedly

**Table 20.1.** Composition of the families in the BSEF

| Family | Adult | | Young | | Infant | Total |
|---|---|---|---|---|---|---|
| | Male | Female | Male | Female | | |
| L | 1 | 1 | - | - | - | 2 |
| P | 1 | 1 | - | - | 1 | 3 |
| R | 1 | 1 | - | - | 1 | 3 |
| Y | 1 | 1 | 1 | - | - | 3 |
| B | 1 | 1 | 1 | 1 | - | 4 |
| G | 1 | 1 | 1 | 1[a] | 1 | 5 |

[a]A young female who was considered to be sexually mature.

larger than her brother but smaller than her parents. In family G, a young female was considered to be almost sexually mature because of her size. Although infants of families P and R were carried by their mothers, the infant of family G was sometimes independent of his/her mother while the mother was resting or feeding.

Figure 20.1 shows the distribution of home ranges of these six families. Each family had a home range of about 25 ha, and these ranges sometimes overlapped with those of the neighbors. In areas which were shared between two families (P, B) or among three families (R, Y, G), inter-family conflicts were observed. All encounters were accompanied by vocalization and sometimes involved chasing. The vocalization was performed by adult females in almost all cases, sometimes with assistance of adult males or young males and females. Chasing behavior was observed between adult males. Trees which were frequently used by gibbons for foraging varied with the seasons, not only in the core area of their home range but also at the boundaries (Fig. 20.2). Furthermore, gibbons often visited trees which would provide foods in the following season.

A solitary male was recorded outside the home range of family B for the first time on January 25, 1996 (Fig. 20.3). Subsequently, he was found to be living in and around the range of family B, and frequently encountered the members of family B. In most cases, he was kept away by their vocalizing. Later, he was observed in a tall tree at the roadside of Jl. Ibu-2. On May 31, he repeatedly climbed up and down the tree before pausing on a branch 40 m high. After sitting on the branch for 20 minutes, he leaped over the main road (Jl. Ibu-2) to a branch of another tree on the opposite side. The width of the road was about 7 m, and the distance between the branches was about 5 m. After this incident, he was observed in the home range of family R for about 1 month, and entered the home range of family G in July 1996. A sexually mature young female of family G appeared to accept him, and they were sometimes observed sitting together on a branch, although the solitary male was often chased off by her father.

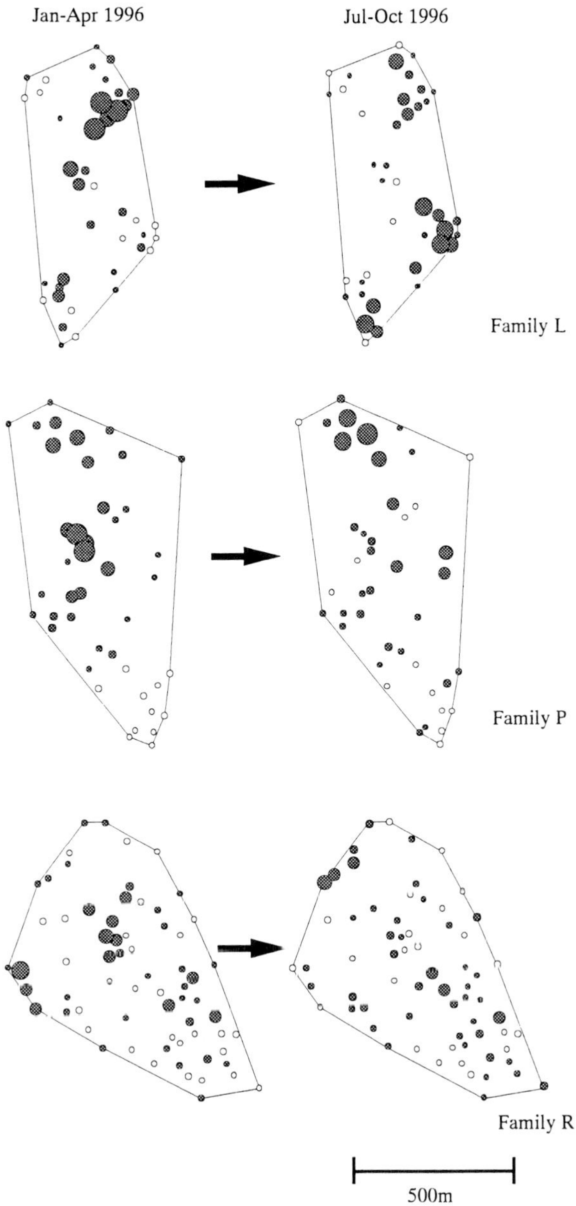


**Fig. 20.2.** Seasonal differences in the center of activity of three families in the BSEF. Gibbons of these families were located once every tracking session (05:00–07:00, 07:00–09:00, 09:00–12:00, 12:00–15:00) for 25 days. *Larger circles* represent trees that were visited more frequently

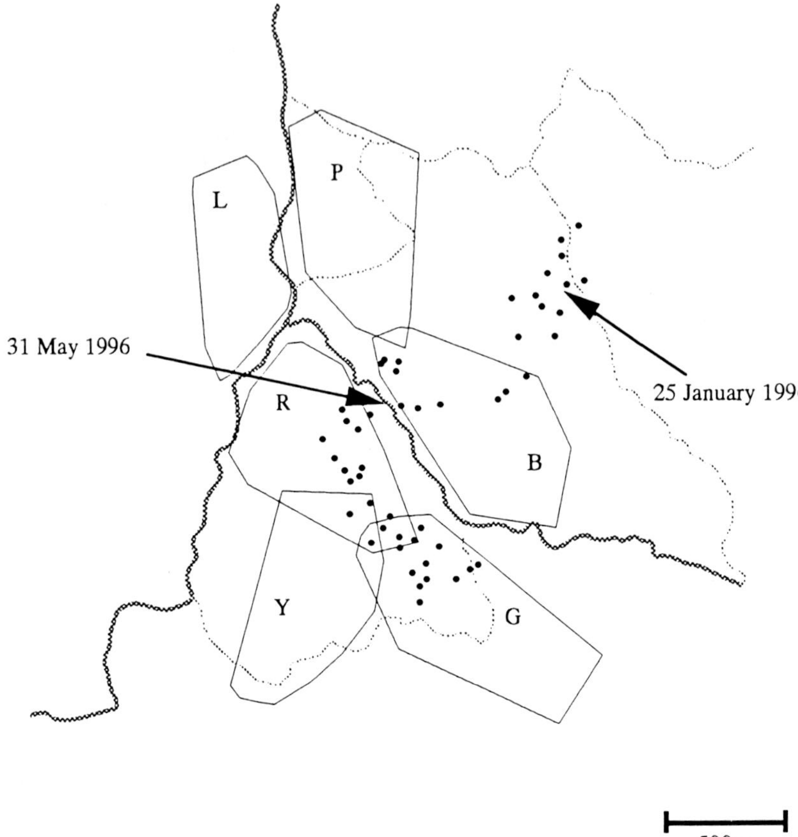


**Figure 20.3.** Movement of a solitary young male. He was found outside the home range of family B on January 25, 1996, for the first time. On May 31, he leaped over the forest road. Then he approached a young female of family G

## 20.4 Gibbon Families in the Isolated Forests

Table 20.2 shows the composition of the families in seven isolated forests around the Tahura Bukit Soeharto. The approximate areas of these fragmented forests were 16, 18, 24, 25, 24, 20 and 20 ha (from top to bottom in Table 20.2). A pair, comprising male and female, in Tanimaju was released in about 1984, but there was no information on the origin of other pairs and families. The family size (4.9 individuals on average) in these forests was higher than in the BSEF (cf. Table 20.1).

Figure 20.4 shows the durations over which individuals stayed on the same bough based on day-long observations at the Amborawang Forest (a) and Tanimakmur-2 Forest (b) in June 1996 and 1997. At Amborawang, in 1996 the father spent more time on a bough with his mate than with his daughter ($U$=13.0, $P$<0.05), and in 1997 he spent more time on a bough with his daughter than with his mate ($U$=8.0, $P$<0.01).

**Table 20.2.** Composition of the families in the isolated forests

| Family (Forests) | Adult Male | Adult Female | Young Male | Young Female | Infant | Total |
|---|---|---|---|---|---|---|
| Amborawang | 1 | 1 | - | 1[a] | - | 3 |
| Tanimakmur-1 | 1 | - | 1 | 1 | | 3 |
| Tanimakmur-2 | 1 | 1 | 1 | 1[a] | 1 | 5 |
| Tanimakmur-3 | - | 1 | 2 | 2 | 1 | 6 |
| Tanijaya | 1 | 1 | 1 | 2 | 1 | 6 |
| Tanimaju | 1 | 1 | 1 | 2 | - | 5 |
| Sumberkarya | 1 | 1 | 2 | 1 | 1 | 6 |

A pair of adult male and female in Tanimaju was released in *c*. 1984.
[a]A young female who was considered to be sexually mature

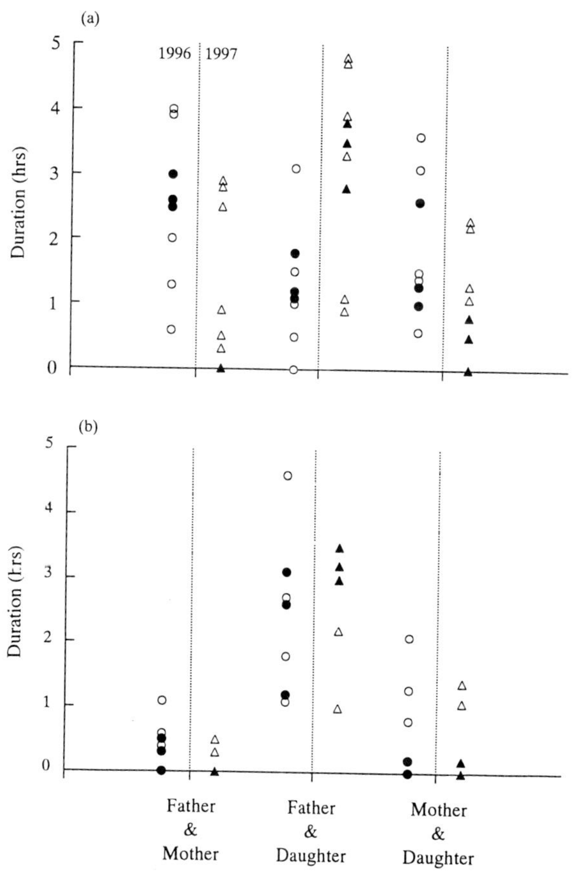


**Figure 20.4a,b.** Duration of an individual staying on the same bough with another per 11 h at Amborawang **(a)** and Tanimakmur-2 **(b)** Forests in June 1996 (*circles*) and June 1997 (*triangles*). *Open symbols* show the estimates from over 6 h (< 11h) tracking data and *solid symbols* from 11 h tracking data

In fact, the daughter generally spent more time with her father in 1997 than in 1996 ($U$=11.5, $P$<0.05). The relationships between the mother and her daughter became estranged ($U$=15.5, $P$<0.05). On the other hand, at Tanimakmur-2 Forest, the father and his daughter were observed frequently on the same bough in both 1996 and 1997 (Kruskal–Wallis $H$=10.7, $P$<0.01 in 1996, $H$=8.6, $P$<0.05 in 1997). Between 1996 and 1997, there were no significant changes in the relationships between the father and the mother ($U$=10.0, $P$>0.1), or between the mother and her daughter ($U$=16.0, $P$>0.1). There were also no significant changes between these years in the amount of time that the father and his daughter spent together on the same bough ($U$=15.0, $P$>0.05). These results showed intimacy between the father and his sexually mature daughter.

## 20.5 Forest Roads as Barriers

This study showed that gibbons might be one of the species most sensitive to forest fragmentation. They can live in home ranges straddling tracks, but their movements appeared to be restricted by main roads (Fig. 20.1). According to Gittins (1983), gibbons spend most of the day in the main canopy (about 15–35 m high in primary forest), not only for foraging and resting but also for traveling between food resources. It is very rare for them to walk on the ground because they are unskillful bipedal walkers on the ground, and become vulnerable to attacks by predators. Gibbons are more likely to travel along the edge of a forest patch, and are less likely to cross the roads when the canopy is discontinuous. This might be due to psychological pressure, as shown by the fact that a solitary male leaped over the road after much hesitation.

Whitington and Treesucon (1991) showed that the white-handed gibbon *H. lar* was largely dependent on seasonal foods. A given food resource for gibbons may be available for only a few weeks in a year, or only once every few years (e.g., Medway 1972, Leighton and Leighton 1983). In this study at the BSEF, the trees where gibbons were observed varied with the seasons (Fig. 20.2), suggesting that the gibbons moved among widely scattered resources all year round. Furthermore, they sometimes visited trees which, though they had no resources available at the time, would provide resources in the following season. They appeared to pay attention to the phenology of trees in their home ranges. Their feeding strategies would probably be markedly affected not only by the partial disappearance of fruiting trees accompanied by construction of roads, but also by appearance of the new roads themselves. We should also note that human activities, such as road maintenance and some works accompanied by planting, are likely to accelerate the restriction of free-ranging behavior of animals. In the BSEF, main roads were frequently bulldozed and widened to accelerate the drying process after heavy rain. Also, *Macaranga* species along the main road were partly barked (by girdling) and cut down in order to accelerate the growth of dipterocarp seedlings which had been planted there. Such treatments have the same effects as road-widening. If these roads were further widened, they would become barriers preventing the passage of gibbons of all ages. It should be remembered that forest fragmentation occurs not only as a result of logging and cultivation, but also by the construction of roads, railroads, fences and firebreaks (Buechner 1987, Stamps et al. 1987, Schonewald-Cox and Buechner 1992, Fahrig and Merriam 1994).

## 20.6 Isolated Forests as Fragile Sanctuaries

Gibbons are one of the most conspicuous animals, since they sometimes take a rest on the tops of tall trees and sing loudly. One week's stay near a forest is sufficient to determine whether gibbons are living there or not. Many local people living around the forests inhabited by gibbons could tell how many gibbons there are. Are these forests really suitable places for gibbons? Isolated forest patches can often supply some food to animals (Wheelwright 1983, Stiles 1988) and are preferable to no forests. We can say quite opportunistically that these isolated forests function as natural "sanctuaries".

However, gibbons in these small forests near the dwellings of local people were frequently influenced by human activities such as logging and agriculture. In a forest at Amborawang, a fig tree of about 40 m in height had been sanctified and spared from felling for many years. Between December and February every year, the tree had provided food and a resting place for a gibbon family. Then, the tree was eventually felled in order to timber a valuable tree, *Eusideroxylon zwageri* (ulin) that had been leaning against it. This event triggered burning for cultivation in the area. Gibbons in isolated forests are always faced with a no-win situation.

Furthermore, larger family size in these isolated forests suggested that the discontinuity of forests prevented natural dispersion of young individuals. Generally, sexually mature gibbons are known to disperse from the range of their natal families (e.g., Carpenter 1940, Tilson 1981). However, in some families of this study, there were females that were similar in size to adults and were considered to be sexually mature (Table 20.2). Brockelman et al. (1998) recently reported that lack of available vacant areas for sub-adults induces their delayed dispersal. In the case of the isolated forests that we observed, it is never possible for mature gibbons to leave their natal territories, because they would be obliged to walk more than 2 km across the fields and/or grasslands to visit the nearest forest.

From a long-term view, the isolated families are likely to be confronted by an inbreeding problem. Monogamy is basically maintained through the exclusion of mature offspring from breeding pairs (e.g., Kleiman 1977, Wittenberger and Tilson 1980) and this is important for the avoidance of inbreeding (e.g., Pusey 1980, see also Shields 1982). Mating among close relatives reduces the level of genetic diversity, which might influence the growth of populations, the ability of populations to adapt to environmental change, and the potential for evolution (Allendorf and Leary 1986, Lande and Barrowclough 1987, Mitton 1993). Such reduced levels of genetic diversity were considered to be a characteristic of species inhabiting oceanic islands or isolated springs (Kilpatric 1981, Schnell and Selander 1981, Echelle et al. 1989). Today, however, concerns about the deterioration of genetic diversity in small isolated populations are growing in conservation biology (summarized by Heywood and Stuart 1992).

Although we did not observe any incestuous mating in these two forests, if gibbons do not behave peremptorily to avoid incest, the problem of inbreeding depression will remain as a future possibility so long as the young do not disperse. Indeed, incestuous pairs consisting of a son and his widowed mother or of siblings were sometimes found even in populations of natural forests (e.g., Tilson 1981). It is certain that it takes a much longer time to detect the evidence of inbreeding depression.

## 20.7 Conclusion

In the process of evolution, the populations of gibbons have been separated by rivers and straits which have accelerated the divergence of the subgenus or subspecies of *Hylobates*. This is why the taxonomy within subgenus *Hylobates* has still been disputable and remains a dynamic field (e.g., Geissmann 1995). Presently, however, the distribution of gibbons has been further fragmented by human activities. Some populations have been confined to small, isolated forests, and the others free-ranging behavior has been limited. The future of these animals depends on the maintenance of non-fragmented forest reserves.

In Indonesia, there have been a number of rehabilitation programs for orangutans. Our objective here, however, is not to argue for such programs for the conservation of gibbon populations. We believe that these programs are certainly useful; however, it is more imperative that we learn what is bringing the gibbons to the point of extinction, what impact such extinction might have on other components of the ecosystem, and how we might apply this knowledge to the maintenance of other species. The implementation of effective conservation measures will depend on an understanding of the ecological information on that species. Primate conservation has been part of a broader program of habitat, floral and faunal preservation and is not specifically focused on the primates themselves. Now we face a turning point and must develop a management system for protected forests so as to shelter other species from the fate of the orangutan. The smartest primate, after all, is not the gibbon, not the orangutan, but we, who have termed ourselves *Homo sapiens*.

## References

Abensperg-Traun M, Smith GT, Arnold GW, Steven DE (1996) The effects of habitat fragmentation and livestock-grazing on animal communities in remnants of gimlet *Eucalyptus salubris* woodland in the Western Australian wheatbelt. I. Arthropods. J Appl Ecol 33:1281–1301

Allendorf FW, Leary RF (1986) Heterozygosity and fitness in natural populations of animals. In: Soulé ME (ed) Conservation biology: the science of scarcity and diversity. Sinauer Associates, Sunderland, pp 57–76

Bellamy PE, Hinsley SA, Newton I (1996) Local extinctions and recolonisations of passerine bird populations in small woods. Oecologia 108:64–71

Brockelman WY, Reichard U, Treesucon U, Raemaekers JJ (1998) Dispersal, pair formation and social structure in gibbons (*Hylobates lar*). Behav Ecol Sociobiol 42:329–339

Buechner M (1987) A geometric model of vertebrate dispersal: tests and implications. Ecology 68:310–318

Carpenter CR (1940) A field study in Siam of the behavior and social relations of the gibbon *Hylobates lar*. Comp Psychol Monogr 16:1–212

Chivers DJ, Raemaekers JJ (1986) Natural and synthetic diets of Malayan gibbons. In: Else JG, Lee PC (eds) Primate ecology and conservation. Cambridge University Press, Cambridge, pp 115–123

Chivers DJ (1991) Species differences in tolerance to environmental change. In: Box HO (ed) Primate responses to environmental change. Chapman and Hall, London, pp 5–37

Dunbar RIM (1988) Primate social systems. Croom Helm, London

Echelle AF, Echelle AA, Edds DR (1989) Conservation genetics of a spring-dwelling desert fish, the Pecos gambusia (*Gambusia nobilis,* Poeciidae). Conserv Biol 3:59–169

Ellefson, JO (1974) A natural history of the white-handed gibbon in the Malayan Peninsula. In: Rumbaugh DM (ed) Gibbon and siamang vol 3, Karger, Basel, pp 1–136

Fahrig L, Merriam G (1994) Conservation of fragmented populations. Conserv Biol 8:50–59

Forys EA, Humphrey R (1996) Home range and movements of the lower keys marsh rabbit in a highly fragmented habitat. J Mammal 77:1042–1048

Geissmann T (1995) Gibbon systematics and species identification. Int Zoo News, 42:467–501

Gittins SP (1980) Territorial behavior in the agile gibbon. Int J Primatol 1:381–399

Gittins SP (1983) Use of the forest canopy by the agile gibbon. Folia Primatol 40:134–144

Gittins SP, Raemaekers JJ (1980) Siamang, lar and agile gibbons. In: Chivers DJ (ed) Malayan forest primates: ten year's study in tropical rainforest. Plenum, New York, pp 63–105

Grand TI (1984) Motion economy within the canopy: four strategies for mobility. In: Rodman PS, Cant JGH (eds) Adaptations for foraging in nonhuman primates. Columbia University Press, New York, pp 54–72

Harris LD (1980) Forest and wildlife dynamics in the southeast. Trans North American Wildlife Natural Resource Conference 45:307–322

Harris LD (1984) The fragmented forest: island biogeography theory and the preservation of biotic diversity. University of Chicago Press, Chicago

Heywood VH, Stuart SN (1992) Species extinctions in tropical forests. In: Whitmore TC, Sayer JA (eds) Tropical deforestation and species extinction. Chapman and Hall, London, pp 91–117

Kilpatric CW (1981) Genetic structure of insular populations. In: Smith MH, Joule JJ (eds) Mammalian population genetics. University of Georgia Press, Athens, pp 28–59

Kiyono Y, Hastaniah (1997) Slash-and-burn agriculture and succeeding vegetation in East Kalimantan. PUSREHUT Special Publication vol 6, Mulawarman University, Samarinda, Indonesia

Kleiman DG (1977) Monogamy in mammals. Q Rev Biol 52:39–69

Lande R, Barrowclough GF (1987) Effective population size, genetic variation, and their use in population management. In: Soulé ME (ed) Viable populations for conservation. Cambridge University Press, Cambridge, pp 87–123

Leighton DR (1987) Gibbons: territoriality and monogamy. In: Smuts BB, Cheney D, Seyfarth R, Wrangham R, Struhsaker T (eds) Primate societies. University of Chicago Press, Chicago, pp 135–145

Leighton M, Leighton DR (1983) Vertebrate responses to fruiting seasonality within a Bornean rainforest. In: Sutton SL, Whitmore TC, Chadwick AC (eds) Tropical rainforest: ecology and management. Blackwell, Oxford, pp 181–196

MacKinnon J, MacKinnon K (1977) The formation of a new gibbon group. Primates 18:701–708

Medway L (1972) Phenology of a tropical rainforest in Malaya. Biol J Linn Soc 4:117–146

Mitani JC (1984) The behavioral regulation of monogamy in gibbons (*Hylobates muelleri*). Behav Ecol Sociobiol 15:225–229

Mitani JC (1990) Demography of agile gibbons (*Hylobates agilis*). Int J Primatol 11:411–424

Mitton JB (1993) Theory and data pertinent to the relationship between heterozygosity and fitness. In: Thornhill NW (ed) The natural history of inbreeding and outbreeding. University of Chicago Press, Chicago, pp 17–41

Pimm SL (1991) The balance of nature? ecological issues in the conservation of species and communities. University of Chicago Press, Chicago

Pusey AE (1980) Inbreeding avoidance in chimpanzees. Anim Behav 28:543–552

Redpath SM (1995) Habitat fragmentation and the individual: tawny owls *Strix aluco* in woodland patches. J Anim Ecol 64:652–661

Rodman P (1978) Diets, densities and distributions of Bornean primates. In: Montgomery GG (ed) The ecology of arboreal folivores, Smithsonian Institution, Washington D.C., pp 465–478

Saunders DA, Hobbs RJ, Margules CR (1991) Biological consequences of ecosystem fragmentation: a review. Conserv Biol 5:18–34

Schnell GD, Selander RK (1981) Environmental and morphological correlates of genetic variation in mammals. In: Smith MH, Joule JJ (eds) Mammalian population genetics. University of Georgia Press, Athens, pp 60–99

Schonewald-Cox CM, Buechner M (1992) Park protection and public roads. In: Fiedler PL, Jain SK (eds) Conservation biology: the theory and practice of nature conservation, preservation, and management. Chapman and Hall, London, pp 373–396

Shafer CL (1990) Nature reserves: island theory and conservation practice. Smithsonian Institution, Washington D.C.

Shields WM (1982) Philopatry, inbreeding, and the evolution of sex. State University of New York, Albany

Smith GT, Arnold GW, Sarre S, Abensperg-Traun M, Steven DE (1996) The effects of habitat fragmentation and livestock grazing on animal communities in remnants of gimlet *Eucalyptus salubris* woodland in the Western Australian wheatbelt. II. Lizards. J Appl Ecol 33:1302–1310

Stamps JA, Buechner M, Krishnan VV (1987) The effects of edge permeability and habitat geometry on emigration from patches of habitat. Am Nat 129:533–552

Stiles FG (1988) Altitudinal movements of birds on the Caribbean slope of Costa Rica: implications for conservation. In: Almeda F, Pringle CM (eds) Tropical rainforests: diversity and conservation. pp 243–258

Temerin LA, Cant JGH (1983) The evolutionary divergence of Old World monkeys and apes. Am Nat 122:335–351

Tilson RI (1981) Family formation strategies of Kloss's gibbons. Folia Primatol 35:259–281

WCMC (1996) Forest statistics: forests for life campaign WWF/WCMC 1996 [http://www.wcmc.org.uk/]

Wheelwright NT (1983) Fruits and the ecology of resplendent quetzals. Auk 100:286–301

Whitington C, Treesucon U (1991) Selection and treatment of food plants by white-handed gibbons (*Hylobates lar*) in Khao Yai National Park, Thailand. Nat Hist Bull Siam Soc 39:111–122

Wilcove DS, MacMillan M, Winston KC (1986) Habitat fragmentation in the temperate zone. In: Soulé ME (ed) Conservation biology: the science scarcity and diversity. Sinauer Associates, Sunderland, pp 237–256
Wittenberger JE, Tilson RL (1980) The evolution of monogamy: hypotheses and evidence. Ann Rev Ecol Syst 11:197–232

# Part 5

# Rehabilitation Techniques and Constraints on Sustainable Forest Management

# 21 Treatment to Develop Mycorrhiza Formation on Dipterocarp Seedlings

SUHARDI

## 21.1 Introduction

Forests in Indonesia extend over approximately 65 million ha with most of the area dominated by dipterocarps, which are one of Indonesia's main exports. Most dipterocarps belong to families in which mutualistic associations with fungi are known to form. There is increasing evidence that mycorrhiza colonization increases the survival and growth of planted dipterocarp seedlings (e.g., Chapter 22, this volume). There is little doubt that one of the functions of mycorrhizal symbiosis is to increase the efficiency of the uptake of water and minerals such as phosphate ions from the soil (e.g., Read 1994). This is especially important in Indonesia where 80% of the land area has acidic soils with an associated shortage of available phosphate ions. Therefore, the release of phosphate into the soil by mycorrhizal symbiosis is particularly important for plant growth in Indonesia. Dipterocarp seedlings naturally establish in forest floors which have moderate light and moisture conditions. However, when dipterocarp seedlings are planted in areas affected by shifting cultivation and fire, they are exposed to particularly dry and exposed conditions. As a result of these harsh conditions, the selection of suitable mycorrhizal fungi for use in inoculation can be important, especially during the establishment of dipterocarp forests in reforestation programs (Julich 1988, Smits 1992).

This chapter presents the experimental results of an investigation into the conditions which optimize the association of fungi with *Shorea* tree species, including *Shorea bracteolata*, *S. acuminata*, *S. smithiana*, *S. johorensis*, and *S. parvifolia*. The experiments were conducted at the Bukit Soeharto Education Forest and Long Nah in East Kalimantan and Jambi Education Forest of Gadjah Mada University in Sumatra. The optimal conditions of many environmental factors, such as the relationship between light levels and temperature, the inoculation of fungi (*Scleroderma columnare*), and the effect of soil conditioners such as rice husk charcoal are discussed with reference to the results of these field and nursery experiments.

Ecological Studies, Vol. 140
Guhardja et al. (eds.) Rainforest Ecosystems of East Kalimantan: El Niño, Drought, Fire and Human Impacts

## 21.2 Shading Effects

Generally, mycorrhizal fungi need relatively low temperatures during the early stages of propagation. Mycorrhizal infection generally occurs when the environmental temperature is lower than 30°C. When the soil surface is completely covered with leaf litter, twigs and other plants, the soil-root systems are kept at a relatively low temperature with increased humidity, thus providing good conditions for mycorrhizal growth. There are however, interspecific differences in the suitability of dipterocarps to different conditions. For example, *Shorea leprosula* is a relatively adaptable dipterocarp species suiting open areas with high light levels and dry conditions (Suhardi et al. 1993, Zipperlin and Press 1996, Ishida et al. 1999). Therefore, this species would be important for planting of timber production, especially in areas degraded by shifting cultivation and fire. In comparison, *Shorea acuminatissima*, *Shorea bracteolata*, *Shorea johorensis*, *Shorea parvifolia*, and *Shorea seminis* require shaded conditions during the early stages of their growth (Suhardi et al. 1996, unpublished).

When these relatively shade-adapted dipterocarp seedlings were planted in environments with different light and phosphate fertilization levels, it was found that the

**Table 21.1.** Effect of shading on the height growth of *Hopea gregaria* seedlings

| Treatment | Height (cm) |
|---|---|
| Glass house | 8.8 a |
| Screen house | 1.5 b |

The letters (a, b) after data show the significant differences ($P < 0.05$).

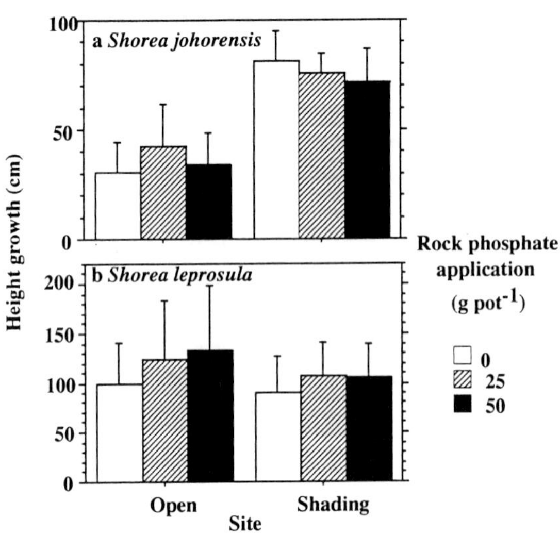


**Fig. 21.1a,b.** Height growth of *Shorea* seedlings in an open and a shaded site under different rates of rock phosphate fertilizer application. **a** *Shorea johorensis* **b** *Shorea leprosula*

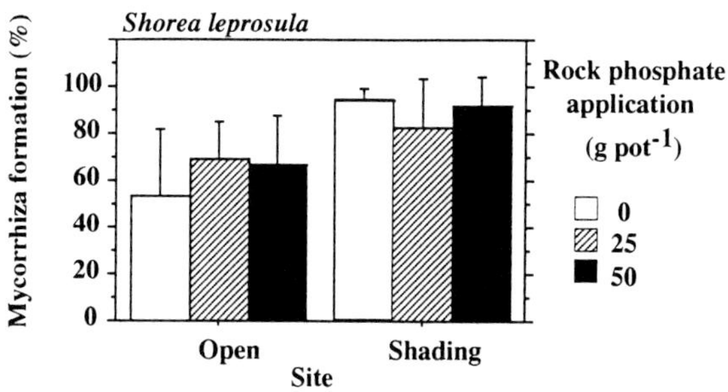


**Fig. 21.2.** Mycorrhiza formation of *Shorea leprosula* seedlings in an open and a shaded site under different rates of rock phosphate fertilizer application

seedling height growth and mycorrhiza formation were very poor in open areas compared to shaded sites. For example, growth in *Shorea johorensis* and *Hopea gregaria* was found to be enhanced in shaded sites compared with open areas (Fig. 21.1a, Table 21.1), i.e., these species favor shaded conditions. Suhardi et al. has observed in another experiment that the average mycorrhiza formation rate in the above group of *Shorea* spp., which favor shade conditions, was 26% in the shaded site compared with 6% in the open site (data not shown). In contrast, the growth of *Shorea leprosula* was slightly enhanced in the open site compared to the shaded site (Fig. 21.1b), although the difference was not significant statistically, i.e., this species grow well in sunlit conditions. The average mycorrhiza formation rate in *Shorea leprosula* was 89.3 ±14.7% in the shaded site and 62.9 ± 22.2% in the open site (Fig. 21.2). A more sun-adapted dipterocarp showed similar height growth and mycorrhiza formation in open and shade sites. Application of phosphorus fertilizer did not affect seedling growth of the dipterocarps used, probably due to the high mycorrhiza formation rates (Fig. 21.1, and 21.2).

## 21.3 Effect of Inoculum and Selection of Appropriate Fungi

In the natural conditions of dipterocarp forests, where dipterocarp seedlings are distributed around a parent tree, the seedlings may be able to acquire suitable fungi for mycorrhiza formation from the parent. Occasionally the seeds disperse to areas away from the parent tree and germinate amongst trees of other families. In this case, they cannot be easily colonized by suitable ectomycorrhizal fungi. Dipterocarp seedlings found in these areas soon disappear.

In order for planted dipterocarp seedlings to successfully establish dipterocarp plantations, a source of suitable mycorrhizal inoculum is important. For instance, inoculation could increase the growth of *Dryobalanops* seedlings (Suhardi 1993). In order to construct a list of mycorrhizal fungi which could be used as a source of inoculum in dipterocarp plantations, observations have been made for approximately six years in Wanariset, East Kalimantan (Yasman 1993). About 172 distinguishable species from 36 genera of fungi were found associated with dipterocarps in this area, approximately 40 have been identified to species level. At Wanariset, the dominant mycorrhizal fungi belonged to the genera; *Amanita*, *Boletus*, *Cantharellus*, *Clavaria*, *Clitopilus*, *Cortinarius*, *Gomphidius*, *Lactarius*, *Leccinum*, *Paxillus*, *Russula*, *Tricholoma*, *Hebeloma*, *Hydnum*, *Hygrophorus*, and *Inocybe* (Yasman 1993). In the Philippines, 32 species of ectomycorrhizal fungi were found in dipterocarp forests with the most abundant fungi belonging to the genera; *Rusulla* and *Lactarius* (Zarate et al. 1993). In mycorrhiza fungi experiments, *Laccaria*, *Pisolithus*, *Rhizopogon*, and *Scleroderma* are the most commonly used species in the tropical regions of Southeast Asia (Supriyanto et al. 1993, Zarate et al. 1993). Although Indonesia has large dipterocarp forests of more than 100 million ha, no *Pisolithus tinctorius* have been found or reported in the country.

It is possible that dipterocarp trees of the same species but at different stages of growth may develop associations with different species of fungi. For example, seedlings initially associate with *Scleroderma*, but after five years *Laccaria* may become dominant. During field observations within well developed dipterocarp forests the author has observed that *Boletus* and *Amanita* and other species are more dominant than *Scleroderma*, which was the most common fungi in the nursery.

## 21.4 Inoculum, Mulching, and Charcoal Treatments to Enhance Mycorrhizal Infection

It is clear that there is a special relationship between certain dipterocarp species and mycorrhizal fungal species (Chapter 22, this volume). However, areas left exposed for long periods following slash-and-burn agriculture and fires may have reduced levels of the suitable mycorrhizal fungi (Read 1994). In such conditions, mycorrhizal fungal infection cannot be simply brought about by fungi associated with one plant quickly infecting the other plants. Therefore, reinoculation is often needed to increase the rate of infection of mycorrhiza fungi amongst planted dipterocarp seedlings. Some methods for the reinoculation of planted dipterocarp seedlings have been developed: 1) seedlings with existing ectomycorrhizal fungi are planted amongst the new seedlings, 2) soil from around parent trees or infected nurseries is used, 3) fruit bodies and spores of suitable fungi are buried and 4) cultured mycelia of the selected fungi are buried.

In nurseries, the source of the soil is especially important. However, the species composition in the area where the soil is collected must be considered. For example, the soil from the Bukit Soeharto Education Forest (a well developed dipterocarp forest) showed increased mycorrhizal formation compared with the soil from Lempake (an extremely degraded dipterocarp forest) in East Kalimantan (Suhardi and Darmawan 1991). In addition, mulching treatment can also be useful for mycorrhizal formation on dipterocarp saplings. There is increased mycorrhizal formation under thick layers of leaf litter on forest floors where conditions are humid and temperatures lower. However, not all types of litter are suitable for mycorrhizal development around dipterocarp seedlings. Although *Imperata cylindrica* (the local name is alang-alang) is easily obtained in East Kalimantan, it is not a suitable material for covering planted seedlings. Apparently, a toxic substance data in alang-alang grass reduces mycorrhizal formation (Suhardi, unpublished data).

In order to enhance mycorrhizal formation, charcoal treatments have been used by some agriculturists and foresters in the tropics. Typically the materials used are woody charcoal, rice husk charcoal, and coconut husk charcoal. Brick fragments are also sometimes used. There is a remarkable promotion of growth in dipterocarp seedlings following the application of rice husk charcoal to the soil, and these are accompanied by changes of leaf color from yellow-green to dark-green (Mori and Marjenah 1994). Therefore, the use of rice husk charcoal should be recommended when dipterocarp seedlings are planted in fields.

## 21.5 Conclusion

On the forest floor of the humid tropics, nutrient constraints rather than water shortage is the dominant factor limiting the growth and establishment of dipterocarp seedlings. Most dipterocarp species growing in Southeast Asia are colonized by microbial mutualists. In particular, mycorrhizal formation enhances plant growth in areas with shaded, wet, and cool conditions. There is mounting evidence that one of the functions of mycorrhizal symbiosis is to enhance the efficiency of the uptake of phosphate ions from the soil. Phosphate is known to be an important limiting nutrient in lowland tropical soils (e.g., Vitousek 1982, Troth et al. 1986). Therefore, developments which increase the infection rate of mycorrhizal fungi are important in the rehabilitation of degraded forests in the tropics. Enhanced mycorrhizal growth increases the chance of seedlings becoming established in natural and planted forests. Also, good mulching, reinoculum and charcoal treatments are effective at enhancing mycorrhizal colonization. The effect of charcoal treatments on plant growth and the photosynthesis process are treated in the following chapter (Chapter 22, this volume).

## References

Ishida A, Nakano T, Matsumoto Y, Sakoda M, Ang LH (1999) Diurnal changes in leaf gas & change and chlorophyll fluorescence in tropical tree species with contrasting light requirements. Ecol Res 14:77–88

Julich W (1988) Dipterocarpaceae and mycorrhizae. GFG Report No 9 Special Issue, GTZ GFG at Mulawarman University, East Kalimantan

Mori S, Marjenah (1994) Effect of charcoaled rice husk on the growth of Dipterocarpaceae seedlings in East Kalimantan with special reference to ectomycorrhiza formation. J Jpn For Soc 76:462–464

Read DJ (1994) Plant-microbe mutualisms and community structure. In: Schulze ED, Mooney HA (eds) Biodiversity and ecosystem function. Springer, Berlin Heidelberg New York, pp 181–209

Smits WTM (1992) Mycorrhizal studies in dipterocarp forests in Indonesia. Mycorrhizas in ecosystems 9. CAB International, pp 283–292

Suhardi (1993) Effect of fertilizer, charcoal and mycorrhiza inoculation to the growth of *Dryobalanops* sp. Proceeding National Seminar on peat BPPT

Suhardi, Darmawan A (1991) Effect on light intensity , soil type and inoculation on mycorrhizal formation and initial growth of *Shorea bracteolata* seedlings. Annual report of PUSREHUT 1:47–62

Suhardi, Faridah E, Iskandar E, Rahayu S (1993) Mycorrhiza formation and growth of *Shorea leprosula* after using charcoal and rockphosphate in Bukit Suharto. Proc Inter Workshop BIO-REFOR 1993, BIO-REFOR/IUFRO/SPDC, pp 196–200

Supriyanto, Setiawan I, Omon M (1993) Effect of *Scleroderma* sp. on the growth of *Shorea mecistopteryx* Ridi seedings. Proc Inter Workshop BIO-REFOR 1993, BIO-REFOR/IUFRO/SPDC, pp 186–188

Troth JL, Campbell JL, Allen HL (1986) Nutrients: Use of forest fertilization and nutrient efficiency genotypes to manage nutrient stress in conifer stands. In: Hennessey TC, Dougherty PM, Kossuth SV, Johnson JD (eds) Stress Physiology and Forest Productivity. Martinus Nijhoff, Dordrecht, pp 9–99

Vitousek P (1982) Nutrient cycling and nutrient use efficiency. Amer Nat 119:553–572

Yasman I (1993) Ectomycorrhizal sporocarp appearance in a Dipterocarp Forest, East Kalimantan, Indonesia. Proc Inter Workshop BIO-REFOR 1993, BIO-REFOR/IUFRO/SPDC, pp 179–181

Zarate JT, Watling R, Jeffries P, Dodd JC, Pampolina NM, Sims K, Lorilla EB, de la CRUZ RE (1993) Survey of ectomycorrhizal fungi associated with pines and dipterocarps in the Philippines. Proc Inter Workshop BIO-REFOR 1993, BIO-REFOR/IUFRO/SPDC, pp 183–185

Zipperlin SW, Press MC (1996) Photosynthesis in relation to growth and seedling ecology of two dipterocarp rain forest tree species. J Ecol 84:863–876

# 22 A Convenient Method for Inoculating Dipterocarp Seedlings with the Ectomycorrhizal Fungus, *Scleroderma columnare*

SHIGETA MORI and MARJENAH

## 22.1 Introduction

Mycorrhizas are symbiotic associations between specialized root-inhabiting fungi and living roots of plants. Ectomycorrhizal colonization alters the niche structure of host plants by increasing their capacity for nutrient acquisition (Allen 1991). It is well known that phosphate uptake of the host trees is enhanced by excretion of organic acids from ectomycorrhizas (Mejstrik and Krause 1973, Ho and Zak 1979, Lapeyrie et al. 1987, Iwase 1991), for instance, and that ectomycorrhizal fungi usually promote nitrogen uptake of the host plants (Plassard et al. 1994). Both phosphate and nitrogen are essential nutrients for plants, which require them for carbon gain, growth and other critical life processes (Evans 1983). Recent studies have revealed that various other microorganisms such as bacteria (Garbaye 1994) and soil fauna also help mobilize nutrients in the mycorrhizosphere (Setala 1995). Furthermore, water uptake of host plants is promoted by mycorrhiza through the huge mycelial surface area that develops (Duddridge et al. 1980). Thus, clear differences should exist in niche nutrition status between mycorrhizal and non-mycorrhizal plants growing in the same place (Allen 1991).

Mycorrhizas are viewed as "biofertilizers" and potential alternatives to chemical fertilizers for infertile tropical soils where reforestation is being carried out (Lee 1998). Most dipterocarp species are obligate mycotrophs that cannot survive without ectomycorrhizas (Smits 1994), since they markedly improve the ability of the dipterocarp trees to take up nutrients from the nutrient-poor tropical soil (Lee and Lim 1989, Turner et al. 1993). *Scleroderma columnare* is known to be one of the most common and most strongly growth-promoting of the ectomycorrhizal fungi associated with dipterocarp species (Santoso 1988). Recently, Kikuchi and Ogawa (1995) confirmed that nitrogen-fixing bacteria co-exist around the ectomycorrhiza of *Scleroderma* spp. associated with dipterocarps. Thus, it is better not only for encouraging mycorrhizal fungi, but also to promote the diversity of soil microbes, to provide suitable conditions for dipterocarp tree symbioses with species such as *Scleroderma columnare*.

In this chapter, we introduce a simple and practical technique for inoculating dipterocarp seedlings with the ectomycorrhizal fungi, *S. columnare*, and describe its

Ecological Studies, Vol. 140
Guhardja et al. (eds.) Rainforest Ecosystems of East Kalimantan: El Niño, Drought, Fire and Human Impacts

effects on the growth and various physiological responses of the host plant. We examined the efficiency of the technique both in nursery and field experiments. The affinity between *S. columnare* and 12 dipterocarp species was also examined. This simple and convenient inoculation method helps maintain high quality nursery stock for several important dipterocarp tree species.

## 22.2 The Inoculation Method and Its Effects

### 22.2.1 A Convenient Inoculation Method

An ectomycorrhizal fungi, *Scleroderma columnare*, was successfully inoculated into dipterocarp seedlings by applying forest top soil as a source of inoculum and rice husk charcoal to induce its colonization (Mori and Marjenah 1994). Forest top soil contains ectomycorrhizal fungi and various other microorganisms, so it can introduce a natural diversity of microorganisms to the rhizosphere. As fruit bodies of *S. columnare* have been observed on wood charcoals, it is known that the species can tolerate neutral or slightly alkaline conditions. Since a large proportion of ectomycorrhizal fungi grow preferentially in acidic conditions (Hung and Trappe 1983), the alkaline charcoal provides a good, cheap selective medium for *S. columnare*. Evenly-sized thin rice husks can be uniformly carbonized by the procedure described in Mori and Marjenah (1994). The method is also ecologically friendly, since it avoids use of autoclaves and fungicides, and biological diversity in the mycorrhizosphere of host plants is promoted by the natural forest soil, which includes various micro-organisms such as nitrogen-fixing bacteria.

### 22.2.2 Effects of Mycorrhizal Colonization on Seedling Growth

A greater degree of ectomycorrhizal colonization was promoted in potted seedlings placed on rice husk charcoal spread on a concrete nursery bed, than in potted seedlings placed directly on the concrete without the charcoal (Fig. 22.1). The abundance of ectomycorrhizal colonization (scored according to the grading of Grand and Hervey 1982) in the pots was further increased by adding forest top soil to the charcoal nursery bed (square symbols in Fig. 22.1). One of the main fungi found in the pots was identified as *Scleroderma columnare,* based on its fruit bodies. The similarity of rhizomorphs found in the pots with and without fruit bodies indicated that this fungus occurred in most of the pots. The frequency of other types of mycorrhiza was negligible. Thus the inoculation method would be effective only for *S. columnare.*

There were close correlations between the degree of ectomycorrhizal colonization and height increment of the seedlings (Fig. 22.1). Height growth was promoted to be 22%–68% above that of the control seedlings due only to the spreading of rice

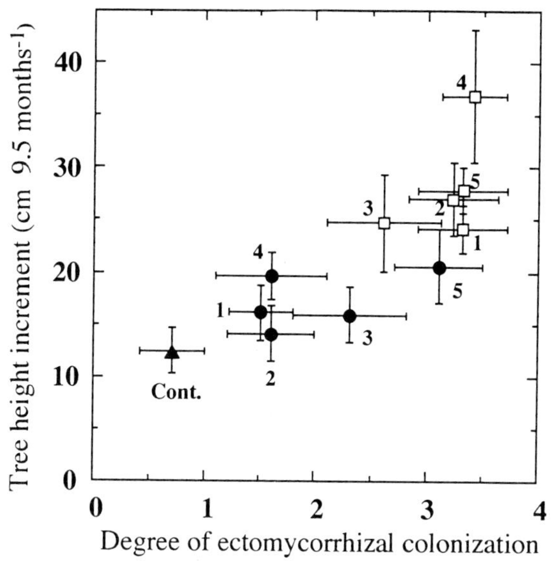


**Fig. 22.1.** Degree of ectomycorrhizal colonization and height increment of *Shorea smithiana* seedlings. For each treatment, 16 seedlings planted in plastic bags were placed on nursery beds with various mixtures of rice husk charcoal and forest top soil. Numbers show the amount of rice husk charcoal on the nursery bed (l $m^{-2}$). *Squares, dots*, and *triangles* represent treatments with forest top soil (15 l $m^{-2}$), without forest top soil, and controls (with neither rice husk charcoal nor forest top soil), respectively. *Solid lines* represent ±1 SE

husk charcoal. Spreading rice husk charcoal alone increased growth of the seedlings to about 1.5 times the heights of controls, and top soil mixed in with the charcoal produced increases of 2–3 times that of controls. Thus, the inoculation method is quite simple but it effectively promotes both *S. columnare* colonization and growth of the seedlings.

## 22.3 Diurnal Changes in Photosynthesis

There were clear differences in diurnal changes of net photosynthetic rates between *Shorea smithiana* seedlings colonized well and poorly by *Scleroderma columnare* (Fig. 22.2). Midday depressions in net photosynthetic rates were restricted to very short periods, and recovery from such depressions was strong and rapid in the well colonized seedlings as compared to the poorly colonized seedlings. Furthermore, both nitrogen (N) and phosphate (P) contents of sample leaves in seedlings with abundant ectomycorrhizal colonization were significantly higher than contents in weakly colonized seedlings. The mean (±1 SD) N contents (%) in leaves (n = 4) of well- and poorly- colonized seedlings was 1.38 (±0.14) and 1.18 (±0.08), respectively, and their P contents (ppm) were 1212 (±118) and 823 (±254), respectively.

In contrast to the positive effect of *S. columnare* colonization on photosynthesis, other ectomycorrhizal fungi such as *Laccaria laccata* and *Heberoma crustuliniforme* do not promote photosynthesis and suppression of midday depression (Dosskey et al. 1990). The inoculation method used here promoted the colonization specifically of *S. columnare* together with various other non-mycorrhizal soil microbes. Thus, the inoculation method is likely to help increase the photosynthetic rate of *Shorea*

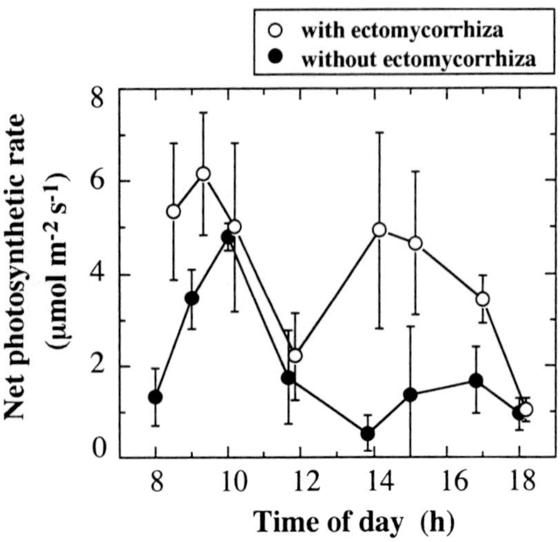


**Fig. 22.2.** Diurnal changes in net photosynthetic rate of *Shorea smithiana* seedlings on a clear day. *Open* and *closed circles* represent the leaves of seedlings with strong and poorly developed ectomycorrhizal colonization, respectively. The second leaf from the terminal bud was used for the measurements. All seedlings were thoroughly irrigated one day before the measurements. *Vertical lines* represent ±1 SE

*smithiana* seedlings. The moderation of midday depression in photosynthetic rate of the colonized seedlings indicates that this inoculation technique could be especially useful for seedlings designated for planting at hot and dry sites.

## 22.4 Application of Ectomycorrhizal Fungi in the Field

The addition of rice husk charcoal to planting holes increased ectomycorrhizal colonization of planted seedlings and promoted their growth (Mori and Marjenah 1994). The addition of 300 $cm^3$ rice husk charcoal per planting hole when planting *Shorea smithiana* and *Shorea laevis* seedlings in a degraded forest site, resulted in height increases of 32% and 38%, respectively, compared to those without the addition (Fig. 22.3). The infection rates of the ectomycorrhizal fungi *Scleroderma columnare* in *S. smithiana* seedlings planted with and without rice husk charcoal were 67.5% and 37.5%, respectively. Thus, amending the surface soil around planting holes by adding rice husk charcoal appears to be effective for promoting colonization by *S. columnare.*

## 22.5 Variation of Affinity Between Different Dipterocarp Species and *S. columnare*

The degree of colonization of *Scleroderma columnare* varied among the 12 dipterocarp species tested (Fig 22.4). Application of rice husk charcoal and forest top soil to the pot medium strongly promoted colonization of *S. columnare* in *Shorea ovalis, Shorea leprosula* and several other species. However, in some tree species, for ex-

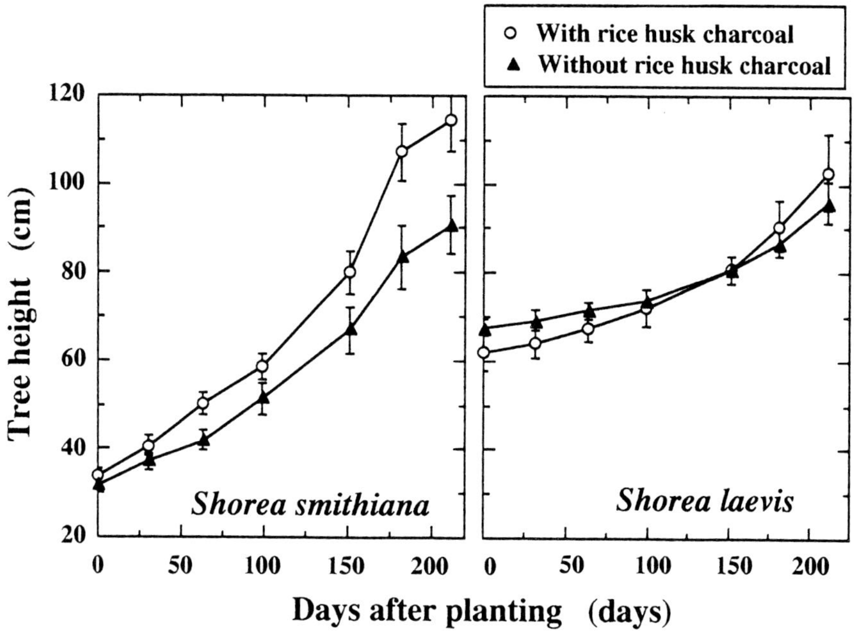


**Fig. 22.3.** Height increment of seedlings in the field with and without rice husk charcoal. The numbers of seedlings of *Shorea smithiana* and *Shorea laevis* used in each treatment were 40 and 50, respectively. *Vertical lines* represent ±1 SE

ample *Dipterocarpus tempehes* and *Shorea acuminatissima*, the application did not enhance the colonization of *Scleroderma columnare*.

It is therefore important to determine the affinity between dipterocarp species and ectomycorrhizal fungus species. Tree species with strong affinity to *S. columnare* (Fig. 22.4) would probably be good choices for planting in dry and nutrient poor habitats (being more likely to be tolerant of these conditions), following application of the ectomycorrhizal inoculation technique described. On the other hand, the technique is not likely to be appropriate for tree species with low affinity to *S. columnare*. Thus, species-specific affinities between dipterocarp trees and ectomycorrhizal fungi should be carefully considered in the application of inoculation procedures. Furthermore the complex relationships between the symbionts require detailed studies to widen the scope of these practices.

## 22.6 Conclusion

In Borneo Island, 267 species of dipterocarp trees have been recorded, and each species has favored habitats, varying from mountain ridges to riversides (Ashton 1983). It has been suggested that the clumped distribution of dipterocarps is due to the ecological adaptation of associations between specific species and compatible fungi at different types of site (Ashton 1983, Smits 1994). Close relationships be-

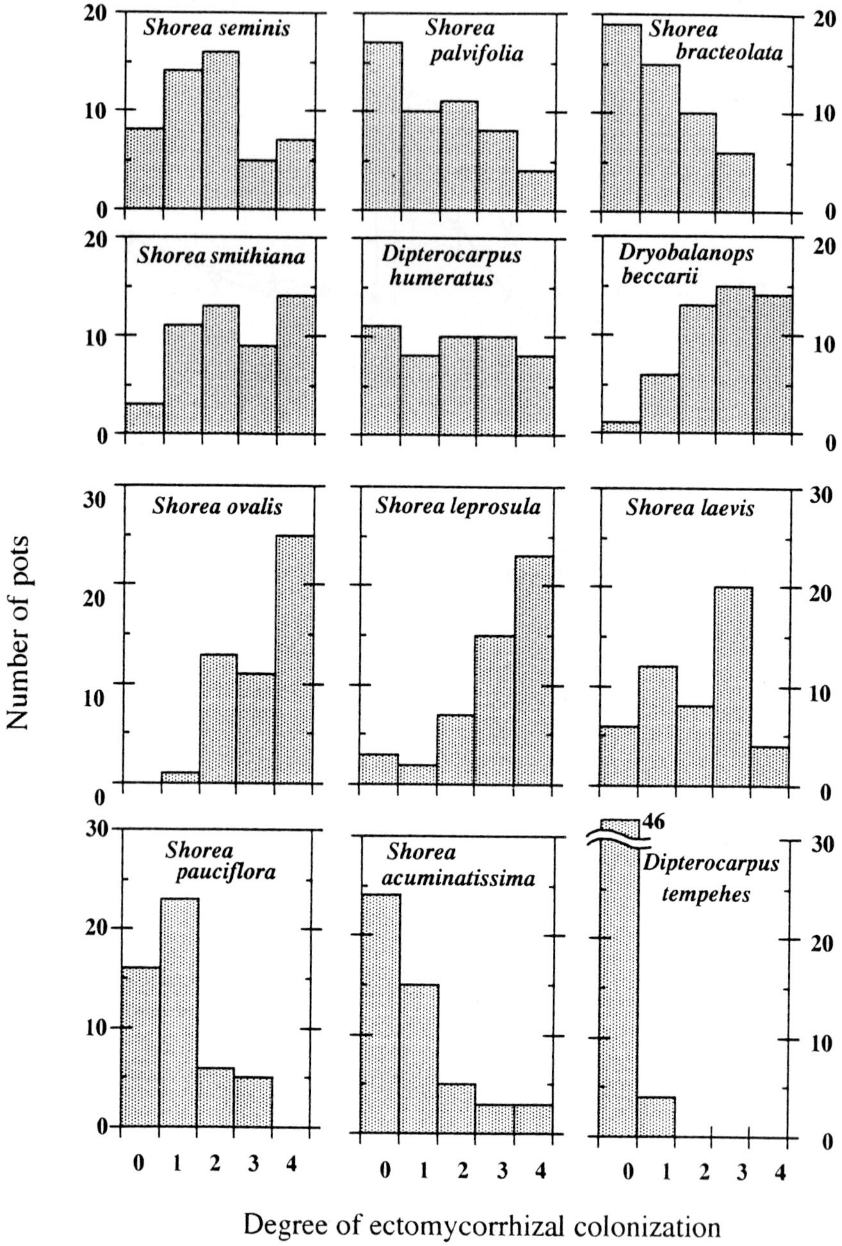


**Fig. 22.4.** Degree of ectomycorrhizal colonization of *Scleroderma columnare* observed in 12 dipterocarp species. For each species 50 one-year-old seedlings were transplanted in 2-liter pots containing 150 $cm^3$ of rice husk charcoal and 200 $cm^3$ of forest top soil. Size and appearance of the seedlings were almost the same within each species and they did not possess ectomycorrhiza at the time of transplanting. The degree of ectomycorrhizal colonization was examined 10 months after the transplanting

tween mycorrhizal fungal diversity and host-plant diversity have also been described (van der Heijden et al. 1998).

Concerning the growing needs for rehabilitation and reforestation, more integrated studies are needed on dipterocarp mycorrhizal associations because the optimization of forest productivity requires full exploitation of the positive relationships found in natural coexistence. Controlled inoculation experiments with clearly identified fungal species are needed for better understanding of dipterocarp- mycorrhizal physiology and to explore their potential for application in forestry (Lee 1998). Nevertheless, the technique for promoting key mycorrhizal fungi introduced in this chapter is already applicable to several dipterocarp species. Furthermore, it can help increase forest productivity and maintain the diversity of other soil microbes, which play important roles in forest health.

## References

Allen M (1991) The ecology of mycorrhiza. Cambridge University Press, Cambridge

Ashton PS (1983) Dipterocarpaceae. In: Van Steenis CGG J (ed) Flora Malesiana. Martinus Nijhoff, The Hague, pp 237–552

Dosskey MG, Linderman RG, Boersma L (1990) Carbon-sink stimulation of photosynthesis in Douglas-fir seedlings by some ectomycorrhizas. New Phytol 115:269–274

Duddridge JA, Malibari A, Read DJ (1980) Structure and function of mycorrhizal rhizomorphs with special reference to their role in water transport. Nature 287:834–836

Evans JR (1983) Nitrogen and photosynthesis in the flag leaf of wheat (*Triticum aestivum* L.). Plant Physiol 72:297–302

Garbaye J (1994) Helper bacteria: a new dimension to the mycorrhizal symbiosis. New Phytol 128:197–210

Grand LF, Harvey AE (1982) Quantitative measurement of ectomycorrhizae on plants roots. In: Schenck NC (ed) Methods and principles of mycorrhizal research. The America Phytopathological Society, St. Paul, pp 157–164

Ho I, Zak B (1979) Acid phosphatase activity of six ectomycorrhizal fungi. Can J Bot 57:1203–1205

Hung L, Trappe JM (1983) Growth variation between and within species of ectomycorrhizal fungi in response to pH in vitro. Mycologia 75:234–241

Iwase K (1991) Gluconic acid synthesis by the ectomycorrhizal fungus *Tricholoma robustum*. Can J Bot 70:84–88

Kikuchi J, Ogawa M (1995) Nitrogen fixing (Acetylene reducing) activity in the mycorrhizas of Dipterocarp seedlings. In: Suzuki K, Sakurai S, Ishii K, Norisada M (eds.) Proc intern workshop of BIO-REFOR, Tokyo, BIO-REFOR IUFOR/SPDC, pp 71–80

Lapeyrie F, Chilvers GA, Bhem CA (1987) Oxalic acid synthesis by the mycorrhizal fungus *Paxillus involutus* (Batsch. Ex Fr.) Fr. New Phytol 106:139–146

Lee SS (1998) Root symbiosis and nutrition. In: Appanah S, Turnbull JM (eds) A review of Dipterocarps taxonomy, ecology and silviculture. CIFOR, Bogor, Indonesia. pp 99–114

Lee SS, Lim KL (1989). Mycorrhizal infection and foliar phosphorus content of seedlings of three dipterocarp species growing in a selectively logged forest and a forest plantation. Plant Soil 117:237–241

Mejstrik VK, Krause HH (1973) Uptake of 32P by *Pinus radiata* roots inoculated with *Suillus luteus* and *Cenococcum graniforme* from different sources of available phosphate. New Phytol 72:137–140

Mori S, Marjenah (1994) Effect of charcoaled rice husks on the growth of Dipterocarpaceae seedlings in East Kalimantan with special reference to ectomycorrhiza formation. J Jpn For Soc 76:462–464

Plassard CD, Barry D, Eltrop L, Mousain D (1994) Nitrate uptake in maritime pine (*Pinus pinaster*) and ectomycorrhizal fungus *Hebeloma cylindrosporum*: effect of ectomycorrhizal symbiosis. Can J Bot 72:189–197

Santoso E (1988) Rengaruh mikoriza terhap diameter batang dan bobot kering anakan *Dipterocarpaceae* (in Indonesian). Bul Pen Hutan (Forest Res Bull) 504:11–21

Setala H (1995) Growth of birch and pine seedlings in relation to grazing by soil fauna on ectomycorrhizal fungi. Ecology 76:1844–1851

Smits WTM (1994) Dipterocarpaceae: mycorrhizae and regeneration. Tropenbos Series 9, The Tropenbos Foundation. III, Wageningen

Turner IM, Brown ND, Newton AC (1993) The effect of fertilizer application on dipterocarp seedling growth and mycorrhizal infection. For Ecol Manag 57:329–337

van der Heijden MGA, Klironmos JN, Ursic M, Moutoglis P, Streitwolf-Engel R, Boller T, Wiemken A, Sanders IR. (1998) Mycorrhizal fungal diversity, ecosystem variability and productivity. Nature 396:69–72

# 23 Photographic Estimation of Light Environments on Forest Floors and Effects of Light on the Growth of Dipterocarp Seedlings

HISASHI YAHATA

## 23.1 Introduction

The light environment in the forest plays an essential role in gap dynamics and seedling growth following enrichment planting designed to rehabilitate degraded forests in the tropics, and light intensity relative to full sunlight (relative light intensity) is often used as an conventional index of light conditions under the forest canopy. However, under clear conditions, the relative light intensity in forest gaps is extremely variable, because direct beams of solar light (sunflecks) reach the forest floor at varying intervals and for varying duration. To avoid sunflecks giving distorted values, relative light intensity is commonly measured under cloudy conditions. Further, some authors recommend taking mean values from many points in a forest stand as the light index, especially when the solar altitude is high (Ando 1985). However, these measures are unsuitable for analyzing the relationships between light conditions and the growth of individual seedlings in the understory of forests precisely because the effect of sunflecks (which may be very significant) is ignored. Thus, direct and continuous measurements of light intensity under the canopy over long periods should ideally be used, but this requires a considerable number of quantum sensors and recorders.

As an expedient means of avoiding these problems, several studies have estimated light intensity using canopy photographs (Anderson 1964, Evans and Coombe 1959, Madgwich and Brumfield 1969, Takenaka 1987). Furthermore, the increasing power and convenience of computerized techniques have significantly expanded the potential for estimating light intensity using photographs (Chazdon and Field 1987). This chapter describes a method for estimating light conditions in a forest understory using photographic images, taken with fish eye hemispherical lenses. This technique was applied in a study of the effects of light intensity on the growth of dipterocarp seedlings planted in a secondary rainforest in Bukit Soeharto Education Forest (BSEF), which is also discussed.

Ecological Studies, Vol. 140
Guhardja et al. (eds.) Rainforest Ecosystems of East Kalimantan: El Niño, Drought, Fire and Human Impacts

## 23.2 Photographic Estimation of Light Intensity on the Forest Floor and Determination of Its Validity

### 23.2.1 Relationships Between Light Intensities Inside and Outside the Forest

The relationships between light intensities inside and outside a forest usually differ considerably depending on whether there is direct solar light or not. (Figure 23.1 shows the diurnal change of light conditions monitored at different locations under a canopy.) In the study described here, diurnal changes of light intensity (photosynthetically active photon flux density, PPFD) were measured in a forest of BSEF using photon flux density sensors (IKS-25, Koito, Tokyo, Japan).

The data obtained on a relatively cloudy day showed a high correlation between outside and inside light intensity, because only diffuse light outside the forest penetrated through the canopy (the left figure in Fig. 23.2). The penetrating PPFD was also highly correlated to the gap size. On the other hand, there was no clear relationship between inside and outside PPFDs on a relatively clear day (the right figure in Fig. 23.2). This may be attributable to the variability in whether or not the canopy intercepts direct solar light. Some of the light intensities measured inside the forest

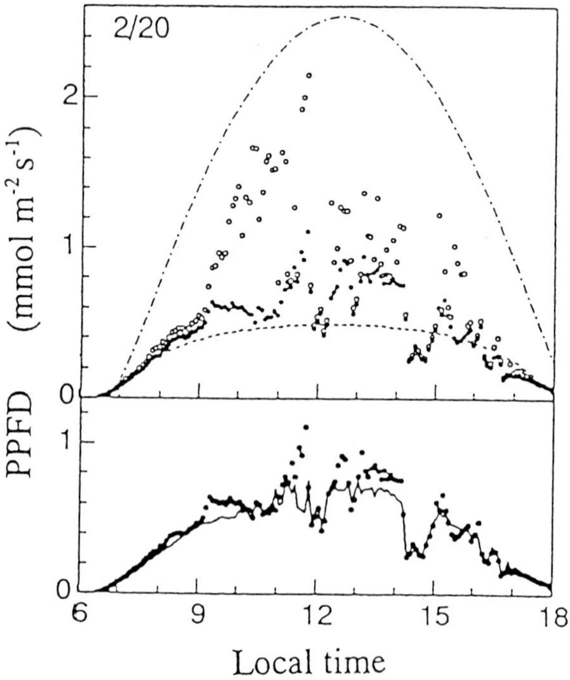


**Fig. 23.1.** Diurnal patterns of total and diffuse photosynthetically active photon flux density (*PPFD*) (*upper figure*) and estimated diurnal patterns of diffuse PPFDs (*lower figure*) in February in BSEF. *Open* and *closed circles* indicate measured values for total and diffuse PPFD, respectively. *Solid* and *broken lines* indicate the predicted values for total and diffuse PPFDs on completely clear days, respectively. In the *lower figure*, *solid lines* indicate the estimated diffuse PPFD, and *closed circles* show measured PPFD

exceeded concurrent intensities measured in the open, which is unexpected since the intensity of forest floor light should be the same or lower than that of open light. These apparently deviant data are due to short time lags that inevitably occur between cloud cover occluding forest and open sites (the interval between the forest and open sites was about 700 m).

Even under cloudy conditions, whenever the outside PPFD exceeded more than 1 mmol photons $m^{-2} s^{-1}$, the PPFD inside the forest tended to be saturated (the left figure in Fig. 23.2). Fig. 23.3 shows an example of the relationships between total PPFD (the sum of direct and diffuse PPFDs) and diffuse PPFD, measured in an open space using a ring shade (Yahata 1991a). It was observed that the diffuse PPFD was almost as strong as total PPFD, when the total PPFD was less than 0.6 mmol photons

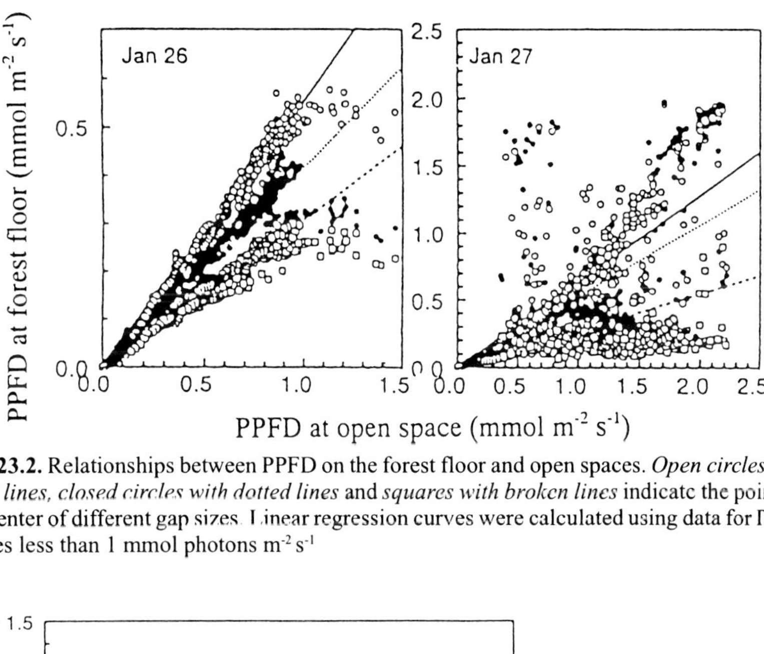


**Fig. 23.2.** Relationships between PPFD on the forest floor and open spaces. *Open circles with solid lines, closed circles with dotted lines* and *squares with broken lines* indicate the points at the center of different gap sizes. Linear regression curves were calculated using data for PPFD values less than 1 mmol photons $m^{-2} s^{-1}$

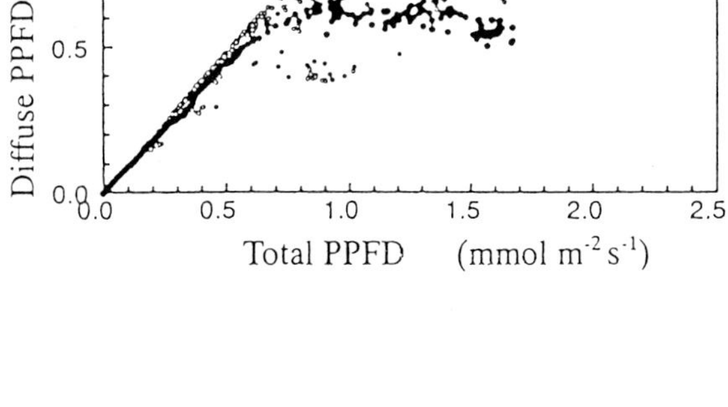


**Fig. 23.3.** Relationship between the PPFD of total light (the sum of diffuse and direct light) and the PPFD of diffuse light measured throughout a day in January with a ring shade. *Solid dots* indicate the data measured in the morning up to 11:00 a.m., when the sky was clear

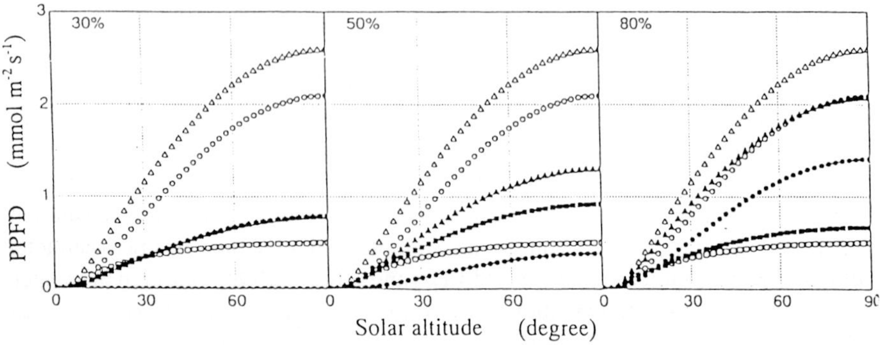


**Fig. 23.4.** Estimated PPFDs of total light (*triangles*), direct light (*circles*), and diffuse light (*squares*) plotted against solar altitude in BSEF. *Open* and *solid marks* indicate estimates for clear and cloudy conditions, respectively. Estimations were made for total PPFD values equivalent to 30%, 50%, and 80% of values predicted for completely clear light conditions

$m^{-2}$ $s^{-1}$, but the diffuse PPFD was saturated whenever total PPFD was more than 1.0 mmol photons $m^{-2}$ $s^{-1}$.

The author also derived estimates of total, direct, and diffuse PPFDs as functions of solar altitude under cloudy conditions (Fig. 23.4). Diffuse PFFD was fairly stable at total PFFD values greater than 2 mmol photons $m^{-2}$ $s^{-1}$ (80% of the level predicted for completely clear conditions), and remained fairly constant even if total PFFD values increased beyond this level. Thus, the PPFD on the forest floor reflects the diffuse PPFD outside the forest, as long as there is no sunfleck. Therefore, there was a good correlation between the open and forest floor light regimes under most cloudy conditions, but the correlation under fine conditions was poor (e.g., on January 27 as shown in Fig. 23.2). However, correlation coefficients for clear days could be calculated using data obtained when total PPFD values were less than 1 mmol $m^{-2}$ $s^{-1}$. There was not a great difference in regression coefficients (slopes) observed between cloudy and fine days, because sufficient numbers of measurements were taken to smooth out the variation. The regression coefficients (slopes) indicate the measured relative PPFD values, which were used to check the validity of the photographic estimates in the following exercise.

## 23.2.2 Estimation of PPFD by Photographic Images

Accurate determination of the PPFD at specific points by the use of photographs of the canopy taken from the forest floor, in conjunction with calculations of the solar track above the canopy, has been facilitated by recent developments in computer techniques. Canopy photographs are commonly taken using a camera with a fish eye lens (e.g., a Nikon FA and Fish eye-Nikkor 8 mm 1:2.8) pointed at the zenith as precisely as possible with the use of a level. Resulting photographs can be orientated by using two small sticks vertically inserted at the outer edge of the lens and measuring the alignment of the sticks with a compass (e.g., SUUNTO, Espoo, Finland). A video camera was used for this research to obtain negative film images which were digitized and fed into a personal computer using a video scanner (98 HVS-10, Hal Laboratory, Tokyo, Japan). The author developed software to analyze the solar track in a photograph at any latitude and longitude, and at any time of year, using the Microsoft Quick Basic Language (Yahata 1991b, 1993). Mitchell and Whitmore (1993) highlighted another potential problem in such analyses, criticizing a previous method to determine the relative light intensity based on calculations of the gap area ratio from a semispherical photograph, since the area distortion of fish eye lenses causes an error. The program used here can eliminate this distortion error.

To estimate PPFD more accurately in a forest, both direct solar and diffuse PPFDs in open conditions should be known. Thus, both of these components were measured using the ring shade, as described above. However, because the solar track changes day by day and manual adjustment of the ring shade is essential, it is difficult to measure diffuse PPFD over long periods at a remote place. In contrast, total PPFD can be measured easily. Yahata (1991a) reported a method for dividing total PPFD into its diffuse and direct beam PPFD components in cases where only total PPFD data are available, even when the sky condition varies from clear to cloudy. Yahata's technique was applied in this study to estimate the diffuse PPFD and compare it with the measured PPFD. The results gave potentially useful information, but systematic errors, discussed below, affected the derived correlation coefficients and, therefore, the utility of the method. If the Yahata's estimation method was correct, data were precisely gathered, and PPFD was correctly estimated, then the slope of the regression between estimated and measured relative light intensities should be 45 degrees (i.e., there should be a perfect 1:1 relationship between them). However, as Fig. 23.5 shows, there was a relatively low correlation coefficient (0.679) and a regression coefficient (gradient) that was, at 1.145, slightly higher than 1.0. This means that the estimated values were higher than the measured values. The loss of

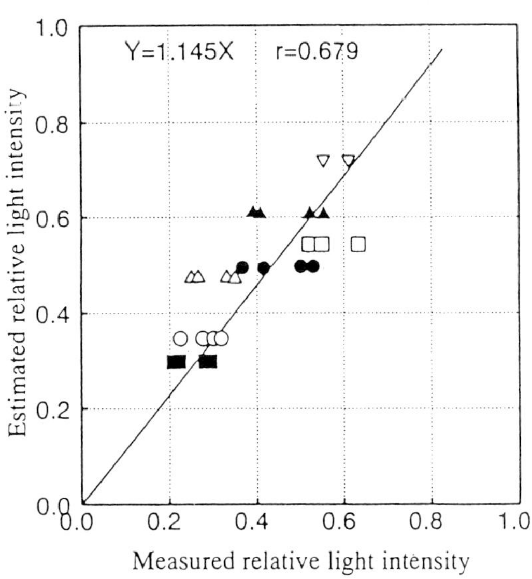


**Fig. 23.5.** Relationships between the estimated and measured relative light intensities using photographic images. The measured relative light intensities are derived from the slopes of the regression curves shown in Fig. 23. Different symbols indicate different sites in a forest

small, thin leaves from the image during processing of the photograph could have caused this overestimation.

PPFD was estimated under different weather conditions and at different points under the same weather conditions. The results were generally satisfactory for measured and estimated PPFD, but some discrepancies were found in the data obtained under fine conditions, (Fig. 23.6). This is largely attributable to the fact that the

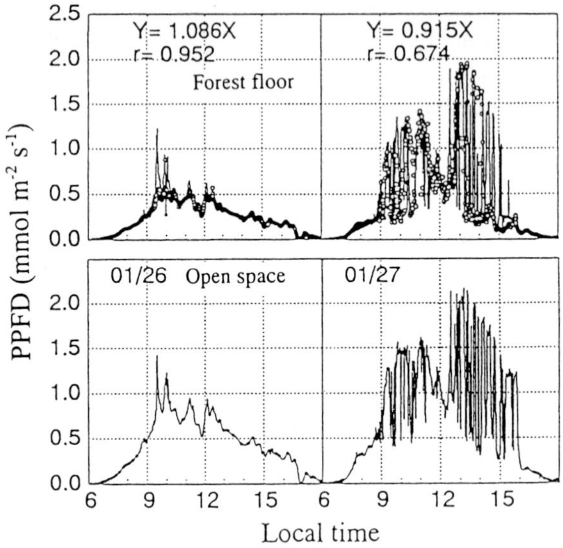


**Fig. 23.6.** Diurnal changes of PPFD inside the forest and in open spaces on a cloudy day (January 26) and a relatively fine day (January 27). In the *upper figures*, *dotted lines* indicate the measured PPFD while the *solid lines* indicate PPFD estimated using photographic images of the canopy

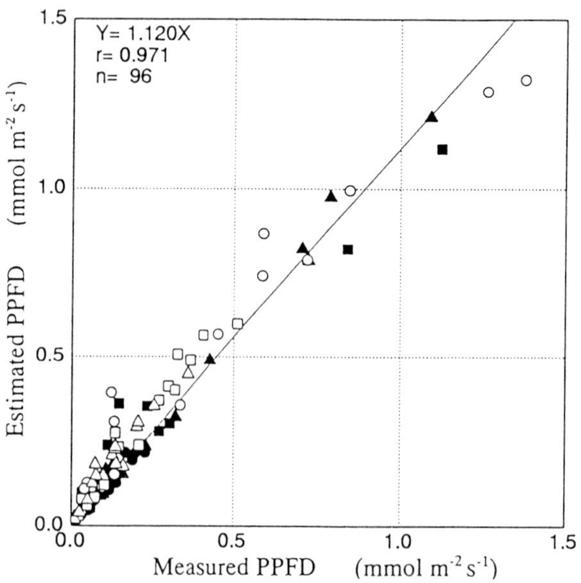


**Fig. 23.7.** Relationships between estimated PPFD and average PPFD every 3 h from 6:00 to 18:00. Different symbols indicate different sites in a forest

distance between the two sensors (inside and outside the forest) caused lags, in terms of cloud coverage, as mentioned earlier.

When instantaneous measurements are made, high correlation coefficients are usually observed under cloudy conditions, whereas poor correlations are generally obtained under fine conditions. The variation caused by random changes during fine conditions could be minimized by the use of mean values integrated over longer periods. Fig. 23.7 shows an example of three-hour means, for instance, in which the correlation coefficient was improved to r=0.971. In addition, estimation of the three-hour means was fairly good even under fine conditions. Furthermore, the correlation coefficient calculated using the diurnal integrated photon flux density was as high as 0.982. This result confirmed the hypothesis that photographic estimation of light environments is useful for determining the light conditions affecting plant growth. When applying the technique it must be noted that the estimated light intensity is about 10% brighter than the true value, because the regression coefficient (slope) was 1.12. Nevertheless, computerized analysis of photographic images of the canopy provides a very convenient and practical technique for estimating light conditions on the forest floor. Furthermore, the method is very efficient for determining long-term light regimes in canopy gaps on the forest floor, as long as the gaps remain stable.

## 23.3 Light Intensity and Seedling Growth

By calculating seasonal estimates of PPFD affecting individual dipterocarp seedlings, we can analyze the integrated effect of light intensity on their growth and physiological processes. In BSEF, the height growth increment was recorded for a

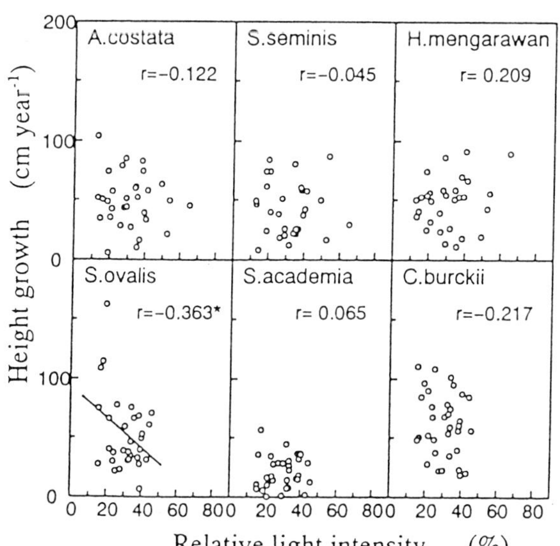


**Fig. 23.8.** Relationships between the height increment of seedlings of six dipterocarp species (*Anisoptera costata*; *Shorea seminis*; *Hopea mengarawan*; *Shorea ovalis*; *Shorea academia*; and *Cotylelobium burckii*) and the relative light intensity under cloudy conditions

full year after planting and was correlated with the light conditions on the forest floor (Fig. 23.8). The results did not generally show a clear dependence of seedling height growth on the relative light intensity, except for *Shorea ovalis*, for which height growth decreased with increasing light intensity. Furthermore, although the overall dependency of height growth on light intensity was weak in dipterocarp seedlings, it has been reported that the maximum height decreased as the relative light intensity increased for several tree species (*Anisoptera costata, Shorea ovalis, S. academia,* and *Cotylelobium burckii*), based on a boundary layer analysis (Webb 1972). Thus, these species may be able to grow faster in shade rather than in full light conditions.

To determine the effect of the time duration of sunlight exposure on seedling growth, the relationships between the height growth of six dipterocarps and a gap light index (Canham et al. 1988) for successive three-hour periods, 06:00–9:00, 9:00–12:00, 12:00–15:00, and 15:00–18:00, was calculated (data not shown, Yahata 1993). The height growth in *Shorea ovalis, S. academia*, and *Cotylelobium burckii* seedlings showed a negative correlation with increasing light intensity in the morning. Both height and diameter growth in the *Shorea ovalis* seedlings were reduced in stronger light conditions, indicating that this species has a relatively strong preference for shade. The overall results suggest that the seedlings in the gap site used mainly the diffuse light in the morning for carbon fixation. Ackerly and Bazzaz (1995) also found that leaves in a tropical forest gap used mainly diffuse light for carbon gain. Similarly, Ishida et al. (1999) observed that for tree seedlings under relatively large gaps in Peninsular Malaysia, well diffused light in the morning was more important for net photosynthesis than light in the afternoon. However, *Hopea mengarawan* seedlings increase in diameter most quickly under strong light (data not shown), indicating that this species is a relatively light demanding species. *Shorea academia, S. seminis, Anisoptera costata,* and *Cotylelobium burckii* appear to be intermediate species for light and shade tolerance.

## 23.4 Conclusion

In conclusion, the use of computerized techniques is very useful for determining the long-term light regimes on the forest floor. The analyses of light environments integrated over time need to examine the linkage between light level and seedling growth. Because of the interspecific variations in light-use among dipterocarps, more work is needed to elucidate the effect of time duration and quality of incident light on seedling growth. It has also been reported that canopy closure causes not only a decrease in the amount of light but also a decrease in the ratio of red to far-red light, and the low red/far-red ratio sometimes promotes shoot elongation (Morikawa, et al. 1976, 1983). Thus, it will be important to examine the diurnal time changes of light quantity and quality for understanding seedling growth and morphology.

# References


Ackerly DD, Bazzaz FA (1995) Seedling crown orientation and interception of diffuse radiation in tropical forest gap. Ecology 76:1134–1146

Anderson MC (1964) Studies of the woodland light climate. 1. The photographic computation of light conditions. J Ecol 52:27–41

Ando T (1985) The management points of multi-storied forest (in Japanese). Ringyo-Kagaku-Shinkosho, Tokyo

Canham CD, Denslow JS, Platt WJ, Runkel JR, Spies TA, White PS (1988) Light regimes beneath closed canopies and tree-fall gaps in temperate and tropical forests. Can J For Res 20:620–631

Chazdon RL, Field CB (1987) Photographic estimation of photosynthetically active radiation: evaluation of a computerized technique. Oecologia 73:525–532

Evans GC, Coombe DE (1959) Hemispherical and woodland canopy photography and the light climate. J Ecol 47:103–113

Ishida A, Nakano T, Matsumoto Y, Sakoda M, Ang LH (1999) Diurnal changes in leaf gas exchange and chlorophyll fluorescence in tropical tree species with contrasting light requirements. Ecol Res 14:77–88

Madgwich HAI, Brumfield GL (1969) The use of hemispherical photographs to assess light climate in the forest. J Ecol 57:537–542

Mitchell and Whitmore (1993) Use of hemispherical photographs in forest ecology; calculation of absolute amount of radiation beneath the canopy. O.F.I. Occasional Papers Oxford Forestry Inst, Oxford University Press, Oxford, pp 1–39

Morikawa Y, Asakawa S, Sasaki S (1976) Growth of pine and birch seedlings under lights with different spectral compositions and intensities. J Jpn For Soc 58:174–178

Morikawa Y, Sasaki S, Asakawa S (1983) Light quality and seedling growth on forest floors (in Japanese). Sinrin-Ricchi XXV:1–7

Takenaka A (1987) Analysis of light transmissity of forest canopies with a telephoto method. Agri For Meteorol 40:359–369

Webb RA (1972) Use of the boundary line in the analysis of biological data. J Hort Sci 47:309–319

Yahata H (1991a) Photographic estimation of light environments on forest floor using computerized technique — Estimation of photon flux densities of the diffuse light and the direct solar direct beam light only with the total light (In Japanese). Bull Ann Meetings Jpn For Soc Kyushu 44:111–112

Yahata H (1991b) Photographic estimation of light environments on forest floor using computerized technique (II) Methodology and test of validity. Annual Report of PUSREHUT 1:32–67

Yahata H (1993) Estimation of light regimes in a forest floor using photographs and the effect on seedling growth of several species of dipterocarps in a tropical rainforest (in Japanese with English summary). Jpn J Forest Environ 35:10–19, 1993

# 24 In Situ Study of the Effects of Elevated Temperature on Photoinhibition in Climax and Pioneer Species

Atsushi Ishida, Takeshi Toma, Dadang I. Ghozali, and Marjenah

## 24.1 Introduction

Understanding the abiotic and biotic factors for seedling growth and the ecophysiological responses against physiological stresses has been of great interest to ecologists, biologists, and foresters. Interactions of multiple environmental factors are relevant to the physiological performance of plants in nature (Ludlow 1987, Gamon and Pearcy 1989, 1990; Ishida et al. 1999a,b). For example, the inhibitory effects of strong light on photochemical capacity in leaves can be exacerbated by other stress conditions, such as cold (e.g., van Kijk and van Hasselt 1993, Leitsch et al. 1994), limited nitrogen supply (e.g., Ferrar and Osmond 1986), and water stress (e.g., Björkman and Powles 1984, Kao and Forseth 1992). In contrast to the large number of studies on the interactive effects of cold and high light intensity, there have been only a few studies on photoinhibition caused by a combination of high light intensity and high temperature (Ludlow 1987, Gamon and Pearcy 1990, Mulkey and Pearcy 1992, Königer et al. 1998).

By measuring chlorophyll fluorescence, we can obtain qualitative and quantitative information about photoinhibition and photoprotective mechanisms in chlorophyll in fields (e.g., Bilger et al. 1995). The term 'photoinhibition' has sometimes been used to include not only sustained-damage to the photosynthetic apparatus but also the reversible midday depression of photochemical capacity. Under sunlit conditions, reductions in ΔF (Fm'–F; variable fluorescence) can result from increased initial fluorescence (F) and/or decreased maximal fluorescence (Fm'). The reduction in Fm' in the light is known as non-photochemical quenching (NPQ), associated with thermal dissipation of excess light energy for protective mechanisms in the photosynthetic apparatus (Björkman and Demmig-Adams 1994). When the reversible depression in Fm' occurs in the light, the efficiency of photosynthetic energy conversion decreases. Such depression in Fm' causes depression of ΔF/Fm' (fluorescence ratio in light or photosystem (PS) II quantum yield), sometimes called 'down-regulation.'

On the other hand, after the leaves are returned to darkness, the fluorescence ratio at a dark (Fv/Fm) increases in three recovery phases (e.g., Somersalo and Krause 1990, Leitsch et al. 1994, Thiele et al. 1998): 1) a very rapid phase (< 10 min. in the

Ecological Studies, Vol. 140
Guhardja et al. (eds.) Rainforest Ecosystems of East Kalimantan: El Niño, Drought, Fire and Human Impacts

dark) associated with the relaxation of the trans-thylakoid proton gradient, 2) a fast phase (< 60 min.) associated with the xanthophyll-cycle, and 3) a slow phase (> 60 min.) associated with the repair of damaged D1 protein. Therefore, sustained-photoinhibition (chronic photoinhibition) can be evaluated from a reduced Fv/Fm, as measured just before dawn, in the leaves of plants which grow in fields.

Recently, open (cleared) sites caused by forest fires (Mori, Chapter 3, this volume) or by abandonment after shifting cultivation (Kiyono and Hastaniah, Chapter 17, this volume) are rapidly increasing in East Kalimantan. The leaves of plants planted in such open sites are exposed to high solar radiation and high-temperature conditions throughout the year. Under such conditions, sustained-photoinhibition or midday depression of photochemical capacity (down-regulation of PSII) can be caused by heat and high light loads, and can be one of the possible factors limiting growth of the planted tree seedlings. This chapter deals with how the interactive effects of high light and high temperature on 'photoinhibition' of PSII differ among pioneer and climax species, and we discuss the possible protective mechanisms in photoinhibition.

## 24.2 High-Temperature Treatment

The interactive effects of high light and elevated temperature on photoinhibition were compared among eight pioneer plants and the seedlings of three dipterocarps with various hardwood densities (Table 24.1). The dipterocarp species can be grouped according to wood density, which is often regarded as being inversely related to growth rate and positively related to shade tolerance (Whitmore 1984). In the dipterocarps, the maximum net photosynthetic rate under light-saturated conditions was higher in plant species with lower hardwood density than in those with higher hardwood densities (Fig. 24.1).

The large forest fires in February 1998 occurred in East Kalimantan during a prolonged drought linked to an El Niño Southern Oscillation (ENSO) event (Toma et al., Chapter 2, this volume). Many pioneer plants naturally invaded open areas caused by the forest fires in Bukit Soeharto Education Forest (BSEF) in East Kalimantan. For rehabilitation of the deforestation, seedlings of the three dipterocarp species, *Shorea ovalis* (Korth.) Bl., *Hopea rudiformis* Ashton, and *Dryobalanops beccarii* Dyer, were planted in an open site in June, 1998. Among the tree species, *Shorea ovalis* has the lowest hardwood density among the three dipterocarps.

To clarify the effect of elevated leaf temperature on photoinhibition in the field, three dipterocarp seedlings and the eight pioneer plants were surrounded and covered with a transparent plastic sheet for fifteen days from 20 October, 1998 (High-temperature treatment) (Fig. 24.2). Air temperature inside the plastic sheet boxes was higher (1–5˚C) than that in ambient air. In the pioneer and climax species, the leaf temperature inside the plastic sheet boxes was about 4ºC higher than that under ambient air conditions (data not shown). Leaf temperatures at a given photosynthetic photon flux density (PPFD) were about 2˚C lower in the pioneer species than in the climax dipterocarp species, under both ambient air and the high-temperature

**Table 24.1.** Species used in the present study

| Species | Abbreviation in figures | Wood density[a] (kg $m^{-3}$) | Maximum height(m) | Maximum net photosynthetic rate[b]($\mu$mol $m^{-2}$ $s^{-1}$) |
|---|---|---|---|---|
| Pioneer plants | | | | |
| *Glochidion obscurum* | Go | - | 15 | 18.5 |
| *Homalanthus populneus* | Hp | - | 9 | 18.5 |
| *Macaranga triloba* | Mt | 300 | 9 | 22.6 |
| *Mallotus paniculatus* | Mp | 480 | 12 | 24.4 |
| *Melastoma malabathricum* | Mm | - | 3 | 19.0 |
| *Piper aduncum* | Pa | - | 5 | 20.0 |
| *Trema cannabina* | Tc | - | 10 | 19.6 |
| *Trema orientalis* | To | 400 | 27 | 18.6 |
| Climax trees (Dipterocarpaceae) | | | | |
| *Hopea rudiformis* | Hr | 750 | 35 | 4.8 |
| *Dryobalanops beccarii* | Db | 720 | 45 | 5.0 |
| *Shorea ovalis* | So | 510 | 60 | 6.2 |

[a]Density is based on 12% moisture content. Data from Choong and Achmadi (1996) and Iwasa et al. (1984).
[b]Net photosynthetic rate was measured at 0900–1100 h under 2000 $\mu$mol $m^{-2}$ $s^{-1}$ PPFD, 350 $\mu$mol $mol^{-1}$ $CO_2$ concentration, and 35°C leaf temperature conditions.

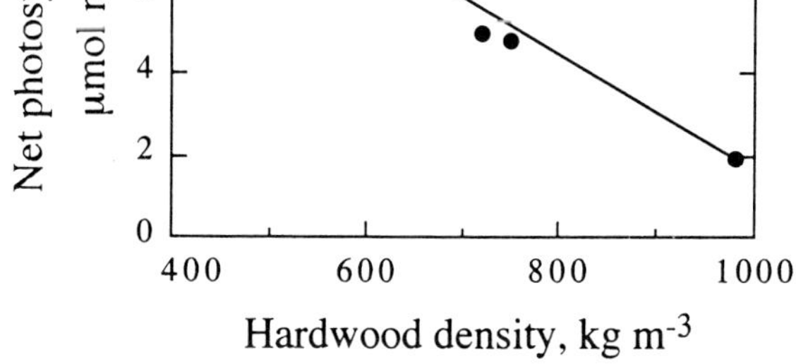


**Fig. 24.1.** The relationship between hardwood density and maximum net photosynthetic rate under high light and ambient $CO_2$ conditions in dipterocarps growing in an open area and large gap sites. Data on species shown is listed in Tables 24.1 and 24.2

**Fig. 24.2.** Photograph of the high-temperature treatment and measurement of chlorophyll fluorescence with a chlorophyll fluorescence meter (MINI-PAM, Walze, Effeltrich, Germany)

treatment conditions. This would be due to the effective cooling effects, i.e., latent heat flux, by generally high transpiration in the leaves of pioneer trees (Bazzaz 1979).

## 24.3 Photoinhibition by Elevated Leaf Temperature

Under ambient air conditions, the differences in dark adapted PSII were small between pioneer and climax dipterocarp trees. The PSII quantum yield at predawn (Fv/Fm) was in the range 0.77–0.83 in all leaves under ambient conditions , except for the leaves of *Dryobalanops beccarii*(Fig. 24.3). Because values of Fv/Fm in unstressed leaves are 0.80–0.83 (Björkman and Demmig, 1987), the values for all leaves except for those of *Dryobalanops beccarii* were optimal or slightly sub-optimal. Since Fv/Fm in the leaves of *Dryobalanops beccarii* was 0.70 ± 0.02 (mean ± 1 SD.), the leaves suffered from low sustained-photoinhibition even under ambient conditions in the open site.

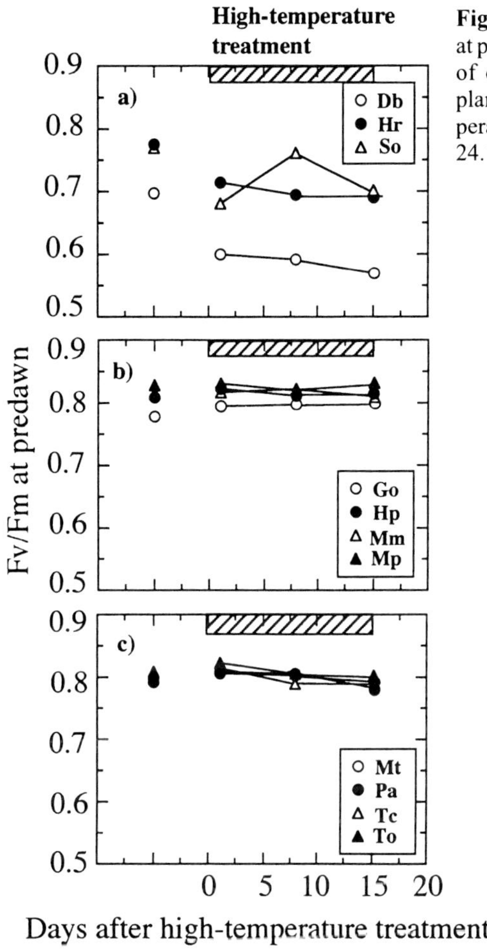


**Fig. 24.3a–c.** Change in PSII quantum yield at predawn (Fv/Fm) in fully expanded leaves of climax tree seedlings **(a)** and pioneer plants **(b,c)**, over fifteen days of high-temperature treatment. Abbreviations as in Table 24.1

The vulnerability of PSII to elevated leaf temperature was lower in the pioneer plants than in the dipterocarps. After fifteen days in the plastic sheet boxes, the values of Fv/Fm in the leaves of *Dryobalanops beccarii* and *Hopea rudiformis* seedlings significantly decreased (*t*-test, $P<0.05$) (Fig. 24.3a). The value of Fv/Fm in *Shorea ovalis* also tended to decrease, but not significantly. On the other hand, the values of Fv/Fm in the leaves of the pioneer plants remained unchanged during the high-temperature treatment (Fig. 24.3b,c).

Under ambient air conditions, PSII quantum yield ($\Delta F/Fm'$) and electron transport rate through PSII (ETR) in the light (800–1200 $\mu mol\ m^{-2}\ s^{-1}$ PPFD) were significantly lower in the dipterocarp species than in the pioneer species (*t*-test, $P<0.01$) (Fig. 24.4). This indicates the usage efficiency of absorbed light for photochemical processes such as carbon gain was higher in pioneer plants than in dipterocarp trees (Demmig-Adams et al. 1996).

In the high-temperature treatment, both $\Delta F/Fm'$ and ETR in the light (800–1200

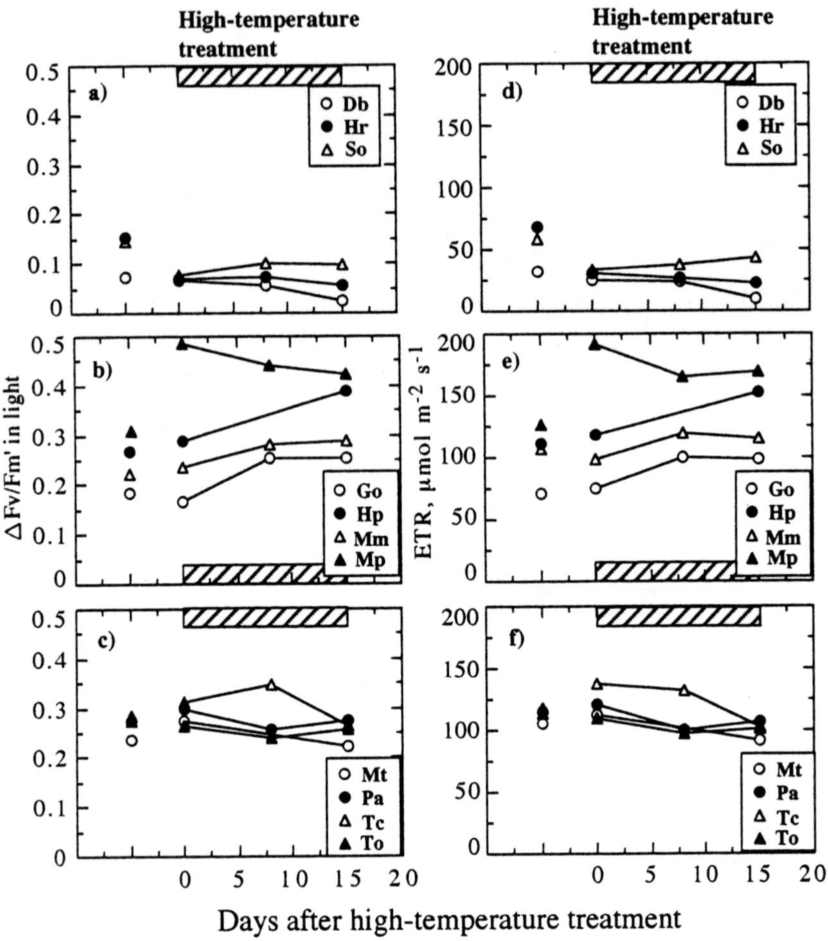


**Fig. 24.4.** Change in PSII quantum yield (ΔF/Fm') **(a,b,c)** and change in electron transport rate **(d,e,f)** through PSII (ETR) in fully expanded leaves of climax tree seedlings **(a,d)** and pioneer plants **(b,c,e,f)** over fifteen days of the high-temperature treatment. Abbreviations as in Table 24.1

μmol $m^{-2}$ $s^{-1}$ PPFD) decreased significantly in the leaves of *Dryobalanops beccarii* and *Hopea rudiformis* seedlings (*t*-test, P<0.01), suggesting that a reversible down-regulation occurred (Fig. 24.4). In the light, ΔF/Fm' and ETR also tended to decrease in *Shorea ovalis*, but not significantly. Decreases in ΔF/Fm' and ETR in the light (800–1200 μmol $m^{-2}$ $s^{-1}$ PPFD) were not found in the leaves of the pioneer plants, suggesting maintenance of the high capacity of PSII in spite of the high-temperature treatment.

The values of NPQ in the light (800–1200 μmol $m^{-2}$ $s^{-1}$ PPFD) after fifteen days in the plastic sheet boxes did not increase compared to those under ambient air conditions for each plant species (Fig. 24.5). Thus, though NPQ is associated with ther-

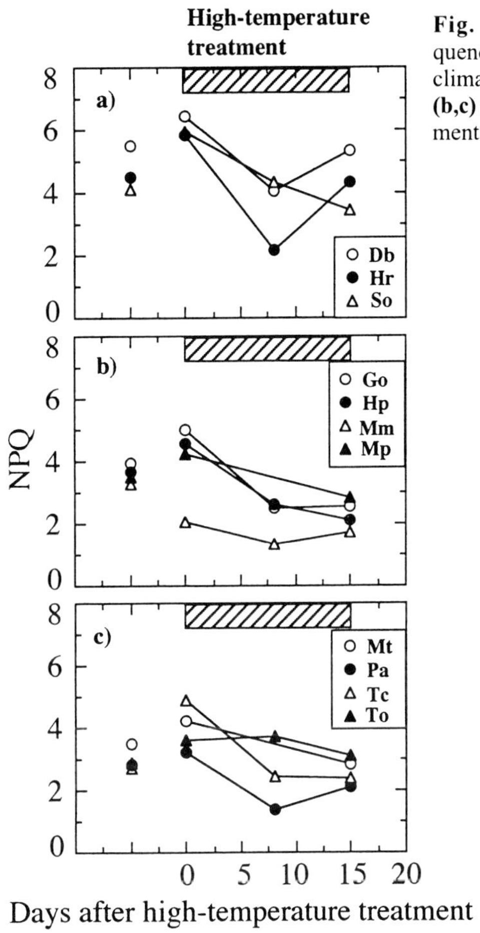


**Fig. 24.5.** Change in non-photochemical quenching (NPQ) in fully expanded leaves of climax tree seedlings **(a)**, and pioneer plants **(b,c)** over fifteen days of high-temperature treatment. Abbreviations as in Table 24.1

mal dissipation of excess light energy (Björkman and Demmig-Adams 1994), the capacity for acclimatization to elevated leaf temperature by increasing NPQ would be low in all the plant species. This result is difficult to interpret and requires further study.

## 24.4 Possible Niche Partitioning

Tropical rainforests contain much of the earth's terrestrial biodiversity. There has been much speculation concerning the factors that allow the co-existence of many plants in tropical rainforests (e.g., Denslow 1980, Lieberman et al. 1995, Hubbell et al. 1999). Hubbell et al. (1999) considered that dipterocarp species can co-exist in natural fields as a result of habitat equilibrium or simply stochastic processes, rather than coexisting simply by niche partitioning. By contrast, as a concept for explaining species diversity, niche partitioning is based on interspecific differences in the response of plants to environmental stresses or resource availability under highly heterogeneous conditions in time and space. Barker et al. (1997) discussed that though there was a possible niche partitioning between dipterocarp climax and pioneer species, there was no niche partitioning within the climax dipterocarp species, based on their data on leaf area-based carbon gain.

The present study indicates that the capacities to tolerate high solar radiation stress differed among dipterocarps. In the saplings of *Shorea leprosula* Miq. (a rapid growing dipterocarp) and *Macaranga gigantea* (Reichb. f. & Zoll.) Muell (a tropical pioneer tree) in Peninsular Malaysia, the photochemical capacity and net photosynthetic rate were higher in a large gap site than in a small gap site. However, in the saplings of *Neobalanocarpus heimii* (King) Ashton (a slow growing dipterocarp), the net photosynthetic rate was lower in a large gap site than in a small gap site (Table 24.2). Kachi et al. (1995) showed that the survival rate of seedlings was higher in *Neobalanocarpus heimii* than in *Shorea leprosula* in a deeply shaded understory. These data support each climax dipterocarp having its own preferred regeneration site which depends on the light environment and on its capacity to acclimate to changing light. *Shorea leprosula* is perhaps adapted to sites with higher rates of tree-fall and multiple-tree (larger) gaps, whereas *Neobalanocarpus heimii* is adapted to sites with lower rates of tree-fall and single-tree (smaller) gaps. Based on the field experiments (elevated leaf temperature) in the present study, *Shorea ovalis* is a sun dipterocarp, and *Dryobalanops beccarii* and *Hopea rudiformis* are shade dipterocarps. Ito et al. (Chapter 13, this volume) also showed that the seedlings of *Shorea ovalis* had relatively high drought tolerance capacity within the dipterocarp species.

**Table 24.2.** Chlorophyll fluorescence and leaf gas exchange characteristics in saplings of a pioneer tree and two dipterocarps with contrasting growth rates[a]

| Species | Fv/Fm at predwn | Max. electron transport rate[b] (μmol $m^{-2}s^{-1}$) | Max. net photosynthetic rate[b] Area basis (μmol $m^{-2}s^{-1}$) | Mass basis (μmol $kg^{-1}s^{-1}$) | Nitrogen basis (μmol N $g^{-1}s^{-1}$) |
|---|---|---|---|---|---|
| *Macaranga gigantea* (a pioneer tree)[c] | | | | | |
| Large gap site | 0.77 a | 125 a | 8.1 a | 155 a | 5.6 a |
| Small gap site | 0.79 a | 85 b | 4.5 b | 112 a | 5.2 a |
| *Shorea leprosula* (a fast-growing dipterocarp)[c] | | | | | |
| Large gap site | 0.78 a | 123 a | 11.0 a | 103 a | 6.2 a |
| Small gap site | 0.77 a | 65 b | 5.4 b | 64 b | 3.8 b |
| *Neobalanocarpus heimii* (a slow-growing dipterocarp)[c] | | | | | |
| Large gap site | 0.76 a | 67 b | 1.9 c | 23 c | 1.3 c |
| Small gap site | 0.79 a | 71 b | 3.5 d | 46 d | 2.7 d |

[a]Modified from Ishida et al. (1999c). Same letters(a–d) after data indicate homogeneous groups in each column according to t-test ($P<0.05$).

[b]Mean values of the diurnal maximum values observed in six fully expanded leaves of each tree species.

[c]Wood density at 12% moisture content is 300 kg $m^{-3}$ in *Macaranga gigantea*, 580 kg $m^{-3}$ in *Shorea leprosula*, and 980 kg $m^{-3}$ in *Neobalanocarpus heimii*.

Within pioneer tree species, no interspecific variations were found in the vulnerability of PSII in the mature leaves. Davies et al. (1998) showed that in 11 pioneer *Macaranga* species in Borneo, seven species were common on clay-rich soils, two preferred sand-rich soils, and two were not strongly affected by soil texture. This suggests that niche partitioning can occur due to variations in soil texture rather than in light environments within pioneer tree species. At least two types of pioneer plants have been recognized, early pioneer and late pioneer species. In south-east Asia, *Macaranga*, *Mallotus*, and *Trema* are examples of early pioneer species, while *Endospermum*, *Octomeles*, and *Scaphium* are examples of late pioneer species (Cheah 1995).

## 24.5 Protective Mechanisms for Photoinhibition in Morphology and Physiology

Various protective mechanisms in photoinhibition have been considered in physiology and morphology. Under drought and high irradiance conditions linked to ENSO events, high photorespiration and high NPQ at midday are be protective mechanisms of leaf physiology (see Chapter 12, this volume), though an increase in NPQ under elevated leaf temperature conditions was not found in the present study. On the other hand, from morphological studies it has been known that vertical leaf inclination angles contribute to maintain high physiological capacity in the top canopy leaves in the tropics (He et al. 1996, Ishida et al. 1999a). In the uppermost canopy leaves of *Macaranga conifera* (Zoll.) Muell. Arg., a pioneer tree, the near-vertical leaf inclination angle helped to maintain a low leaf temperature, a high ΔF/Fm', and a high net photosynthetic rate at midday (see Chapter 12, this volume).

**Table 24.3.** Quantum yield of PSII (ΔF/Fm') and non-photochemical quenching (NPQ) at 400–600 μmol $m^{-2}$ $s^{-1}$ PPFD and leaf inclination angles at the top canopy of the current-year fully expanded leaves in *Planchonella obovata* var *dubia* and *Hibiscus glabera*[a]

| Leaf properties | *Planchonella obovata* var *dubia* Current-year leaves | *Planchonella obovata* var *dubia* Horizontally-fixed leaves[b] | *Hibiscus glaber* Current-year leaves |
|---|---|---|---|
| ΔF/Fm' | 0.30 a | 0.26 b | 0.37 c |
| NPQ | 4.0 a | 4.9 b | 2.9 c |
| Leaf inclination angle (degrees) | 41 a | 0 b | 23 c |

[a]Ishida and Nakano unpublished data.
[b]Artificially excluded hairs on the adaxial leaf surface and horizontally-fixed top canopy leaves (current-year old leaves).
[c]Same letters after data indicate homogeneous groups in each leaf property, according to t-test (P<0.05).

Even within co-existing climax shrubs in a sunlit environment, interspecific variations in photochemical capacity of PSII are found. *Planchonella obovata* var *dubia* and *Hibiscus glaber* co-exist at sunlit ridge parts in the Bonin islands in the subtropical region of Japan. The canopy leaves in *Planchonella obovata* showed lower ΔFv/Fm' and higher NPQ at a given PPFD than those in *Hibiscus glaber* (Table. 24.3). The canopy leaves of *Planchonella obovata* had dense, red hairs on the adaxial leaf surface and more vertical leaf inclination angles, whereas the canopy leaves of *Hibiscus glaber* had no leaf hairs and more horizontal leaf angles. By artificially excluding leaf hairs and fixing the inclination angles at a horizontal position in the top canopy of leaves of *Planchonella obovata*, ΔFv/Fm' at a given PPFD decreased and NPQ at a given PPFD increased. Thus, the pubescence on the adaxial leaf surface and the leaf inclination angles may act as protective mechanisms of PSII in morphology. Canopy leaves with lower photochemical capacity have greater vertical inclination angles, as in *Planchonella obovata*, and those with higher photochemical capacity have more horizontal inclination angles, as in *Hibiscus glaber*. Differences in canopy structure will contribute to enhance total carbon gain in the crown level in each species according to the interspecific differences in photochemical capacity. This data suggests that the linkage between photochemical capacity in leaves and canopy structure is an important concept in understanding light-use at the whole canopy level.

## 24.6 Conclusion

Sustained-photoinhibition was not accelerated by the high-temperature treatment (elevated about 4°C in leaf temperature) in pioneer plants. Thus, global temperature increments might possibly spread the vegetation area of pioneers. On the other hand, interspecific differences in the susceptibility of PSII were found among the three dipterocarps. This suggests a possible niche partitioning within the climax dipterocarps, depending on the variations of light environments according to canopy gap size. Variations in acclimation capacity were found even in co-existing shrubs in a sunlit site. Canopy leaves with lower photochemical capacity had more vertical inclination angles, and those with higher photochemical capacity had more horizontal inclination angles. This contributes to enhance total carbon gain at the crown level in each plant. These data suggest a linkage between photochemical capacity in leaves and canopy structure. Information on the interactive environmental factors limiting leaf carbon gain processes and understanding the protective mechanisms against such environmental stresses in morphological and physiological properties can be useful for improving rehabilitation and conservation techniques in tropical rainforests.

# References

Barker MG, Press MC, Brown ND (1997) Photosynthetic characteristics of dipterocarp seedlings in three tropical rain forest light environments: a basis for niche partitioning? Oecologia 112:453–463

Bazzaz FA (1979) The physiological ecology of plant succession. Ann Rev Ecol Syst 10:351–371

Bilger W, Schreiber U, Bock M (1995) Determination of the quantum efficiency of photosystem II and of non-photochemical quenching of chlorophyll fluorescence in the field. Oecologia 102:425–432

Björkman O, Powles SP (1984) Inhibition of photosynthetic reactions under water stress: interaction with light level. Planta 161:490–504

Björkman O, Demmig B (1987) Photon yield of $O_2$ evolution and chlorophyll fluorescence characteristics at 77 k among vascular plants of diverse origin. Planta 170:489–504

Björkman O, Demmig-Adams B (1994) Regulation of photosynthetic light energy capture, conversion, and dissipation in leaves of higher plants. In: Schulze E–D, Caldwell MM (eds) Ecophysiology of Photosynthesis. (Ecological Studies, vol 100) Springer, Berlin Heidelberg New York, pp 17–47

Cheah LC (1995) Pioneer species for fast growing tree plantations in Malaysia-An evaluation. FRIM Technical Information No. 53. Forest Research Institute Malaysia (FRIM), Kepong, Kuala Lumpur

Choong ET and Achmadi SS (1996) Utilization potential of the dipterocarp resource in international trade. In: Schlute A, Schöne D (eds) Dipterocarp Forest Ecosystems. Towards Sustainable Management. World Scientific, Singapore, pp 481–525

Davies SJ, Palmiotto PA, Ashton PS, Lee HS, Lafrankie JV (1998) Comparative ecology of 11 sympatric species of *Macaranga* in Borneo: tree distribution in relation to horizontal and vertical resource heterogeneity. J Ecol 86:662–673

Demmig-Adams B, Adams III WW, Baker DH, Logan BA, Bowling DR, Verhoeven AS (1996) Using chlorophyll fluorescence to assess the fraction of absorbed light allocated to thermal dissipation of excess excitation. Physiol Plant 98:253–264

Denslow JS (1980) Gap partitioning among tropical rainforest trees. Biotropica (Suppl) 12:47–55

Ferrar PJ and Osmond CB (1986) Nitrogen supply as a factor influencing photoinhibition and photosynthetic acclimation after transfer of shade-grown *Solanum dulcamara* to bright light. Planta 168:563–570

Gamon JA, Pearcy RW (1989) Leaf movement, stress avoidance and photosynthesis in *Vitis californica*. Oecologia 79:475–481

Gamon JA, Pearcy RW (1990) Photoinhibition in *Vitis californica*: interactive effects of sunlight, temperature and water status. Plant Cell Environ 13:267–275

He J, Cee CW, Goh CJ (1996) 'Photoinhibition' of *Heliconia* under natural tropical conditions: the importance of leaf orientation for light interception and leaf temperature. Plant Cell Environ 19:1238–1248

Hubbell, SP, Foster RB, O'Brien ST, Harms KE, Condit R, Wechsler B, Wright SJ, Loo de Lao S (1999) Light-gap disturbances, recruitment limitation, and tree diversity in a Neotropical forest. Science 283:554–557

Ishida A, Toma T, Marjenah (1999a) Leaf gas exchange and chlorophyll fluorescence in relation to leaf angle, azimuth, and canopy position in the tropical pioneer tree, *Macaranga conifera*. Tree Physiol 19:117–124

Ishida A, Toma T, Marjenah (1999b) Limitation of leaf carbon gain by stomatal and photochemical processes in the top canopy of *Macaranga conifera*, a tropical pioneer tree. Tree Physiol 19:467–473

Ishida A, Nakano T, Matsumoto Y, Sakoda M, Ang LH (1999c) Diurnal changes in leaf gas exchange and chlorophyll fluorescence in tropical tree species with contrasting light requirements. Ecol Res 14:77–88

Iwasa T, Kikata Y, Kitano S, Sasaki N, Suzuki K, Hara K (1984) Handbook of Tropical Plants and Trees (In Japanese). Yokendo, Tokyo

Kao W-Y, Forseth IN (1992) Diurnal leaf movement, chlorophyll fluorescence and carbon assimilation in soybean grown under different nitrogen and water availabilities. Plant Cell Environ 15:703–710

Kachi N, Okuda T, Yap SK and Manokaran N (1995) Biodiversity and regeneration of canopy tree species in a tropical rain forest in Southeast Asia. J Environ Sci 9:17–36

Königer M, Harris GC, Pearcy RW (1998) Interaction between photon flux density and elevated temperatures on photoinhibition in *Acacia macrorrhiza*. Planta 205:214–222

Leitsch J, Schnettger B, Critchley C, Krause GH (1994) Two mechanisms of recovery from photoinhibition in vivo: reactivation of photosystem II related and unrelated to D1-protein turnover. Planta 194:15–21

Lieberman M, Lieberman D, Peralta R, Hartshorn GS (1995) Canopy closure and the distribution of forest tree species at La Selva, Costa Rica. J Trop Ecol 11:161–178

Ludlow MM (1987) Light stress at high temperature. In: Hyle DJ, Osmond CB, Arntzen CJ (eds) Photoinhibition. Elsevier, Amsterdam, pp 89–109

Mulkey SS, Pearcy RW (1992) Interactions between acclimation and photoinhibition of photosynthesis of a tropical forest understorey herb, *Alocasia macrorrhiza*, during simulated canopy gap formation. Funct Ecol 6:719–729

Somersalo S, Krause GH (1990) Effects of freezing and subsequent light stress on photosynthesis of spinach leaves. Plant Physiol Biochem 28:467–475

Thiele A, Krause GH, Winter K (1998) *In situ* study of photoinhibition of photosynthesis and xanthophyll cycle activity in plants growing in natural gaps of the tropical forest. Aust J Plant Physiol 25:189–195

van Kijk KJ, van Hasselt PR (1993) Kinetic resolution of different recovery phases of photoinhibited photosystem II in cold-acclimated and non-acclimated spinach. Physiol Plant 87:187–198

Whitmore TC (1984) Tropical rainforests of the Far East, 2nd edn. Clarendon, Oxford

## 25.2 Seedling Growth in Contrasting Light Environments

### 25.2.1 The Study Sites

The forests in BSEF were initially logged commercially, and then local people logged the area again and began slash-and-burn farming in the 1970s. A prolonged drought and large fires occurred in 1982–1983, linked with an El-NiÒo southern oscillation event (Goldammer and Seibert 1990). However, fertile *ulin* trees are still fairly abundant. The seedlings are usually restricted to the area near mother trees, because the fruit is heavy. To examine the effects of contrasting light environments on seedling growth, two research plots were established in existing seedling banks on a slope with sandy-loam topsoil, in July 1992 (Plots 1 and 2 in Table 25.1). The dominant overstory trees were *ulin*, *Endospermum diadenum*, *Dipterocarpus tempehes*, and others. Most of the *ulin* seedlings were derived from the seeds from mast fruiting that occurred around 1988. In Plot 1, called the gap site, a canopy gap was made, and weeding was started in May 1993. The canopy was left intact in Plot 2, called the closed site.

To examine the effects of soil texture on seedling growth, Plot 3 was established on a hilltop with loam-clay topsoil in BSEF in November 1996.*Ulin* were planted under an overstory of pioneer trees, mainly *Macaranga triloba*. The characteristics of the 0–1 m soil layers were almost the same at the closed and gap sites. Although there were no large differences in N contents in the soils among the three plots (Plots 1, 2, and 3), the soil in Plot 3 contained only 23% in available phosphorus, 6% in exchangeable calcium, 3% in exchangeable magnesium, and 26% in exchangeable sodium, of these in the soil in Plot 2.

### 25.2.2 Growth of Seedlings with Changing Light

The death rate of the seedlings at the closed site was 2.8% $year^{-1}$ (Table 25.2). This value is lower than the 9.8–16.0% $year^{-1}$ for red meranti-*Shorea* seedlings in Sabah, Borneo (Whitmore 1984), indicating that *ulin* is a shade-tolerant species (Whitmore 1998). The sudden opening of the canopy increased the seedling death rate. The death rate in the gap site was about twice as high as the death rate in the closed site. The highest death rate (12.2% $year^{-1}$) was recorded in the first five months after opening the canopy. One month after opening the canopy, there was a five-month period of low rainfall (see Chapter 2, this volume). Therefore, this very high death rate was probably due to the drought and sun exposure. Itoh et al. (see Chapter 13, this volume) showed that the growth of climax dipterocarp seedlings decreases with drought stress, and Ishida et al. (see Chapter 24, this volume) showed that the vulnerability of the leaves to heat and brightness stress is higher in slow-growing dipterocarp species than in fast-growing dipterocarp species. The death rate after the sudden opening of the canopy was dependent on the age of the *ulin* seedlings. In the low-rainfall period in 1994, some cohorts of seedlings less than three years old had higher death rates than older seedlings. Jong (1932), Endert (1937), and Tuyt (1939) showed that three- to eight-year-old seedlings grow well when the canopy is opened.

Although *ulin* is considered a shade-tolerant species, the seedlings require an appropriate canopy opening to accelerate growth. The relative growth rates in terms of the height and diameter of the seedlings were higher in the gap site than in the closed site (Table 25.2). The diameter of the ten tallest seedlings increased by an average of 0.389 cm $year^{-1}$ in the gap site, while the increase was 0.123 cm $year^{-1}$ in the closed site. Assuming that these growth rates remain unchanged, opening the canopy and weeding in the gap site should decrease the time required for an *ulin* trunk to reach 10 cm in DBH from 78 to 26 years.

The seedlings grown under bright conditions developed thick leaves and large root systems. The canopy opening also results in relatively low humidity conditions on the forest floor. Forty-two months after the gap formation, the total dry mass of the seedlings in the gap site was 5.9 times that of the seedlings in the closed site (Table 25.3). The thick roots in the gap site were 2.5 times longer horizontally and 1.7 times deeper than in the closed site. The total leaf area (LA)/fine-root mass ratios of individual seedlings were smaller (0.85 times) in the gap site than in the closed site. On the other hand, the leaf mass/fine-root mass ratios were 1.23 times larger in the gap site than in the closed site. Such morphological features seem common to different tree seedlings grown under different light conditions (e.g., Sasaki and Mori 1981) (for dipterocarps, see Chapter 13, this volume).

**Table 25.2.** Growth of *Eusideroxylon zwageri* seedlings in 1993–1996

| Plot | Age[a] (years) | Mean height[a] (cm) | Height growth rate ($cm\ cm^{-1}\ year^{-1}$) | Diameter growth rate ($cm\ cm^{-1}\ year^{-1}$) | Death rate (% $year^{-1}$) |
|---|---|---|---|---|---|
| 1 (gap site) | About 5 | 92±36 | 1.210 | 1.228 | 5.7 |
| 2 (closed site) | About 5 | 85±30 | 1.080 | 1.084 | 2.8 |
| 3 (poor soil site) | 5[b] | 61±22[c] | 1.035[c] | - | 7.9[c] |

[a]In 1993. [b]Seeds were planted and germinated in 1988. [c]The growth data in Marjenah (1994) were used for the values in 1993.

**Table 25.3.** Organs of *Eusideroxylon zwageri* seedlings in November 1996.

| Plot | Mean height (cm) | Total mass (g) | Leaves: SLA[a] ($cm^2\ g^{-1}$) | Leaves: LA[b] ($m^2$) | Fine roots: Mass (g) | Fine roots: Surface area ($m^2$) | LA/fine-root mass ($m^2\ g^{-1}$) | Thick root ranges: Horizon (cm) | Thick root ranges: Vertical (cm) |
|---|---|---|---|---|---|---|---|---|---|
| 1 (gap site) | 174 | 769 | 86.5 | 1.34 | 23.5 | 18.9 | 0.0570 | 116 | 77 |
| 2 (closed site) | 110 | 130 | 126 | 0.265 | 3.95 | 2.70 | 0.0671 | 35 | 36 |
| 3 (poor soil site) | 70.0 | 42.2 | 129 | 0.0608 | 2.42 | 1.42 | 0.0251 | 45 | 54 |

All data are mean values of five seedlings. [a]Leaf area/mass. [b]Total leaf area.

## 25.3 Growth of Seedlings Seeded Artificially on an Unsuitable Soil

Low nutrient levels in the soil seem to affect seedling growth and survival. Although the degree of canopy openness in the *Macaranga triloba* forest (Plot 3, the poor soil site) was intermediate for the three study sites (Table 25.1), the seedlings in this site had the lowest height growth rate and the highest death rate (Table 25.2). This indicates that the slow growth of seedlings resulted from factors such as soil nutrient content rather than the light environment.

The total dry mass of the seedlings in 1996 was about one-third that in the closed site, but the thick roots extended further horizontally and deeper than those in the closed site (Table 25.3). Both the leaf area/fine-root mass and the leaf mass/fine-root mass ratios of seedlings were lowest at the poor soil site. Since development of the root system is favored over shoot development in dry, nutrient-poor soils (Chapin 1980, Fitter and Hay 1981), these results suggest that the seedlings could adapt to soil-nutrient stresses, although the aboveground organs grow slowly. Alphen (1954) showed that *ulin* seedlings do not grow well on sand-rich soils that are nutrient-poor and dry readily. *Ulin* seedling requires moist, but well drained, fertile soils to grow well.

## 25.4 Influence of Fire on Seedling Survival and Regeneration

Relatively large *ulin* trees have a fire-resistance capacity (Goldammer and Seibert 1990), and are also less vulnerable to cutting owing to effective sprouting. Of 26 *ulin* trees from 44 to 92 cm in DBH, 24 (92%) sprouted new shoots from stumps in fallow land after slash-and-burn agriculture (Kiyono and Hastaniah 1997). The ground fires in March 1998 burned the research sites (see Chapter 3, this volume). All four mature *ulin* (30–60 cm in DBH) in the closed and gap sites survived the fires, while only 30 (7.6%) seedlings in the gap site survived. Thus, smaller seedlings were more vulnerable to fire than larger ones at the gap site. After the fire, the survivors in the gap site were 302±90 cm tall and 2.37±0.83 cm in diameter at 0.3 m height, while the dead seedlings were 208±73 cm tall and 1.52±0.77 cm in diameter. Considering that all the seedlings (110±55 cm tall and 0.90±0.40 cm in diameter) died in the closed site, our release cutting helped some *ulin* seedlings to escape burning to death.

## 25.5 Conclusion

The *ulin* (*Eusideroxylon zwageri*) is an ironwood species whose growing stock has decreased due to over-exploitation. Appropriate post-felling treatment is needed to maintain an *ulin* wood resource in the forest. The death of *ulin* seedlings was rare, even in a closed rainforest. Since they grow slowly, natural regeneration in terms of timber production seems poor. Canopy opening and weeding (release cutting) will accelerate seedling growth by improving light conditions. Release cutting might decrease the time required for the *ulin* stem to reach 10 cm in DBH from 78 to 26 years. However, sudden felling increased the death rate of seedlings. Since seedlings less than three years old were more sensitive to sunlight and dry conditions than older seedlings, release cutting should only be performed when existing seedlings are at least three years old. Compared with seedlings that regenerated naturally in their own habitat, the seedlings artificially established on dry, nutrient-poor soil grew very slowly and had a higher death rate, even in moderate light conditions. Only suitable, moist soils should be selected for planting *ulin*. Large *ulin* saplings are very fire-resistant. Therefore, appropriate release cutting will shorten the time required for *ulin* seedlings to gain fire-resistance.

## References


Alphen de Veer EJ van (1954) *Eusideroxylon zwageri*, a remarkable tree species (*Eusideroxylon zwageri*, een merkwaardige boomsoort) (in Dutch). De Tropische Natuur. pp 37–41

Brown WH (1919) Vegetation of Philippine mountains. Manila. (quoted from Richards PW (1952) The tropical rainforest-an ecological study. Cambridge UP, New York)

Chapin FS III (1980) The mineral nutrition of wild plants. Ann Rev Ecol Syst 11: 233–260

Endert FH (1937) Report of an official trip to Belitung from 25 to 30 August 1937 by the adjunct adviser Dr. F. H. Endert (Verslag van een dienstreis naar Billiton van 25 t/m 30 Augustus 1937 van den Adjunct-Adviseur Dr. F. H. Endert) (in Dutch) (Unpublished)

Fitter AH, Hay RKM (1981) Environment physiology of plants. Academic Press

Goldammer JG, Seibert B (1990) The impact of droughts and forest fires on tropical lowland rainforest of East Kalimantan. In: Goldammer JG (Ed) Fire in the tropical biota: Ecosystem processes and global challenges. Ecological studies 84, Springer-Verlag, pp 11–31

Hastaniah, Kiyono Y (1995) Growth and habitat of *ulin* (*Eusideroxylon zwageri*) in a lowland area of East Kalimantan (Keadaan tempat tumbuh dan pertumbuhan *ulin* (*Eusideroxylon zwageri*) di hutan dataran rendah Kalimantan Timur) (in Indonesian). Proc PUSREHUT Research Seminar. Tropical Rain Forest Research Center, Research Institute of Univ Mulawarman, Samarinda, pp 46–52

Jong B de (1932) Experience with the cultivation of ironwood (*Eusideroxylon zwageri*) (Ervaringen met ijzerhout-culturen) (in Dutch). Tectona 25: 37–49

Kesser PJA, Sidiyasa K (1994) Trees of the Balikpapan-Samarinda area, East Kalimantan, Indonesia: A manual to 280 selected species. Tropenbos Series 7, The Tropenbos Foundation, Wageningen

Kiyono Y, Hastaniah (1997) Slash-and-burn agriculture and succeeding vegetation in East Kalimantan. PUSREHUT special publication 6, Univ Mulawarman, Samarinda

MacKinnon K, Hatta G, Halim H, Mangalik A (1996) The ecology of Kalimantan. Periplus Editions, Singapore

MacKinnon J, MacKinnon K (1986) Review of the protected area system in the Indo-Malayan Realm. IUCN, Gland

Marjenah (1994) Data of Arboretum measurements at Bukit Soeharto Research Forest in East Kalimantan, Indonesia. Annual Report of PUSREHUT 4, Univ Mulawarman, Samarinda

Richards PW (1952) The tropical rainforest-an ecological study. Cambridge UP, New York
Sasaki S, Mori T (1981) Growth responses of dipterocarp seedlings to light. Mal For 44: 319–345
Tuyt P (1939) Strip planting of ironwood in Palembang residency (Schaduwrijen cultuur van ijzerhout in de residentie Palembang) (in Dutch). Tectona 32: 805–828
Whitmore TC (1984) Tropical rainforests of the Far East (2nd edn). Clarendon Press, Oxford
Whitmore TC (1998) An introduction to tropical rainforests (2nd edn). Oxford UP, New York
Whitten AJ, Damanik SJ, Anwar J, Nazaruddin H (1987) The ecology of Sumatra (2nd edn). Gadjah Mada UP, Yogyakarta
Witkamp H (1925) De ijzerhout als geologische indicator. Trop Natuur 14: 97–103 (quoted from Richards PW (1952) The tropical rainforest-an ecological study. Cambridge UP, New York)

# 26 Diseases of Dipterocarp Saplings Planted in Bukit Soeharto Education Forest, East Kalimantan

Djumali Mardji

## 26.1 Introduction

Dipterocarp species have long been used as sources of timber for plywood, fiberboard, particle-board, molding and other building materials. To maintain production of dipterocarp timbers, enrichment planting programs have recently been started, and extensive plantations of dipterocarp species have been established (see Chapter 1, this volume). However, studies on diseases of trees in tropical plantations have mainly focused on fast-growing exotic species. Far less work has been done on diseases affecting dipterocarps, and more information is required (Elouard 1998) to ensure the success of the dipterocarp plantations and enrichment planting projects.

Nevertheless, a range of diseases affecting dipterocarp seedlings and saplings in natural and plantation forests in Southeast Asia have been identified. *Dipterocarpus grandiflorus* seedlings killed by *Botryodiplodia theobromae* often occur under their mother trees, for instance, and up to 40% of saplings were found to be killed by species of *Polyporus*, *Humicola* and *Macrophoma* in a natural forest in the Philippines (Quiniones 1980). *Cylindrocladium scoparium* can cause the death of *Dipterocarpus oblongifolius* seedlings in Peninsular Malaysia (Lee and Ahmad 1982). Various bacterial, viral and fungal diseases have been reported on seedlings and saplings of *Shorea* spp. in natural and plantation forests in East Kalimantan, Malaysia, Thailand and the Philippines (Smits et al. 1991). Also, Soeyamto and Mardji (1995) recently observed root-rot and stem cankers caused by an unidentified fungi on one-year-old *Shorea leprosula* seedlings planted in open sites in a plantation at Batu Ampar, East Kalimantan.

Most of the studies cited above focused on a specific species under a narrow range of growth conditions. This chapter, however, deals with diseases found in a systematic analysis of 5-year old saplings representing 15 dipterocarp species, planted using a range of methods, in the Bukit Soeharto Education Forest (BSEF) of Mulawarman University. The aim of the investigation was to determine whether different plant species and site conditions are correlated with differences in incidence and severity of disease. Six diseases (leaf spot, leaf blight, chlorosis, stem canker, dieback, and stunt) were monitored amongst a total of 2603 saplings, representing 15 dipterocarp species, planted in seven blocks of different planting systems

Ecological Studies, Vol. 140
Guhardja et al. (eds.) Rainforest Ecosystems of East Kalimantan: El Niño, Drought, Fire and Human Impacts

(Table 26.1). The diseases on the planted trees were recorded according to their visible symptoms, and disease severity was also recorded (see severity scores listed in Table 26.2) for each infected tree.

**Table 26.1.** Number of observed 5-year-old dipterocarp saplings in the demonstration and experimental forest in Bukit Soeharto Education Forest of Mulawarman University

| | Single planting | | | Group planting | | | |
|---|---|---|---|---|---|---|---|
| Block | A | C | E | B5 | B8 | B9 | D |
| Sapling distance(m) | 3 × 5 | 3 × 3 | 3 × 3 | 1 × 1 | 0.5 × 0.5 | 1 × 1 | 1 × 1 |
| Group distance(m) | | | | 6 × 6 | 6 × 6 | 6 × 6 | 2 × 2 |
| No. of saplings within a group | | | | 5 | 8 | 9 | 4[a] |
| Species | | | | | | | |
| *Cotylelobium lanceolatum* | | | 71 | | | | 15 |
| *Dipterocarpus confertus* | | | 74 | | | | |
| *Dipterocarpus cornutus* | | | 66 | | | | |
| *Dipterocarpus gracilis* | | | 82 | | | | 30 |
| *Dipterocarpus humeratus* | 15 | | 76 | | | | 11 |
| *Dryobalanops beccarii* | 35 | 40 | 86 | 184 | 133 | 117 | 39 |
| *Dryobalanops lanceolata* | | | 81 | | | | 11 |
| *Dryobalanops rappa* | | | 66 | | | | |
| *Shorea acuminatissima* | | 40 | 90 | | | | 11 |
| *Shorea bracteolata* | | | 121 | | | | 24 |
| *Shorea laevis* | 35 | | 76 | 59 | 105 | | 20 |
| *Shorea leprosula* | | | 30 | 219 | 131 | 129 | 11 |
| *Shorea parvifolia* | 30 | 69 | 57 | | | | 16 |
| *Shorea pauciflora* | | | 49 | | | | |
| *Shorea seminis* | | | 67 | | | | 22 |

[a]Mixed with four species.

**Table 26.2.** Scoring system used to assess disease severity in each plant.

| Score | Status | Condition of plant |
|---|---|---|
| 0 | Nil | No infection symptom present or the number of infected leaves very low and width of lesions narrow compared to the width of all leaves. Plant remained sound. |
| 1 | Low | Few infected leaves with few lesions on each leaf, or little defoliation or chlorosis occured, or plant looked sound but other symptom such as stem canker presented. |
| 2 | Medium | Many infected leaves and many lesions on each leaf, or defoliation or chlorosis occured, or other symptom such as stem canker present |
| 3 | Severe | Abundant infected leaves and abundant lesions on each leaf, or abundant defoliation or chlorosis occured, or other symptom such as stem canker presented. |
| 4 | Highly severe | Extensive numbers of infected leaves and extensive lesions on each leaf, or extensive defoliation or chlorosis occured, or other symptom such as stem canker presented. |
| 5 | Dead | All leaves were wilted or defoliated or no indication of life |

## 26.2 Notes on Diseases

Incidence of attack (F) was calculated by the formula of James (1974) as follows:

$$F\ (\%) = \{(\text{Number of infected saplings}) / (\text{Total number of observed saplings})\}\ 100 \qquad (1)$$

Severity of attack (I) was calculated by the formula of de Guzman (1985) and Singh and Mishra (1992), modified as follows:

$$I(\%) = \{(1X_1 + 2X_2 + 3X_3 + 4X_4 + 5X_5) / 5X\}\ 100 \qquad (2)$$

where, $X_i$ is the number of plants of each species with score i, and X is the total number of observed plants of each species.

The incidence of disease was found to differ between species, and even within species it differed between planting sites (Table 26.3). Leaf spot disease was the most frequent disease in the area. The severity of all the diseases was low for each of the species and blocks (data not shown) and many of the trees were sound.

### 26.2.1 Leaf Spot

The major symptom of leaf spot, as the name suggests, is necrotic spots on infected leaves. Leaf spot was the most frequently found disease in all the blocks and all the species except *Shorea pauciflora*. The highest incidence of leaf spot disease occurred in *Dipterocarpus gracilis*. Symptoms and the causal agents of leaf spot differed amongst the plant species. Causal agents observed included species of the fungal genera *Colletotrichum* in *Dipterocarpus humeratus* and *Dryobalanops lanceolata, Heterosporium* in *Shorea laevis*, *Hendersonia* in *Shorea leprosula* and unidentified fungi in *Dryobalanops beccarii*, *Dryobalanops rappa* and *Shorea acuminatissima*. In the Philippines, leaf spots on *Anacardium occidentale* have been found to be caused by *Pestalotiopsis adusta* and those on *Calliandra haematocephala* by *P. langloisii* (Kobayashi and de Guzman 1988).

### 26.2.2 Leaf Blight

Symptoms of leaf blight are necroses on the leaves, which expand to cover broader areas than the ones associated with leaf spot. The necroses, generally of irregular shape, originate at the edge or top of the leaves and then spread to the middle. The disease was found on *Dry. beccarii*, *Dipt. humeratus*, *Shorea bracteolata* and *S. pauciflora*. The highest incidence of leaf blight disease occurred on *S. pauciflora*. The leaves of *Dipt. humeratus* were infected with *Colletotrichum*, and the causal agents of the leaf blight of *Dry. beccarii* were species of *Colletotrichum* and *Pestalotiopsis*. The fungi infecting *S. bracteolata* and *S. pauciflora* trees were not identified.

According to Stevens (1966), the genus *Pestalotiopsis* consists of more than 200

**Table 26.3.** Incidence of diseases (%) on 5-year-old dipterocarp plant species growing on the demonstration and experimental forest in Bukit Soeharto Education Forest

| Species | Block | n | Leaf spot | Leaf blight | Chlorosis | Stem canker | Dieback | Stunt |
|---|---|---|---|---|---|---|---|---|
| *Dipterocarpus humeratus* | A | 15 | 47.2 | 5.6 | - | 2.7 | - | - |
| *Dryobalanops beccarii* | A | 35 | 26.9 | - | 7.5 | - | - | - |
| *Shorea laevis* | A | 35 | 97.2 | - | 2.8 | 8.3 | - | 5.6 |
| *Shorea parvifolia* | A | 30 | - | - | - | 1.7 | - | - |
| *Dryobalanops beccarii* | C | 40 | 22.4 | - | - | - | - | - |
| *Shorea parvifolia* | C | 69 | - | - | - | 2.6 | 5.3 | 3.9 |
| *Cotylelobium lanceolatum* | E | 71 | 1.2 | - | - | - | 12.5 | 5.0 |
| *Dipterocarpus confertus* | E | 74 | 19.8 | - | - | - | - | - |
| *Dipterocarpus cornutus* | E | 66 | 10.3 | - | - | - | - | - |
| *Dipterocarpus gracilis* | E | 82 | 57.8 | - | - | - | - | - |
| *Dipterocarpus humeratus* | E | 76 | 31.0 | - | - | - | - | - |
| *Dryobalanops beccarii* | E | 86 | 1.6 | 6.5 | - | 8.9 | - | - |
| *Dryobalanops lanceolata* | E | 81 | 42.2 | - | - | - | - | - |
| *Dryobalanops lappa* | E | 66 | 34.5 | - | - | - | - | - |
| *Shorea acuminatissima* | E | 90 | - | - | 2.8 | - | 2.8 | 4.9 |
| *Shorea bracteolata* | E | 121 | 0.7 | 0.7 | 0.7 | 0.7 | 2.7 | - |
| *Shorea laevis* | E | 76 | 42.4 | - | 8.5 | 5.1 | - | 2.5 |
| *Shorea leprosula* | E | 30 | - | - | - | 1.0 | - | - |
| *Shorea parvifolia* | E | 57 | - | - | 2.3 | 1.6 | 0.8 | 1.6 |
| *Shorea pauciflora* | E | 49 | - | 0.6 | - | - | 0.6 | 0.6 |
| *Shorea seminis* | E | 67 | - | - | - | 0.7 | - | 0.7 |
| *Dryobalanops beccarii* | B5 | 184 | 24.5 | - | 5.5 | 2.9 | - | - |
| *Dryobalanops beccarii* | B8 | 133 | 14.4 | - | 3.4 | - | 0.3 | 5.6 |
| *Dryobalanops beccarii* | B9 | 117 | 9.2 | - | 3.6 | 0.8 | 0.3 | - |
| *Shorea laevis* | B5 | 59 | 22.5 | - | - | - | 1.0 | - |
| *Shorea laevis* | B8 | 105 | 18.7 | - | 4.4 | 1.6 | 3.4 | 2.5 |
| *Shorea leprosula* | B5 | 219 | 73.0 | - | - | - | 0.5 | - |
| *Shorea leprosula* | B8 | 131 | - | - | - | 2.2 | 0.7 | 2.5 |
| *Shorea leprosula* | B9 | 129 | 15.0 | - | - | 2.5 | - | - |
| *Cotylelobium lanceolatum* | D | 15 | 23.3 | - | - | - | - | - |
| *Dipterocarpus gracilis* | D | 30 | 56.7 | - | - | - | 4.7 | - |
| *Dipterocarpus humeratus* | D | 11 | 3.3 | - | - | - | - | - |
| *Dryobalanops beccarii* | D | 39 | 41.7 | - | - | - | - | - |
| *Dryobalanops lanceolata* | D | 11 | 3.3 | - | - | - | - | - |
| *Shorea acuminatissima* | D | 11 | 6.7 | - | - | - | - | - |
| *Shorea bracteolata* | D | 24 | 16.7 | - | - | - | - | - |
| *Shorea laevis* | D | 20 | 10.0 | - | - | - | - | - |
| *Shorea leprosula* | D | 11 | - | - | - | - | - | - |
| *Shorea parvifolia* | D | 16 | 13.3 | - | - | - | - | - |
| *Shorea seminis* | D | 22 | 13.3 | - | - | - | - | - |

species, most of which are pathogens that may infect many commercial plants in nurseries and plantations, causing leaf blight disease. Many such pathogens are now regarded as *Pestalotiopsis*, and the genus has been found in leaves, stems, flowers and fruits of plants. Butin (1989) stated that *Pestalotiopsis* is a weak parasite that often attacks seedlings affected by heat stress, and is often found on the shoot of such plants. Gibson (1979) noted that *Pestalotiopsis* is a fungus often found on diseased stems and leaves of *Pinus* and considered it to be the main causal agent. However, since various species of *Pestalotiopsis* often attack diseased parts of plants already infected by other pathogens, the true significance of this fungal group as a primary parasite seems doubtful.

### 26.2.3 Chlorosis

Chlorosis appears as a yellowing of some or all the leaves of affected trees. This is generally caused by failure of chlorophyll formation, destruction of chlorophyll as a result of disease, or the lack of certain elements required for chlorophyll formation. This disease was found on *Dry. beccarii*, *Shorea acuminatissima*, *S. bracteolata*, *S. laevis* and *S. parvifolia*. The highest incidence of the disease occurred in block A and the species most affected was *Dry. beccarii*.

Almost all the leaves of saplings at the research site with these symptoms turned yellow. These saplings were suppressed in growth, and had fewer leaves than disease-free saplings, although they were taller than those affected by stunt disease (see below). Chlorosis of the leaves was often followed by insect attack and infection by leaf spot fungi. The causal agent of the chlorosis was suspected to be a virus, because other plants growing close to diseased specimens often remained healthy.

### 26.2.4 Stem Canker

Stem canker occurs on tree stems, near the ground. Wounds on recently-damaged branch stumps are the main sites of infection of the fungus, because they are moist and not covered by callus immediately. In this disease, stem cankers (necroses of bark and cambium, often with sunken lesions, followed by formation of callus, eventually causing swelling and death of the shoot, or 'dieback') were observed on infected trees after the pathogen had girdled the stem. The infected parts of the stem were hypertrophic (i.e. slightly thicker) than the sound parts. Severe infection resulted in dieback, and epicormic branches sprouted from the stem under the infection site. This disease was found on *Dipt. humeratus*, *Dryo. beccarii*, *Shorea leprosula*, *S. seminis, S. bracteolata*, *S. laevis*, and *S. parvifolia* . The fungi causing the stem canker on these dipterocarp species were from the genus *Stereum* (Basidiomycetes). The causal fungus of stem canker on these dipterocarp saplings were from the genus *Stereum* (Basidiomycetes), and the symptoms were similar to those reported for other stem canker-causing fungi.

## 26.2.5 Dieback

The main symptom of dieback is similar to that of stem canker. The difference is that in dieback there is no clear wound or canker, but the shoot dies following the spread of necrotic symptoms, starting from the apex and progressing down towards the main stem. The color of the infected shoot changes from bright to dark green. Infected shoots also clearly differ from sound stems since they defoliate, and sometimes epicormic branches appear on them. The epicormic branches die by the time the pathogenic attack has extended to the basal part of the shoots. This disease is probably caused by damage to vascular cells in the infected area, resulting in failure to transport water and nutrients to the shoot tip. At the study site, this disease was found on *Cotylelobium lanceolatum*, *Dry. beccarii*, *S. acuminatissima*, *S. bracteolata*, *S. laevis*, *S. leprosula*, *S. parvifolia* and *S. pauciflora*. The causal fungus or pathogen responsible for dieback on the dipterocarp saplings has not yet been identified.

One of the most common pathogens linked with dieback disease among fast-growing species in India is *Cylindrocladium quinqueseptatum* (Singh and Chaturvedi 1993), which also causes this disease in Brazil (Gibson, 1975). It has also been observed to infect *Eucalyptus urophylla* in a plantation in Sumatra (Nuhamara, 1991). However, in the Philippines, dieback in *Cassia fructicosa* was reportedly caused by *Diatrypella favacea* and *Valsa kitajimana* (Kobayashi and de Guzman, 1988), and *Pestalotiopsis* sp. was found to be the causal agent of dieback on seedlings of *Swietenia macrophylla* in Batu Ampar, East Kalimantan (Mardji 1995). The latter agent only attacked the younger parts of seedlings: older parts were not susceptible, so the seedlings could sprout again.

Sharma et al. (1985, 1986) reported that dieback on *Eucalyptus* spp. infected by *Corticium salmonicolor* (which also causes stem canker/pink disease) was the effect of toxic substance(s) that damaged wood cells and resulted in leaf wilting. Wilting diseases of other tree species are primarily caused by damage to water transport cells, resulting from infection by various microorganisms (Gäumann 1951, Butin 1989). Epicormic branches that sprout from the sound part of infected tree stems can not grow continuously, but eventually die when the pathogen spreads to basal parts of the stem, from where the epicormic shoots originate. Thus, the pathogen eventually causes the death of the whole tree. Such epicormic branches have also been observed in various other tree species infected by *C. salmonicolor*, including *Eucalyptus* spp. (Seth et al. 1978, Sharma et al. 1985), *Paraserianthes falcataria* (Eusebio et al. 1979), *Cordia alliodora* (Neil 1987), *Acacia auriculiformis* (Florence and Balasundaran 1991) and *A. mangium* (Mardji 1994).

## 26.2.6 Stunt

Stunting of plants is usually caused by light, nutrient or water limitations, resulting in small, weakly-growing plants, often due to intense competition from weeds. Weeds are potentially dangerous for planted trees, because they can grow faster than dipterocarp trees and may smother them. There were many suppressed and dead saplings covered with weeds in our study site. If suppressed saplings are not released in time, they will die. Stunt disease was found on *C. lanceolatum*, *Dry. beccarii*, *S. accuminatissima*, *S. laevis*, *S. leprosula*, *S. parvifolia*, *S. pauciflora* and *S. seminis*.

## 26.3 Disease Incidence on Different Dipterocarp Species

Mixed species planting (block D) clearly reduced the incidence of most diseases other than leaf spot, which was found at almost the same level in both single and mixed species plots (Table 26.3). The incidence of leaf spot disease in *Dipt. humeratus*, *Dry. lanceolata*, *S. laevis* was suppressed by the mixed species planting, while the incidence in *Dipterocarpus gracilis* and *Dry. beccarii* was not. The only other disease recorded in the mixed species planting block was dieback, in *Dipt. gracilis*.

Block A was very bushy and conditions under the bush canopy were very humid when the observations were taken. This environment seemed to promote the development of many diseases. The growth of saplings was also suppressed by the bushiness, and low sapling vitality would be expected to contribute to high susceptibility to leaf spot disease. Indeed, the incidence of leaf spot disease in Block A was very high, especially in *S. laevis*. However, *S. parvifolia* saplings were not infected by leaf spot disease in this block, and it infected few specimens of the species in the other blocks. Thus, saplings of *Shorea parvifolia* seem to be less susceptible to leaf spot disease than *S. laevis*. *S. leprosula*, *S. accuminatissima*, *S. pauciflora*. *S. seminis* also showed high levels of resistance against leaf spot disease.

In the single species blocks (blocks A, B, C, and E), species with low susceptibility towards most of the diseases, except leaf spot, included *Dipterocarpus confertus*, *Dipterocarpus cornutus*, *Dipt. gracilis*, *Dry. lanceolata*, *Dry. rappa*, *S. pauciflora*, and *S. seminis*. *Dipt. gracilis* was resistant towards many diseases but had very high susceptibility (greater than 50%) towards leaf spot disease. *Dry. beccarii*, *S. laevis* and *S. leprosula* appeared to be relatively susceptible towards many diseases, but the rates of severity of these diseases were low (less than 5%).

Leaf spot was the most frequently found disease on the dipterocarp saplings, followed by chlorosis, stem canker, stunt, dieback and leaf blight. However, the severity of the diseases in all species was very low (usually less than 5% except for leaf spot). Serious epidemic diseases were not found at this study site. One reason for this may be that these stands were formed by enrichment planting in secondary forest, rather than by establishment of plantations, in which single species are planted across large cleared areas. This hypothesis is strongly supported by the low incidence of disease found in the mixed species block.

## 26.4 Conclusion

Since enrichment planting and forest plantation involve considerable financial investment, failure of establishment can be economically devastating. Thus, risks associated with diseases of dipterocarp seedlings and saplings should be carefully investigated before embarking on plantation or enrichment planting programs (Elouard 1998). Generally, healthy seedlings should be selected for planting, for all species of dipterocarp, and the planting site should be appropriate for the species chosen. Planting in unsuitable sites will result in poor growth, and weakened or stunted trees tend to be more susceptible to disease. After planting, intensive maintenance is necessary because dipterocarps generally have low tolerance of weed cover, especially in the seedling stage. According to the results of this survey, *S. pauciflora* is the most suitable dipterocarp species for planting on a large scale at Bukit Soeharto, at least with respect to disease tolerance. Several other species could also be recommended, including *S. acuminatissima*, *S. parvifolia* and *S. seminis*. Prevention of disease can also be assisted by mixed species planting.

## References


Butin H (1989) Krankheiten der Wald- und Parkbäume. Georg Thieme Verlag, Stuttgart

de Guzman ED (1985) Field diagnosis, assessment and monitoring tree diseases. Inst For Conserv UPLB, College of Forestry, Laguna, the Philippines

Elouard C (1998) Pests and diseases of Dipterocarpaceae. In Appanah S, Turnbull JM (eds) A review of Dipterocarps. CIFOR, Bogor, Indonesia. pp 115–131

Eusebio MA, Quimio JrMJ, Ilagan FP (1979) Canker of Moluccan Sau (*Albizia falcataria* (L.) Back.) in Bislig. Sylvatrop Philipp For Res J 4: 191–214

Florence EJM, Balasundaran M (1991) Occurrence of pink disease on *Acacia auriculiformis* in Kerala. Indian Forester 117: 494–496

Gäumann E (1951) Pflanzliche Infektionslehre. 2.Auflage. Verlag Birkhäuser, Basel

Gibson IAS (1975) Diseases of forest trees widely planted as exotics in the tropics and southern hemisphere. Part I. Important members of the Myrtaceae, Leguminosae, Verbenaceae and Meliaceae. Commonwealth Mycological Institute, Kew, Surrey, England

Gibson IAS (1979) Diseases of forest trees widely planted as exotics in the tropics and southern hemisphere. Part II. The Genus Pinus. Commonwealth Mycological Institute, Kew, Surrey, England

James WC (1974) Assessment of plant diseases and losses. Ann Rev Phytopath 12: 27–48

Kobayashi T, de Guzman ED (1988) Monograph of tree diseases in the Philippines with taxonomic notes on their associated microorganisms. Bull For For Prod Res Inst 351: 99–200

Lee SS, Ahmad AM (1982) *Cylindrocladium scoparium* Morgan-A new pathogen of some forest trees species in Peninsular Malaysia. Pertanika 5: 72–75

Mardji D (1994) Epidemiologische Untersuchungen von *Corticium salmonicolor* Berk. & Br. an *Acacia mangium* Willd. in PT. ITCI, Kenangan, Ostkalimantan (Indonesian). Ph.D. Dissertation, Universität Georg-August, Göttingen, Germany

Mardji D (1995) Penyakit hutan tanaman industri. Bahan Kuliah Fakultas Kehutanan Unmul, Samarinda (in Indonesian, Unpublished report)

Neil PE (1987) *Corticium salmonicolor* (pink disease) on *Cordia alliodora*. Vanuatu Forest Service Report 3

Nuhamara ST (1991) Berbagai penyakit pada Hutan Tanaman Industri (HTI) di Indonesia. Makalah Seminar Nasional "Peningkatan Produktivitas Hutan Tanaman Industri (HTI) Melalui Upaya Pengendalian Hama dan Penyakit Secara Terpadu" (in Indonesian). Kerjasama Fak. Kehutanan IPB dengan Dep. Kehutanan Republic. Indonesia, Bogor

Quiniones SS (1980) Notes on the diseases of forest trees in the Philippines. Sylvatrop Philipps For Res J 5: 263–271

Seth SK, Bakshi BK, Reddy MAR, Singh S (1978) Pink disease of *Eucalyptus* in India. Eur J For Path 8: 200–216

Sharma JK, Mohanan C, Florence EJM (1985) Disease survey in nurseries and plantations of forest tree species grown in Kerala. Kerala Forest Research Institute, Research Report 36

Sharma JK, Florence EJM, Sankaran KV, Mohanan C (1986) Toxin bioassay for assessing relative susceptibility of Eucalyptus to pink disease. Proc National Seminar on "Eucalypts in India, Past, Present and Future". Kerala Forest Research Institute, pp 400–403

Singh UP, Mishra GD (1992) Effect of powdery mildew (*Erysiphe pisi*) on nodulation and nitrogenase activity in pea (*Pisum sativum*). Plant Pathology 41: 262–264

Singh S, Chaturvedi AN (1993) Stagnating disease of Eucalypts. J Trop For 9: 364–367

Smits WTM, Yasman I, Leppe D, Noor M (1991) Some observations on diseases of Dipterocarpaceae. Biotrop Special Publication 41: 147–163

Soeyamto Ch, Mardji D (1995) Inventarisasi hama dan penyakit pada Hutan Tanaman Industri (*Eucalyptus deglupta* dan *Paraserianthes falcataria*) PT(in Indonesian). Kiani Hutani Lestari, Batu Ampar. Laporan Penelitian (Research Report) II. Faculty of Forestry, Mulawarman Univ Samarinda

Stevens FL (1966) The fungi which cause plant disease. Johnson Reprint Corporation, New York

# 27 Participatory Forest Management

Makoto Inoue

## 27.1 Introduction

Since the late 1970s many participatory forest management systems have been practiced in the tropics because professional foresters noticed that they could not manage the forests sustainably under the principle of conventional and industrial forestry. A participatory forest management system is considered to be an indispensable policy concept to achieve sustainable forest management at present. According to initial definitions, social forestry and community forestry (FAO 1978) were defined similarly as any situation which intimately involves local people in a forestry activity for rural development. These days, however, it seems that the term social forestry involves a wider range of comprehensive participatory activities such as farm forestry or individual forestry, and the term community forestry implies collective activities rather than individual activities.

It is important that local resource management systems, based on local knowledge, are identified and examined as to whether they are sustainable both ecologically and socio-culturally before the introduction of participatory forest management systems at the local level. It is reasonable to transfer management initiative to the local people when their resource management systems are identified to be sustainable. On the contrary, professionals such as social foresters (Gelder and O'Keefe 1995) and development planners may play an important role in promoting participatory forest management when the systems are not so sustainable.

In this chapter, the sustainability and fragility of the Kenyah's resource management systems will be examined first, where the framework used to evaluate the local resource management systems will be demonstrated. Next, a basic course to introduce participatory forest management will be recommended. And finally, present Indonesian forest policy will be examined in terms of participation of the local people.

Ecological Studies, Vol. 140
Guhardja et al. (eds.) Rainforest Ecosystems of East
Kalimantan: El Niño, Drought, Fire and Human Impacts

## 27.2 The Kenyah's Resource Management Systems and Their Changes

In chapter 15 the Kenyah's swidden agriculture was clarified as quite reasonable from an ecological point of view, but so fragile that it had changed drastically from the traditional, via quasi-traditional or transitional, to non-traditional swidden agriculture.

Our range of vision, however, should not be limited to swidden agriculture since indigenous people have various ways of using their environment. Local resource management system by the Kenyah people seems to consist of three sub-systems as follows;

1) Swidden system: having a function to provide them with foodstuffs such as rice, cassava, and various kinds of vegetables at present swiddens and fuelwood at former swiddens.
2) Communal forest system: having a function to provide timber for buildings, and logs for canoes and coffins, and other non-wood forest products such as rattan, resin, fruits, birds and animals from the area where the practice of swidden agriculture is prohibited.
3) Other systems: collecting non-wood forest products outside swidden areas and communal forests, fishing in rivers, and so on.

Since changes in the swidden system were already mentioned in chapter 15, the communal forest system will be focused on here. In the Apo Kayan region or the isolated central plateau of Borneo island, most land other than swidden areas is covered with forests where people hunt birds and animals, collect wood for housing materials, canoes, shingles, and coffins, and collect non-wood forest products such as aloe-wood, rattan, and resin.

There are special forests, relatively near the villages, where swidden making is prohibited in order to reserve big trees and non-wood forest products. These areas are called "Tanaq Ulen (Mulen)", or customary communal forests. Most villages have one or a few Tanaq Ulen which are reached at most within a two hours' walk from the settlements. Basically outsiders are eliminated from the use of resources in Tanaq Ulen. On the other hand, all the members of the village can use the Tanaq Ulen resources without restriction. However, they should not fell or utilize trees that are marked with a special sign, called "Atep (Tip)", made by other members to show they will use the tree. The configuration of Atep depends on the social stratum (see Fig. 27.1).

In some villages, however, customary communal forests have been degraded. I found such phenomena to be marked in the seat of the region. This was especially so in a village resided in by a great chief of the Kenyah, where a part of Tanaq Ulen was canceled and opened to practice swidden agriculture at the beginning of the 1990s as a result of discussion by the villagers. It is obvious that the significance of Tanaq Ulen in their livelihood has deteriorated in the village. On the contrary, splendid Tanaq Ulen covered with rich forests can be seen in the most isolated villages.

**Fig. 27.1.** "Atep" of commoners at Long Sungai Barang village in Apo Kayan

## 27.3 Evaluation of the Resource Management System by the Kenyah

### 27.3.1 Framework for Evaluation

The discussion of "commons" is useful in examining the sustainability and fragility of local resource management systems. Here we define the "local commons" as collective resource management institutions for local resources, principally for the "common pool resources (CPRs)". CPRs are natural or man-made resources sufficiently large that it is costly to exclude users from obtaining subtractable resource-units which are produced or appropriated (Ostrom 1992).

The local commons can be classified into two in terms of the existence of regulation (Inoue 1995).

1) "Tight local commons", where certain regulations are established in the group, and rights and duties are clearly defined.

2) "Loose local commons", where the members of the group have relatively free access without specified regulation.

To examine ecological functions, the concept of sustainable utilization is categorized into three groups (Inoue 1997).

1) "Haphazard sustainable use" is defined as a mode of utilization where unconscious action brings upon sustainable use of the resources.
2) "Incidental sustainable use" is defined as a mode of utilization where conscious actions for other purposes achieve sustainable use.
3) "Intended sustainable use" is defined where sustainable resource management is achieved as intended. Some regulations are then incorporated into customary law. For example, there were regulations on tools, season, and species in utilizing "Iriai-rin" or communal forests in Japan.

Under local commons with sufficient ecological function, "intended" sustainable use can be observed.

At the time, local commons satisfying most of the following seven conditions (Ostrom 1990) are endowed with sufficient socio-cultural function;

1) Boundaries of CPRs are clearly defined.
2) Appropriation rules are congruent with local conditions.
3) Members can participate in modifying the operational rules.
4) Monitors are accountable to the appropriators or are the appropriators.
5) Graduated sanctions, depending on the seriousness and context of the offense, exist.
6) Conflict-resolution mechanisms exist.
7) Local society has a certain extent of self-governing.

## 27.3.2 Swidden Sub-System

An example of "haphazard" sustainable use are unburned trees remaining in swiddens which prevent soil erosion to some extent. This is just an extensive way of land readjustment, which results in this situation occurring.

One of the examples of "incidental" sustainable use is the rotation system of the swiddens shown in chapter 15. Secondary vegetation is not cut until the trees grow as large as a human thigh or waist, because it is easier to weed the swiddens. As a result enough biomass for sustainable cultivation can be secured.

Another example is the shifting of swiddens every year, because consecutive cultivation would increase weeds in the swiddens. As a result, the potential of germination of the seeds mixed with the topsoil is preserved, and forests can recover quickly.

No examples of "intended" sustainable use are found in the swidden systems. The system is then characterized as "loose local commons" from the viewpoint of ecological function.

On the other hand, conflict over swidden land among villagers could be avoided or settled owing to the existence of customary land tenure systems (Inoue and Lahjie 1990) satisfying the seven conditions. This means that the swidden system achieves sufficient socio-cultural functions. The system can then be characterized as "tight local commons" from the viewpoint of socio-cultural function.

### 27.3.3 Communal Forest Sub-System and the other Sub-Systems

Description of the present state of Tanaq Ulen, a kind of CPR, in Apo Kayan shows that the communal forest sub-system does not have a sufficient ecological function, but has a socio-cultural function by means of using Atep to avoid conflict. The sub-system is then regarded as "loose local commons" from the view point of ecological function and as "tight local commons" from the view point of socio-cultural function.

On the contrary, natural resources outside communal forests and swidden areas are utilized freely. Even outsiders can make use of primary forest other than Tanaq Ulen. Actually the village borderlines do not make sense, because everyone, both the villagers and outsiders, can access the resources. Other sub-systems are not then regarded as "local commons". If anything, the nature of the systems are similar to "global commons" where a tragedy of the commons (Hardin 1968) would easily occur.

## 27.4 Basic Course to Introduce Participatory Forest Management

The framework to evaluate local resource management systems as demonstrated above must also be useful in setting up participatory forest management at the local level, such as for social forestry and community forestry.

Firstly, we can judge whether a technical application of local knowledge and local technology is reasonable or not, by examining the ecological function of the commons. In the case of the Kenyah system, for example, technical cooperation by outsiders such as NGOs, the local government, and other professionals may be necessary, because their resource management systems are sustainable but they are not endowed with sufficient ecological function.

Secondly, evaluation of the socio-cultural function of the commons is helpful to examine modes of formal institutional arrangements for participatory forest management. To set up a participatory forest management system, levels of participation in local development planning should be specifically taken into consideration . The term "participation", however, can not tell us the actual level of participation because the meaning of the term varies too widely. The wide spectrum of participation can be categorized into three areas as follows;

A) "Participatory top-down approach": a blueprint approach where residents are considered to be wage laborers, volunteers, fund providers, etc.
B) "Professional-guided participatory approach": relatively flexible blueprint approach where drafts of a plan made by professional planners are examined by the residents and citizens and modified through discussion, workshops, etc.
C) "Endogenous bottom-up approach": learning process approach where professionals just act as facilitators.

We have the possibility to make formal institutional arrangements for the "endogenous bottom-up approach" when the systems are endowed with a sufficient socio-cultural function of the commons, such as the Kenyah's system, because the residents can manage their resources well by themselves. On the other hand, application of a "professional-guided participatory approach" seems to be reasonable when the systems are not endowed with a sufficient socio-cultural function.

In any case the "participatory top-down approach" should be avoided because this approach is usually considered to be nominal and fake. As a matter of fact, this approach should not be included as a participatory approach.

## 27.5 National Forest Policy in Terms of Participation

In Indonesia the term "social forestry" is an umbrella terminology for governmental participatory forest management programs. Here, the social forestry programs promoted by the government on national forest land and non-forest land will be reviewed first. Then, NGO-supported community-based forest management will be discussed.

### 27.5.1 Social Forestry Programs on National Forest Land

#### Programs on Production Forests Controlled by Companies

1) Tumpang Sari in Java: The "Tumpang Sari" system, or intercropping agroforestry, where peasants intercrop agricultural products between the rows of planted trees such as Teak trees has existed from more than 100 years ago. The peasants, however, do not have any rights to the trees and cannot be the main players in forest management.
2) Forest village social development program in forestry concession areas: Forestry concessions with forest village development (HPH Bina Desa) was introduced in 1991 and which evolved into the "Forest village social development program (Pembinaan Masyarakat Desa Hutan)" in 1995. Logging companies are put under obligations to construct roads, elementary schools, clinics, churches, mosques, to help the people to introduce agroforestry systems, to buy their agricultural produce, and to plant fast growing species on degraded land and alang-alang grassland. Under the program, however, the local people cannot be involved in forest management and are rather excluded from management.
3) Tumpang Sari and Transmigration with industrial tree plantation (HTI-Trans) in industrial tree plantation areas: Even though the tumpang sari systems were introduced for the people, they are no more than temporary measures. Under the HTI-Trans program, the people are just considered to be low cost laborers. The local people cannot be involved in forest management under these two programs.

**Community Forest (Hutan Kemasyarakatan) Program**

The "Community forest" program was first launched in 1995 based on the decree of the Minister of Forestry No.622. A revised version of the "community forest" program was proclaimed in October 1998 by decree of the Minister of Forestry and Plantation No. 677 (Keputusan Menteri Kehutanan dan Perkebunan Nomor 677 / Kpts-II /1998).

Community forestry under the program is practiced by the people, specifically cooperatives of people living within and near the forests who obtain the "community forestry concession (Hak Pengusahaan Hutan Kemasyarakatan)". The concession, for which the period is 35 years, is granted on production forests (hutan produksi), protection forests (hutan lindung), and conservation areas (kawasan pelestarian alam) such as national parks, in the national forest which are free of other rights. All the activities of producing wood and non-wood forest products, including planting, tending, protecting, harvesting, and marketing, for the purpose of self consumption and sale, are regarded as community forestry activities.

It is obvious that forest management by local people is not approved on concession areas of forestry and industrial tree plantations. In order to avoid and settle conflicts between the companies and the people, it is recommended to convert certain parts of the areas under forestry concession and industrial tree plantation concession to that under the community forestry concession.

## 27.5.2 Social Forestry Programs on Non-Forest Land

An "Individual forest (Hutan Rakyat)" program is practiced on private land for which the people hold a certification of landownership, outside national forest. The main activity of individual forestry is regreening or afforestation and *Paraserianthes falcataria* is planted by many people subsidized by a reforestation fund. In order to introduce the program, land registration should go ahead. It seems then that the expansion of the individual forest program may be more difficult in outer Indonesia, such as in East Kalimantan, than in Java.

## 27.5.3 Community-Based Forest Management (Sistem Hutan Kerakyatan:SHK)

The NGO-supported SHK is defined as customary forest management systems by the local people, especially the indigenous peoples. The people depending on the forests have initiative, form organizations, and develop customary regulations to manage forests.

The SHK can be considered to be a local commons. The existence of the system shows that they have rather sufficient socio-cultural functions institutionally. The SHK can then be practiced through the "endogenous bottom up approach".

If the system is endowed with sufficient ecological function, a local resource management system can be fully applied technically, too. On the contrary, professional technical assistance by foresters is necessary when they do not have sufficient ecological function.

The government, however, has hesitated to permit the SHK. Even in the latest community forest program, the government gives priority to rights other than community forestry concessions, such as forestry concessions, industrial tree plantation concessions, tree felling rights (Ijin Pẹmungutan Kayu) to develop oil palm plantations and transmigration areas, etc. However, it is proven that large-scale forest development projects such as logging, industrial tree plantations, and conversion to agricultural land, are not so sustainable and if anything, can often be causes of deforestation.

Most of the people practicing SHK must live in areas covered with relatively rich forests, which are designated as conservation areas and production forests. In the conservation areas, however, a community forestry concession can be approved only in a utilization zone (zona pemanfaatan) and the main activity acceptable is confined to recreation. This is because the government does not rate the SHK correctly.

It is thus recommended that the areas where the local people practice SHK are designated as a traditional use zone (zona pemanfaatan tradisional) or socio-cultural zone (zona sosio/membudaya) under the condition that the SHK is sustainable. It may be easier for the government to accept this recommendation because the cost to change policy is very low.

## 27.6 Closing Remark

The author does not intend to underestimate the local resource management system. Local resource management systems, such as the Kenyah's system in which the communal forest system is regarded as SHK, should be duly appreciated. Provided that the community forest program is given priority in national forest policy, the SHK is not so contradictory to the governmental community forest program . All of the existing SHK in Indonesia should be investigated objectively from the viewpoints of both ecological function and socio-cultural function, in cooperation with NGOs and the government.

## References

FAO (1978) Reappraisal of forestry education and training needs in the Asia/Far East Region. FAO Advisory Committee on Forestry Education, Rome

Gelder BV, O'Keefe P (1995) The New Forester. Intermediate Technology Publications, London

Hardin G (1968) The Tragedy of the Commons. Science 162: 1243–1248

Inoue M (1995) Swidden agriculture and tropical forest: changes in traditional swidden system in Kalimantan (in Japanese). Kobundo, Tokyo

Inoue M (1997) Examination of Tropical Forests as Commons: based on positive survey in Kalimantan, Indonesia (in Japanese with English summary). Journal of Environmental Sociology 3:15–32

Inoue M, Lahjie AM (1990) Dynamics of swidden agriculture in East Kalimantan. Agrofor Sys 12: 269–284

Ostrom E (1990) Governing the Commons: The Evolution of Institutions for Collective Action. Cambridge University Press, Cambridge

Ostrom E (1992) The Rudiments of a Theory of the Origins, Survival, and Performance of Common-Property Institutions. In: Bromley DW (ed) Making the Commons Work: Theory, Practice, and Policy. ICS Press, San Francisco, pp 293–318

# 28 Overview of the Changing Forest Ecosystems in East Kalimantan

TOKUNORI MORI, SEIICHI OHTA, ATSUSHI ISHIDA, TAKESHI TOMA, and TERUKI OKA

## 28.1 Introduction

Increasingly, attention is being focused on tropical forests. This interest has been motivated by three main concerns: 1) Tropical forests harbor the greatest biological diversity on Earth; 2) they are a major component of the global carbon cycle; and 3) they provide forest goods and services to all sectors, from the local to the global scale. These three concerns often cause conflicting problems in forest management, and forest scientists and administrators are seeking a harmonious solution. Despite recent advances in the understanding of tropical forest ecosystems, knowledge is still insignificant in relation to the three concerns mentioned above. To establish verifiable forest management methods that are ecologically sound, economically viable, socially responsible, and environmentally acceptable, both administrators and scientists including those in the fields of socio-economics, plant ecology, eco-physiology, soil science, animal ecology, entomology, mycology, pathology, silviculture, and forestry have collected accurate and detailed information about the changing forest ecosystems in the tropics.

Currently, the typical Southeast Asian tropical rainforests that cover about 80% of the state of East Kalimantan are seriously threatened by human activities (MacKinnon et al. 1996). Borneo is in the humid tropics, but there have been periodic long dry spells linked to El Niño-Southern Oscillation (ENSO) events since the prehistoric era (Goldammer and Seibert 1990, Walsh 1996). The recent increase in human activities led to extensive and repeated wildfires during the most recent dry spells, and the wildfires caused serious forest degradation (Chapter 3, this volume). This is changing the forest ecosystems which are our study target.

## 28.2 Implications and Misconceptions Concerning Soil Change

Soil is one of the most basic components of a tropical rainforest ecosystem. Soil change with vegetation and land-use changes is a global concern for decision mak-

Ecological Studies, Vol. 140
Guhardja et al. (eds.) Rainforest Ecosystems of East Kalimantan: El Niño, Drought, Fire and Human Impacts

ers, because it is directly related to strategies for returning land from degraded to productive land. Evaluating the changing processes of vegetation is important in regard to the influence on the global environment of the budget of greenhouse gases and change in land productivity. Nevertheless, information on the soils under the degraded ecosystems, such as in secondary forest, *Imperata cylindrica* grassland, and newly established tree plantations, is very scarce. Furthermore, there are still some misconceptions about soil changes through transformation processes in vegetation.

Have soils under degraded ecosystems deteriorated? Ohta et al. (Chapter 4, this volume) revealed that the commonly held view that forest degradation causes soil deterioration is not always true. In fact, the opposite can be true: surface soils under some degraded ecosystems, especially *Imperata* grasslands, maintain higher levels of exchangeable bases in the earlier stages (10–20 years) after selective logging or forest fire than do primary forests, because of ash input from the burned biomass. Substantial amounts of nutrients are still held in the soils. It is of prime importance to restore forest ecosystems where nutrients become fixed in biomass and are conserved more efficiently. Otherwise, the soils will gradually lose nutrients and will be deteriorated irreversibly.

Another commonly held view is that forest degradation is accompanied by accelerated emission of $CO_2$ from the soil. In Chapter 4 in this volume, Ohta et al. demonstrated that soil carbon stock does not decrease with progressing forest degradation; in fact, it increases somewhat with conversion into grasslands. When a forest is lost, the carbon in the topsoil is temporarily mineralised and emitted. However, the climate in the humid tropics and the less infertile soil of the Asian tropics enable the immediate regeneration of secondary vegetation and recovery of the soil carbon stock. Thus, the hypothetical rapid decrease in soil carbon with forest loss and its slow recovery with secondary vegetation build-up, are examples of the misconceptions concerning soil changes in the tropics. Nevertheless, it must be emphasised that in degraded ecosystems the soils are often deteriorated in terms of available nitrogen (N) and phosphate (P) (Ohta 1988), and this will constrain future land-use.

Tropical rainforests grow on highly weathered nutrient-poor substrates (Vitousek and Sanford 1986; Ohta and Effendi 1992a, b; 1996), and are believed to develop mechanisms to conserve nutrients to minimize their leaching loss (Jordan 1985). In Chapter 5, Tanaka et al. concluded that *Imperata* grassland might have well-developed nutrient conservation mechanisms, comparable to those of natural forests, and this might be why the grassland soils have not deteriorated, at least in terms of base cations (Chapter 4, this volume). On the other hand, plantations of *Acacia mangium*, a fast growing timber-tree, are less able to conserve nutrients than natural forest or grassland, because the larger $NO_3^-$ flux in the system that results from high N-fixation capacity by individual trees may accelerate leaching loss of base cations from the soils. This might be a serious problem for sustainable *Acacia* plantations, and other leguminous plantation tree species might also have the potential to accelerate the leaching loss of base elements. Studies focusing on the impact of N-fixing tree species on nutrient conservation are therefore required to avoid irreversible soil deterioration.

Another important concern is the soil change with swidden agriculture, which is still the major source of livelihood for the indigenous people of inland Borneo. Recently, the fallow period of swidden agriculture is becoming shorter and shorter with the increasing population and monetary economy (Chapter 15, this volume). Is this altered swidden agriculture system sustainable in terms of soil productivity? Morisada et al. indicate that the present 10-year fallow period is not sufficient for nutrient buildup in sandy Ultisols (Chapter 6, this volume). Thus, developing tactics which enables rational land-use, is necessary for the sustainable use of soil resources.

Although the major soils in lowland East Kalimantan are Ultisols, a diverse range of soils were found in the province. Effendi et al. (Chapter 7, this volume) illustrated the catenary distribution of soils developed on serpentinite, and revealed their unique characteristics. Aksa et al. (Chapter 8, this volume) also analyzed soils derived from volcanic parent materials in central East Kalimantan and suggested that in the humid tropics the carbon accumulated in the dark-colored volcanic soils may function as a stable carbon pool, even under intense disturbance. This information is fundamental to an evaluation of the effects on soils of changes in the forest and the consequent impact on forest ecosystems.

## 28.3 Fire and Human Impact on Vegetation Changes

The lowland dipterocarp forest in Borneo is one of the most species-rich ecosystems in the world (e.g., Whitmore 1984). The tall forest canopy, which exceeds 60 m, consisting mainly of dipterocarp trees, provides a large number of spatial and temporal niches. Consequently, the dipterocarp ecosystem is home to hundreds of thousands of animal and plant species, which form an intrinsic network of interactions (Schulte 1996). However, the species-rich forests in this region have been widely replaced by degraded plant communities, such as small-tree forests, shrub land, and grasslands. Signs of the degradation include a small standing biomass, species conversion, and the invasion of exotic plants such as *Imperata cylindrica*. Human impact, usually by road construction and selective logging, has accelerated fragmentation in various vegetation types. There is a possibility that such fragmentation reduces not only species richness but also genetic variations within a species (Tsumura personal communications).

The degradation of tropical rainforests involves a common process: road construction for commercial logging, followed by the slash-and-burn agriculture of newly migrant farmers (Chapter 17, this volume). As a result, primary and logged-over forests were changed to slash-and-burn agricultural land and its fallow lands. In addition to this process, the occurrence of drought-linked wildfires has accelerated forest degradation in East Kalimantan. Recently, two large wildfires occurred within a 14-year interval in East Kalimantan during the prolonged droughts in 1982–83 and 1997–98, linked to severe ENSO events (Chapter 3, this volume). Between these two large fires, lesser droughts occurred in 1987, 1991, and 1994, which were also accompanied by wildfires, even in the rainforest region (Chapter 3, this volume).

The large fires inhibited regeneration of canopy trees after selective logging. The damage in the 1982–83 fires was directly related to the intensity of past disturbance, such as logging or forest fires (Chapter 10, this volume). The higher the intensity of disturbance was, the higher the tree mortality became and the larger the canopy openings formed. Heavily disturbed sites were dominated by a few pioneer species, primarily *Macaranga gigantea* and *Macaranga triloba*, with a low top canopy height and a small biomass. After the fires in 1982–83, the 14-year interval between the large wildfires in 1982–83 and 1997–98 was not sufficient duration of time for recovery from the burned forests to near-primary forests in forest structure and biomass. Subsequently, tree mortality in the 1997–98 fires was higher in these *Macaranga*-dominated stands, where the logging intensity and fire disturbance had been high.

Recently, *Imperata* grasslands widely extend in rainforest regions in East Kalimantan. The grassland formation is associated with slash-and-burn agriculture, but the fallow of slash-and-burn agriculture alone rarely forms the grasslands (Chapter 17, this volume). The fallow land usually becomes secondary forest, and ultimately develops into a mixed dipterocarp forest after a fallow period longer than 70 years, if the seeds of primary forest species are provided (Chapter 16, this volume). The establishment of *Imperata* grassland is actually triggered by the abandonment of plantation in perennial crops such as pepper. Land abandoned after the production of cash crops mostly becomes a vast grassland (Chapter 17, this volume). Although tree seedlings can grow in the grasslands, recovery to forests is hopeless. The dormant buds of *Imperata* grass can tolerate fires, whereas the aboveground parts of this plant burn easily. Wildfires from slash-and-burn fields readily spread into *Imperata* grasslands and prevent the survivorship of the growing trees. Thus, repeated-wildfires linked to drought and slash-and-burn agriculture serve to maintain and expand *Imperata* grassland areas.

A cessation of fires will help to stop the developing forest degradation. The aboveground biomass of primary dipterocarp forests exceeds 400 t $ha^{-1}$ (Yamakura et al. 1986), which is much greater than that of degraded vegetation such as *Macaranga*-dominated stands (ca 100 t $ha^{-1}$; Chapter 10, this volume) and *Imperata* grassland (ca 10 t $ha^{-1}$; Chapter 16, this volume). Thus, if we prevent wildfires, the degraded stands will develop to primary dipterocarp forests, and will act as a carbon sink.

## 28.4 Drought and Fire Impact on Forest Functions

During the ENSO years 1997–98, which brought hot and dry weather to the East Kalimantan region (Chapter 2, this volume), it was not only the forest structure but also the leaf gas exchange processes that were affected. With no precipitation for 1–2 months, there was a remarkable decrease in the net photosynthetic rate ($P_n$) and stomatal conductance ($g_s$) in the top canopy leaves in pioneer *Macaranga* and climax dipterocarp trees (Chapter 12, this volume). Furthermore, drought-linked wildfires affected about 3.6 million ha in East Kalimantan, and decreased terrestrial carbon storage in this area (Chapter 3, this volume). These events suggest that the ecosystems in East Kalimantan act as a source of carbon to the atmosphere in ENSO years. Tian et al. (1998) showed that the ecosystems in the Amazon also act as a source of carbon in ENSO years, whereas these ecosystems are globally important carbon sinks in normal wet years. Both high leaf fall and low $g_s$ reduce water loss at the whole-plant level during droughts. The reduced $g_s$ effectively limited $P_n$ in the top canopy leaves of *Macaranga conifera* (Ishida et al. 1999b). Thus, prolonged drought will affect global carbon and water cycles between terrestrial regions and the atmosphere.

The midday depression of $g_s$ was found even in rain forest trees. The stomata in the top canopy leaves of a pioneer species *Macaranga conifera* began to close at a smaller leaf-to-air vapor pressure deficit ($\Delta W$) in drier conditions than in wetter conditions (Chapter 12, this volume). Ishida et al. (Chapter 12, this volume) discussed how variation in the sensitivity of $g_s$ to $\Delta W$ suggests the hydraulic limitation hypothesis (Ryan and Yoder 1997), which states that old trees increase in height less rapidly as they become taller because of an increase in soil-to-leaf hydraulic resistance. Becker et al. (1998) showed that size-dependent mortality in dipterocarp trees in Borneo during the prolonged drought in the 1982–83 ENSO years increased, probably because of the high frequency of drought-induced embolisms in xylem tissue (Tyree et al. 1998) or small and shallow root systems (Newbery 1991) in the trees with higher mortality. The size-dependent vulnerability to drought of rainforest canopy trees might also support the hydraulic limitation hypothesis. During the prolonged drought, mortality was higher in pioneer trees than in climax trees (Condit et al. 1995; Chapter 10, this volume), indicating that pioneer trees are more vulnerable to drought than dipterocarp trees. The top canopy heights are generally lower in pioneer *Macaranga* trees than in climax dipterocarp trees, probably due to the variations of vulnerability to drought. More work will be needed to elucidate the linkage between hydraulic limitation and the canopy height of rainforest trees.

Understanding the impact of drought on whole-plant performance through stomatal behavior, leaf fall, and carbon allocation is important, even in rainforest trees. In the tropics, the inclination angle of the top canopy leaves is a critical determinant of leaf temperature, carbon gain, and water use within the canopy (He et al. 1996, Ishida et al. 1999a). However, functional understanding of water and nutrient uptake through root systems is still poor. Itoh et al. (Chapter 13, this volume) and Hastaniah and Kiyono (Chapter 25, this volume) showed that the ratio of the root mass to the total leaf area of the whole plant increased in dipterocarp seedlings in response to drought. This would contribute to improving belowground water absorption, and to maintaining a relatively high $g_s$ and $P_n$. Although mycorrhizal associations in the root systems enhance the growth of dipterocarp seedlings (Chapters 21 and 22, this volume), we know very little about the quantitative effects of mycorrhizae on water and nutrient uptake. Ishida et al. (Chapter 24, this volume) suggest the linkage between canopy structure and physiological tolerance capacity against strong sun-light at the single leaf level. However, the complex canopy structure and the linkage between carbon gain within the canopy and the use of resources such as light, water, and nutrients, remain poorly understood.

## 28.5 Fire and Human Impact on Animals

The recent repeated fires are critical factors in the forest ecosystem, affecting animals both directly by killing them and indirectly by eliminating their resources. Since the 1997–98 fires in forests were surface fires that burned understory vegetation, litter, and fallen logs and limbs on the soil surface (Chapter 3, this volume), animals that depend on such resources for food and shelter (e.g., Carabidae) are more severely affected than others (e.g., Bostrychidae). Both the ecological and social behavior of each species and the extent of the damage to resources, affect the recovery of the animal populations in the burned area (e.g., Chapter 14, this volume). In the case of primates, two species of leaf monkeys (*Presbytis rubicunda* and *P. frontata*) were not found within Bukit Soeharto Education Forest of Mulawarman University (BSEF) during or after the fire, whereas gibbons were observed finding refuge in the unburned forest stands during and just after the fire, although following the fire their loud morning calls could not be heard for at least 3 months. According to the list of tree species for a 9 ha plot in the BSEF, the fire affected about 57% of the fruiting trees on which gibbons rely. Our survey between December 1998 and March 1999 found that the home ranges of gibbon families were more than double their pre-fire size, and that some nuclear families shared ranges.

The tropical rainforest is the most species-rich environment on Earth. According to Whitmore (1990), the tropical rainforests, which occupy only 7% of the total land area of the Earth, nourish more than half the species in the world. Even a restricted area within BSEF contains more than half the number of different longicorn beetles found in the whole of North America (Makihara 1999). At the species level, the interaction between plants and animals maintains the biological diversity of tropical rainforest.

During the course of this century, however, various human activities and man-made fires have degraded tropical rainforest habitats in East Kalimantan. One of the most notable changes is that a vast forest has been divided into many small pieces with various vegetation types. The fragmentation of habitats causes a slow, but sure, increase in the number of species facing extinction through a decrease in the genetic diversity that enables adaptation to environmental change, although the effects are not apparent immediately (Chapter 20, this volume). Man's actions have disrupted the intricate web connecting the numerous species in the tropical rainforest, and threatens biological diversity not only at the species level, but also at the genetic level, and probably at community and ecosystem levels. Whereas, our knowledge of the process of extinction is limited to a few species, the imbalance of the intricate webs connecting species that is created by the extinction of other species needs to be considered now. The reports on coleopteran insects and the Bornean gibbon in this volume provide good examples (Chapters 14 and 20, this volume).

## 28.6 Community Forestry and Rehabilitation

Swidden agriculture and forest gardens (*Lambo* in Dayak, Sardjono 1996) are still the major livelihood activity of the indigenous people of inland Borneo, and the traditional agriculture is characterized by a long fallow period between short periods of crop production (Whitten et al. 1987; Chapter 15, this volume). The ancient Maya, in Central America, conducted a similar system of forest utilization (Gómez-Pompa and Bainbridge 1995). In Borneo, however, the recent increase in population pressure and the infiltration of a monetary economy have altered the traditional manner of land-use, with shortening of the fallow period and the cultivation of cash crops.

Traditional forest management is usually sophisticated and sustainable (Gómez-Pompa and Bainbridge 1995). There are attempts to learn methods of forest management and utilization from traditional systems. One such activity in Indonesia is 'Community-based Forest Management', which is practiced mainly by the local people with the assistance of experts in socio-culture and the support of an NGO (Chapter 27, this volume). However, mere imitation of traditional systems cannot overcome the pressure from a rising population and a monetary economy. Thus, recent achievements in biological science should be combined with traditional knowledge. The success of such combined systems requires long-term support by administrators and technical experts in ecology, agriculture, forestry, and socio-economics.

Forest gardens fully utilize features of plant resources and are ecologically sustainable. We could re-discover forestry techniques from the gardens which have sophisticated utilization of tree growth behavior and ecological rules. One example is the utilization of symbiosis for raising tree seedlings. Most dipterocarp trees have ectomycorrhizae, and are obligate mycotrophs, which cannot survive without ectomycorrhiza (Smits 1994). Large and tall dipterocarp trees on shallow and rather infertile tropical soils may perish without their symbiotic ectomycorrhiza. A simple inoculation method using a special mycorrhiza, *Scleroderma columnare*, has been developed in our studies (Chapter 22, this volume).

Despite the accumulation of much data on the growth of dipterocarp seedlings (see Chapters 13 and 24, this volume), the rehabilitation and reforestation of dipterocarp species is not yet common. This is because of the diversity in dipterocarp species and their interspecific variations in physiological features. Many dipterocarps are slow growing and require mild environmental conditions in their early stages. Generally, young dipterocarp trees dislike open land and strong sunlight (Chapter 24, this volume) and are intolerant of wildfire (Chapter 3, this volume). Thus, global temperature increments and increased forest fires would spread the vegetation area of pioneers in tropics. Already, a burned forest contains a large number of pioneer trees (Chapter 9, this volume), and enrichment planting is indispensable for sustaining the dipterocarp forests in the burned areas in East Kalimantan. Silvicultural tending of slow-growing trees (Chapter 25, this volume), planting in small canopy-gaps according to interspecific variations of light-requirements (Ishida et al. 1999c), and the cessation of wildfires (Chapter 3, this volume) seem to be essential for restoring a large dipterocarp forest.

## 28.7 Concluding Remark

Along the roadsides in East Kalimantan, we can see a mosaic of vegetation, including farmlands, *Imperata* grasslands, shrub-like savanna, pyrophytic woods, secondary *Macaranga* forests with tall-standing dead-dipterocarp-trees due to the large forest fires, and selectively-logged-over dipterocarp forests. All of these ecosystems, which were once thick primary tropical rainforests of dipterocarps, are the products of human activities and drought-linked wildfires in severe ENSO years. What we have lost in our generations, which have important functions for the world's people, has to be restored as much as possible by our own efforts so as to hand over a sound environment and worthy forest resources to our succeeding generations. The adequate understanding introduced in this book on the process of changing forest ecosystems and its influences on soils, plants, and animals, will become of great value for society to develop a rational policy for achieving sustainable management of tropical rainforests.

## References

Becker P, Ong CL, Goh F (1998) Selective drought mortality of dipterocarp trees: no correlation with timber group distributions in Borneo. Biotropica 30: 666–671

Condit R, Hubbell SP, Foster RB (1995) Mortality rates of 205 neotropical tree and shrub species and the impact of a severe drought. Ecol Monog 65: 419–439

Goldammer JG, Seibert B (1990) The impact of droughts and forest fires on tropical lowland rain forest of East Kalimantan. In: Goldammer JG (ed) Fire in the Tropical Biota. Ecological Studies 84. Springer, Berlin Heidelberg New York, pp 11–31

Gómez-Pompa A, Bainbridge DA (1995) Tropical forestry as if people mattered. In: Lugo AE, Lowe C (Eds) Tropical forest: management and ecology. Ecological Studies 112, Springer, Berlin Heidelberg New York, pp 408–422

He J, Chee CW, Goh CJ (1996) 'Photoinhibition' of *Heliconia* under natural tropical conditions: the importance of leaf orientation for light interception and leaf temperature. Plant Cell Environ 19: 1238–1248

Ishida A, Toma T, Marjenah (1999a) Leaf gas exchange and chlorophyll fluorescence in relation to leaf angle, azimuth, and canopy position in a tropical pioneer tree, *Macaranga conifera*. Tree Physiol 19: 117–124

Ishida A, Toma T, Marjenah (1999b) Limitation of leaf carbon gain by stomatal and photochemical processes in the top of the canopy of *Macaranga conifera*, a tropical pioneer tree. Tree Physiol 19: 467–473

Ishida A, Nakano T, Matsumoto Y, Sakoda M, Ang LH (1999c) Diurnal changes in leaf gas exchange and chlorophyll fluorescence in tropical tree species with contrasting light requirements. Ecol Res 14: 77–88

Jordan CF (1985) Nutrient Cycling in Tropical Forest Ecosystems. John Wiley & Sons, Chichester pp 23–48

MacKinnon K, Hatta G, Halim H, Mangalik A (1996) The ecology of Kalimantan. Periplus Edition (HK), Singapore

Makihara H (1999) Atlas of longicorn beetles in Bukit Soeharto Education Forest. PUSREHUT Special Publication No. 7, Univ Mulawarman, Samarinda, Indonesia
Newbery D McC (1991) Floristic variation within kerangas (heath) forest: re-evaluation of data from Sarawak and Brunei. Vegetatio 96: 43–86
Ohta S (1988) Influence of deforestation on the soils of the Pantabangan area, Central Luzon, the Philippines. Soil Sci Plant Nutr 36: 561–573
Ohta S and Effendi S (1992a) Ultisols of "Lowland Dipterocarp Forest" in East Kalimantan, Indonesia, I. Morphology and physical properties. Soil Sci Plant Nutr 38: 197–206
Ohta S and Effendi S (1992b) Ultisols of "Lowland Dipterocarp Forest" in East Kalimantan, Indonesia, II. Status of carbon, nitrogen and phosphorus. Soil Sci Plant Nutr 38: 207–216
Ohta S and Effendi S (1996) Soils under Lowland Dipterocarp Forests—Characteristics and Classification. In: Schulte A, Schöne D (eds) Dipterocarp Forest Ecosystems, Towards Sustainable Management. World Scientific Publishing, Singapore, pp 29–51
Ryan and Yoder (1997) Hydraulic limits to tree height and tree growth. BioScience 47: 235–242
Sardjono A (1996) The Lembo system: a model for agroforestry in dipterocarp forest ecosystems of East Kalimantan, Indonesia. In: Schulte A, Schöne D (Eds) Dipterocarp Forest Ecosystems, Towards Sustainable Management. World Scientific Publishing, Singapore, pp 354–368
Schulte A (1996) Dipterocarp forest ecosystem theory based on matter balance and biodiversity. In: Schulte A, Schöne D (eds) Dipterocarp Forest Ecosystems, Towards Sustainable Management. World Scientific Publishing, Singapore, pp 3–28
Smits WTM (1994) Dipterocarpaceae: Mycorrhizae and regeneration. Tropenbos Series 9. The Tropenbos Foundation, Wageningen, The Netherlands
Tian H, Melillo JM, Kicklighter DW, McGuire AD, Helfrich III JVK, Moore III B, Vörösmarty CJ (1998) Effect of interannual climate variability on carbon storage in Amazonian ecosystems. Nature 396: 664–667
Tyree MT, Pantiño S, Becker P (1998) Vulnerability to drought-induced embolism of Bornean heath and dipterocarp forest trees. Tree Physiol 18: 583–588
Vitousek PM, Sanford Jr RL (1986) Nutrient cycling in moist tropical forest. Ann Rev Ecol Syst 17: 137–167
Walsh RPD (1996) Drought frequency changes in Sabah and adjacent parts of northern Borneo since the late nineteenth century and possible implications for tropical rain forest dynamics. J Trop Ecol 12: 385–407
Whitten AJ, Mustafa M, Henderson GS (1987) The ecology of Sulawesi. Gadjah Mada Univ Press, Yogyakarta, pp 574–607
Whitmore TC (1984) Tropical rain forests of the Far East, 2nd edn. Clarendon, Oxford
Whitmore TC (1990) An introduction to tropical rain forests. Clarendon Press, Oxford
Yamakura T, Hagihara A, Sukardjo S, Ogawa H (1986) Above ground biomass of tropical forest stands in Indonesian Borneo. Vegetatio 68: 71–82

# Subject Index

## G



## H



## I

## T



## U



## V



## W

# Species Index

## Plants

## N



## O



## P



## R



## S

# Animals

## Fungi

# Ecological Studies

Volumes published since 1993

Volume 101
**Effects of Land Use Change on Atmospheric $CO_2$ Concentrations: South and South East Asia as a Case Study** (1993)
V.H. Dale (Ed.)

Volume 102
**Coral Reef Ecology** (1993)
Y.I. Sorokin

Volume 103
**Rocky Shores: Exploitation in Chile and South Africa** (1994)
W.R. Siegfried (Ed.)

Volume 104
**Long-Term Experiments With Acid Rain in Norwegian Forest Ecosystems** (1993)
G. Abrahamsen et al. (Eds.)

Volume 105
**Microbial Ecology of Lake Plußsee** (1994)
J. Overbeck and R.J. Chrost (Eds.)

Volume 106
**Minimum Animal Populations** (1994)
H. Remmert (Ed.)

Volume 107
**The Role of Fire in Mediterranean-Type Ecosystems** (1994)
J.M. Moreno and W.C. Oechel (Eds.)

Volume 108
**Ecology and Biogeography of Mediterranean Ecosystems in Chile, California and Australia** (1995)
M.T.K. Arroyo, P.H. Zedler, and M. D. Fox (Eds.)

Volume 109
**Mediterranean-Type Ecosystems. The Function of Biodiversity** (1995)
G.W. Davis and D.M. Richardson (Eds.)

Volume 110
**Tropical Montane Cloud Forests** (1995)
L.S. Hamilton, J.O. Juvik, and F.N. Scatena (Eds.)

Volume 111
**Peatland Forestry. Ecology and Principles** (1995)
E. Paavilainen and J. Päivänen

Volume 112
**Tropical Forests: Management and Ecology** (1995)
A.E. Lugo and C. Lowe (Eds.)

Volume 113
**Arctic and Alpine Biodiversity. Patterns, Causes and Ecosystem Consequences** (1995)
F.S. Chapin III and C. Körner (Eds.)

Volume 114
**Crassulacean Acid Metabolism. Biochemistry, Ecophysiology and Evolution** (1996)
K. Winter and J.A.C. Smith (Eds.)

Volume 115
**Islands. Biological Diversity and Ecosystem Function** (1995)
P.M. Vitousek, L.L. Loope, and H. Adsersen (Eds.)

Volume 116
**High Latitude Rainforests and Associated Ecosystems of the West Coast of the Americas: Climate, Hydrology, Ecology and Conservation** (1996)
R.G. Lawford, P. Alaback, and E. Fuentes (Eds.)

Volume 117
**Global Change and Mediterranean-Type Ecosystems** (1995)
J. Moreno and W.C. Oechel (Eds.)

Volume 118
**Impact of Air Pollutants on Southern Pine Forests** (1996)
S. Fox and R.A. Mickler (Eds.)

Volume 119
**Freshwater Ecosystems of Alaska. Ecological Syntheses** (1997)
A.M. Milner and M.W Oswood (Eds.)

Volume 120
**Landscape Function and Disturbance in Arctic Tundra** (1996)
J.F. Reynolds and J.D. Tenhunen (Eds.)

Volume 121
**Biodiversity and Savanna Ecosystem Processes. A Global Perspective** (1996)
O.T. Solbrig, E. Medina, and J.F. Silva (Eds.)

Volume 122
**Biodiversity and Ecosystem Processes in Tropical Forests** (1996)
G.H. Orians, R. Dirzo, and J.H. Cushman (Eds.)

Volume 123
**Marine Benthic Vegetation. Recent Changes and the Effects of Eutrophication** (1996)
W. Schramm and P.H. Nienhuis (Eds.)

Volume 124
**Global Change and Arctic Terrestrial Ecosystems** (1997)
W.C. Oechel et al. (Eds.)

Volume 125
**Ecology and Conservation of Great Plains Vertebrates** (1997)
F.L. Knopf and F.B. Samson (Eds.)

Volume 126
**The Central Amazon Floodplain: Ecology of a Pulsing System** (1997)
W.J. Junk (Ed.)

Volume 127
**Forest Decline and Ozone: A Comparison of Controlled Chamber and Field Experiments** (1997)
H. Sandermann, A.R. Wellburn, and R.L. Heath (Eds.)

Volume 128
**The Productivity and Sustainability of Southern Forest Ecosystems in a Changing Environment** (1998)
R.A. Mickler and S. Fox (Eds.)

Voiume 129
**Pelagic Nutrient Cycles: Herbivores as Sources and Sinks** (1997)
T. Andersen

Volume 130
**Vertical Food Web Interactions: Evolutionary Patterns and Driving Forces** (1997)
K. Dettner, G. Bauer, and W. Völkl (Eds.)

Volume 131
**The Structuring Role of Submerged Macrophytes in Lakes** (1998)
E. Jeppesen et al. (Eds.)

Volume 132
**Vegetation of the Tropical Pacific Islands** (1998)
D. Mueller-Dombois and F.R. Fosberg

Volume 133
**Aquatic Humic Substances: Ecology and Biogeochemistry** (1998)
D.O. Hessen and L.J. Tranvik (Eds.)

Volume 134
**Oxidant Air Pollution Impacts in the Montane Forests of Southern California** (1999)
P.R. Miller and J.R. McBride (Eds.)

Volume 135
**Predation in Vertebrate Communities: The Bialowieża Primeval Forest as a Case Study** (1998)
B. Jedrzejewska and W. Jedrzejewski

Volume 136
**Landscape Disturbance and Biodiversity in Mediterranean-Type Ecosystems** (1998)
P.W. Rundel, G. Montenegro, and F.M. Jaksic (Eds.)

Volume 137
**Ecology of Mediterranean Evergreen Oak Forests** (1999)
F. Rodà et al. (Eds.)

Volume 138
**Fire, Climate Change and Carbon Cycling in the North American Boreal Forest** (2000)
E.S. Kasischke and B. Stocks (Eds.)

Volume 139
**Responses of Northern U.S. Forests to Environmental Change** (2000)
R. Mickler, R.A. Birdsey, and J. Hom (Eds.)

Volume 140
**Rainforest Ecosystems of East Kalimantan: El Niño, Drought, Fire and Human Impacts** (2000)
E. Guhardja et al. (Eds.)